Ai6X6 RESEARCH LAB

Transformation Blueprint Series

Published by EDGE Publish, www.EdgePublish.com

Data Transformation for the AI Era, First Edition, 2026
ISBN-13: 978-1-971148-02-1, Printed in the United States of America.

Ai6X6
RESEARCH LAB

Transformation Blueprint Series

DATA TRANSFORMATION FOR THE AI ERA

Building the Intelligence Fabric of the Enterprise.
The 6x6 Blueprint for Data Sovereignty and Trusted Analytics.

SUBRATA MUKHERJEE

EDGE PUBLISH
www.EdgePublish.com

"For the grace of God, who lights the path.
And, in loving memory of my mother Rupali and my father Dilip, who taught me the importance of kindness and perseverance. This work is born from the love you gave, the sacrifices you made, and the quiet wisdom that still echoes in my life. This one is for you."

Personal message from our Founder & Executive Director

SUBRATA MUKHERJEE

For over twenty-five years, I have sat at the heart of global digital transformations—from shifting a 180-year-old institution like *The Economist* into a digital powerhouse, to architecting hyper-automation for Fortune 500 leaders. I have seen the rise of the internet, the mobile revolution, and the cloud. But none of these shifts compare to the era we have just entered.

The Ai6x6 Research Lab was born out of a singular, urgent observation: Most organizations are trying to solve tomorrow's problems with yesterday's mental models. Traditional transformation is often treated as a "project"—a linear path with a start and an end. But in an era of Agentic AI and self-directing systems, transformation is a living organism. It is a system of behaviors, decisions, and capabilities that must evolve in unison. I founded this Lab to move beyond the "hype" and the "tools." My motivation is to provide leaders with the Cohesion, Schema, and Custody required to navigate a world where the boundary between human intuition and machine intelligence is disappearing. People often ask why "6×6." Throughout history, the number six has been the sacred geometry of our existence—from the sexagesimal systems of ancient Sumerian astronomy to the hexagonal patterns found in the farthest reaches of our cosmos. Mathematically, six is the first "perfect number" because its divisors (1, 2, and 3) add up to itself, making it uniquely balanced. In my journey across the U.S., Europe, and Asia, I realized that this ancient symmetry holds the key to modern enterprise. By distilling organizational complexity into six fundamental pillars and viewing them through six transformative lenses, we activate a collective "sixth sense" for the organization—the ability to perceive and act on what others miss. This architecture forms the foundation of our three flagship methodologies—the "*guided navigation system every transformation needs*".

My journey to this blueprint was forged in the engine rooms of the world's most complex organizations. I have led wholesale business transformations for *The Economist*, reinvented Bloomberg's digital presence following the *Businessweek* acquisition, revitalized the iconic *Dummies* brand within Wiley, and optimized the global ratings processes within S&P Global and McGraw-Hill. This strategic work was balanced by deep technical execution: implementing RPA, AI, IoT, and Digital Twins within Raytheon, and architecting CRM, BPM, and digital workplace automations for the global clients of Cognizant and Unisys.

This perspective was further refined through rigorous academic pursuit, drawing on my MBA in Marketing from Jadavpur University, my Master's in Technology Management from Columbia University, and executive certifications from Harvard University and the Indian Institute of Management (IIM-C) to bridge the gap between high-level theory and industrial-scale execution.

Today, through the Ai6x6 Research Lab, I lead the development of the frameworks that are now being adopted by executive teams globally. At the center of this work is three methodologies: 6X6 Blueprint, IGNITE Methodology and REVEAL Methodology our multidimensional navigation system that unifies architecture, execution, and evaluation across the enterprise.

However, my background as a Tech Journalist for *The Telegraph* for over a decade and a columnist for *CIO Magazine* has taught me that true insight requires total independence. I have built this Lab as a space where my team and I can express our opinions freely, without the interference of corporate sponsors. Our goal is not just to report the news, but to "fold" it into the Ai6x6 Blueprints—giving you the critical "so-what" and the "what-next" to make global shifts actionable for your organization.

We are democratizing this knowledge because I believe that the ability to thrive in the AI era should not be reserved only for those who can afford expensive global consulting firms. Our mission is to empower businesses of all sizes with the targeted, practical, and battle-tested information that I often wish I had at my disposal during my own leadership assignments. Through the Ai6x6 Transformation Blueprints and our Executive Mentor Network, we are scaling AI literacy to ensure that as systems become more autonomous, leaders become more intentional. We are not just adopting new tools; we are reimagining the nature of the enterprise itself.

Ai6X6 RESEARCH LAB

NEW YORK | LONDON | KOLKATA

To our esteemed Industry Thought Leaders: Thank you!

Your generous time and profound insights have been the bedrock of this book. Each interview has been converted into sections and one-page cards that have been sprinkled across six books. You have provided a vital perspective, transforming our work from theory into a deeply practical guide rooted in real-world leadership in the AI-Era. Your willingness to share your hard-won wisdom, navigating complex challenges and revealing the future of your industries, enriches every page. You all have successfully navigated the AI transition in your careers, your companies and your industries. This book is a testament to your expertise. We are immensely grateful for your unwavering support and belief in this project. Thanks a lot, from me and my team at Ai6x6 and Executive Mentor Network!

TABLE OF CONTENTS

guided navigation for transformation

BLUEPRINT SERIES

SECTION A

Introduction

Ai6x6™ Transformation Blueprint

Ai6x6 Research Lab is a global intellectual institution dedicated to advancing the science and practice of organizational transformation. We develop original frameworks, diagnostic systems, capability playbooks, and research models that help leaders navigate complexity with clarity, precision, and confidence. Our work spans **Leadership, Workforce, Data, Customer, Product, and Operations**—the six pillars that define modern enterprise maturity. We are based in New York, London & Kolkata.

OUR MISSION FOR ORGANIZATIONAL TRANSFORMATION

To build a globally recognized body of knowledge that helps organizations think clearly, act decisively, and transform responsibly in an age of AI acceleration. We essentially play **six roles** for the industry:

THE CATALYST BRING ALIGNMENT TO LEADERSHIP	**THE STRATEGIST** BRING STRUCTURE TO AMBIGUITY	**THE ARCHITECT** BRING CLARITY TO COMPLEXITY
THE ENGINEER BRING DISCIPLINE TO TRANSFORMATION	**THE ETHICIST** BRING INTEGRITY TO OUTCOMES	**THE GUARDIAN** BRING OVERSIGHT TO AUTONOMY

WHAT WE DO?

1. Build Transformation Frameworks: We architect the Ai6x6 Transformation Blueprints— a structured, multi-lens transformation model that helps orgs decode patterns, risks, behaviors, and decision pathways.
2. Develop Diagnostic Systems: Our maturity diagnostics and radar models translate organizational signals into actionable insights, identifying Growth, Risk, Kill, and Shift zones.
3. Publish Research & Methodologies: We codify best practices in enterprise design through multi-volume series and research playbooks, available in bookstores or on our website ai6x6.com.
4. Create Executive Tools: From RACI matrices to transformation roadmaps, we build high-utility models for immediate executive application. We have many accelerated off-the-shelf offerings on our website.
5. Support Global Organizations: Our frameworks empower enterprises, consulting teams, and innovation labs. Our advisory team worldwide including our Executive Mentor Network is here to help.

ACCOMPLISHMENTS

Subrata Mukherjee has delivered enterprise-scale transformation across the Americas, Europe, and Asia, shaping modernization programs for some of the world's most influential organizations. His work spans AI/ML, automation, digital product strategy, enterprise architecture, and large-scale organizational transformation. He has led digital reinvention in global publishing, driven automation and AI adoption in aerospace and defense, modernized ratings and compliance processes in financial services, and guided product and data transformation across multiple industries. In 2016, he was recognized as one of Folio's Top 100 People in Publishing for his impact on digital innovation.

INDUSTRIES

His career spans aerospace and defense, financial services, publishing, healthcare, consulting, education, manufacturing, and technology services. He has delivered transformation in highly regulated environments, mission-critical engineering contexts, global media organizations, and fast-moving digital businesses. This cross-industry exposure gives him a multidimensional perspective on how AI reshapes strategy, operations, and organizational identity.

EXPERTISE

Subrata's expertise spans the full arc of enterprise transformation, beginning with early work in CAD/CAM for steel plants and smart-city infrastructure, which shaped his systems-thinking foundation. He then became a specialist in enterprise CRM and BPM transformation, leading large-scale modernization programs that re-architected customer operations, workflow orchestration, and enterprise process design. Over time, he transitioned into a strategist, product thinker, and transformation architect, guiding organizations on designing, adopting, and scaling next-generation digital capabilities. Today, he is recognized as a leading voice in AI and Emerging Technologies, with deep expertise across AI/ML, Generative AI, LLMs, RAG architectures, intelligent automation, hyper-automation, conversational AI, chatbots, agentic systems, and decision-automation frameworks. His work is rooted in strategy, product design, operating-model transformation, and enterprise-scale adoption—not coding or engineering execution—and focuses on translating emerging AI capabilities into practical, scalable, enterprise-grade systems that reshape how organizations operate and compete.

EDUCATION

Subrata holds a Master's in Technology Management from Columbia University, an MBA in Marketing from Jadavpur University, and a Business Communications certification from Harvard University. He also holds a Management by Values certification from the Indian Institute of Management, Calcutta, deepening his grounding in leadership behavior, organizational culture, and values-driven transformation. He is a Certified Product Owner. His academic foundation spans technology, strategy, communication, and organizational leadership, informing his multidisciplinary approach to AI-era transformation.

COMPANIES

Across his career, Subrata has delivered transformation programs for organizations including The Economist, Raytheon Technologies, S&P Global, Bloomberg, Wiley, Cognizant, Unisys, and Gujarat Ambuja Cement. His work has ranged from leading The Economist's digital transformation and modernizing Bloomberg's subscription and data infrastructure, to launching Raytheon's enterprise-wide Citizen Development program for Automation and AI, transforming the Dummies brand at Wiley, modernizing S&P's Ratings Process after the sub-prime mortgage crisis, scaling CRM and Automation/AI consulting portfolios at Cognizant and Unisys, and establishing the technology backbone and smart-building systems for the Gujarat Ambuja conglomerate.

THOUGHT LEADERSHIP

Subrata is a prolific speaker who has presented at more than 100 global conferences and events. His writing has appeared in CIO Magazine, and he served as a weekly technology columnist for The Telegraph. Through the Ai6x6 Research Lab, he publishes frameworks, diagnostics, and transformation blueprints that help organizations navigate the AI economy. In 2026–27, he will launch the Executive Mentor Network to develop a global community of AI Strategy–certified coaches and expand AI literacy among senior leaders. Beyond his professional work, he is an avid traveler who has explored more than 40 countries and an accomplished portrait artist.

www.linkedin.com/in/sm360

1. GIVE **VALUE**, REPEATEDLY, BECOME A GOTO BOOK

The 18-book series brings in a lot of content from a practitioner perspective. The advice we are providing have worked for us in our past projects across industries; so we are preaching what we practiced before

2. BE **BRIEF**, PROVIDE CONTENT IN SMALL NUGGETS

I have never read a book cover to cover. I have short attention span, and more and more people are suffering from the same syndrome. Hence each micro-topic we have introduced is covered in two to three pages

3. BE **COMPREHENSIVE** FOR REAL WORLD IMPACT

Transformation is not easy. In the real world there are a lot of inter-connected challenges. We have proposed detailed frameworks that will connect the dots and provide comprehensive solutions to your team/client

4. BE **PRACTICAL**, PROVIDE EXAMPLES TO DEMYSTIFY

Most companies and leaders know they and their companies need to transform, but don't know where to start. I wanted to provide a step-by-step guide or a MOSCOW priority list on getting started

5. PUSH **BOUNDARIES** AND DARE LEADERS TO DREAM

Real transformation is not tinkering around the edges and playing safe; its often about self disrupting and pushing boundaries and reinventing oneself. Applies to individuals, teams and organizations.

6. PUSH LEADERS TO **INTROSPECT** & ASK QUESTIONS

Since the world around us is moving so fast, leaders often have a fear of been exposed about their lack of knowledge and ask "stupid" questions. These book educates you enough, so you can ask intelligent questions

7. BE **EXCITING**, DON'T BE BORING

I have tried to keep the book interesting, small text, lot of structured tables, quite a few images, bold formatting, case studies, Tips and Tricks, Interviews, RACI matrix, buying guides, assignment questions etc.

8. BRING **OTHERS**, HEAR FROM MY COMMUNITY

Tech sector worldwide is very collaborative. We learn from each other every day and we use the knowledge to impact lives positively. This book attempts to bring a few voices from my network and hear their perspective

9. PROMOTE **AI ECONOMY**, USE AI TOOLS YOURSELF

Generative AI has been used extensively for formatting content in a repeatable format & page size control. Without GenAI such an extensive book series would not have been possible in a short time.

10. BE **DIFFERENT**, CHARTER YOUR OWN COURSE

Most management self-help books are pages and pages of long text with guidelines provided by publishers that are a bin antiquated. Modern learning happens byte sized and is multi-modal and multi-channel..

The **Guided Navigation for Transformation** framework is a comprehensive, end-to-end methodology designed to de-risk and accelerate AI adoption within an organization. It moves away from "random acts of digital" by providing a disciplined, three-stage approach that balances human capability with technical execution. The core of the system is the **6x6 Lens Sequence**, which ensures that every strategic move is evaluated through six critical dimensions: **Leadership, Workforce, Data, Customer, Product, and Operations**.

By applying these lenses consistently across the Blueprint, Reveal, and Ignite phases, the framework creates a "Golden Thread" that links high-level strategy to on-the-ground deliverables. This is not just a consulting model, but an educational ecosystem with an 18-book series that ensures that AI transformation is both measurable and sustainable.

"Guided Navigation for Transformation" in the AI Era

We provide guided navigation through three distinct phases: First, we establish your **Transformation Blueprint** to align skills. Second, we run the **Reveal Diagnostic** to assess your current maturity. Finally, we execute the **Consulting Playbook** to turn the 'Art of the Possible' into a 'Science of Deliverables'.

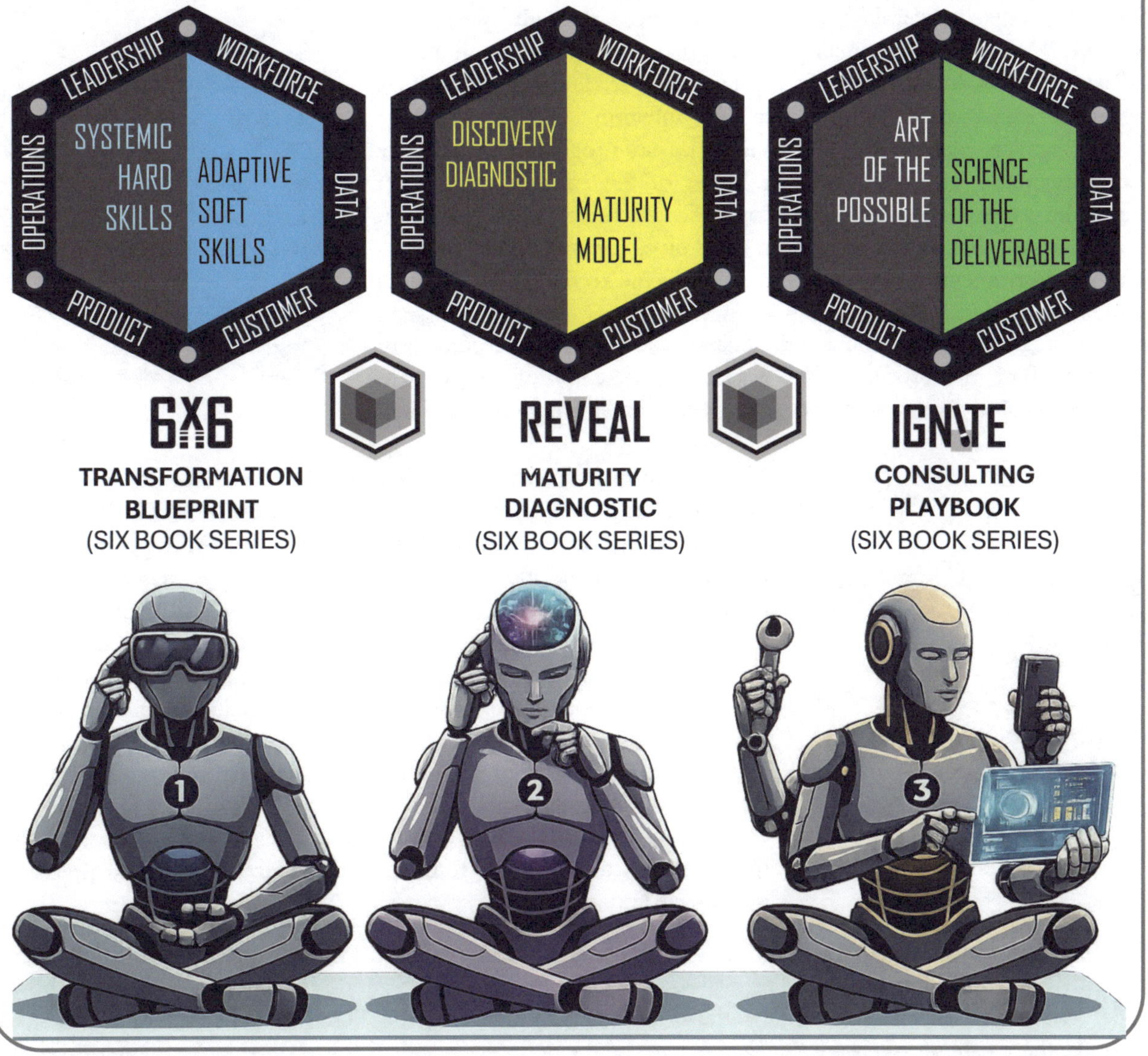

The **AI6x6 Transformation Engine™** is the flagship construct of the AI6x6 Research Lab—a precision-engineered ecosystem that serves as the definitive "Operating System" for the modern enterprise. Built on a proprietary 6×6 geometry, it unifies architecture, execution, and calibration into a single, high-performance unit. The Engine provides leaders with the high-torque machinery and **Guided Navigation** required to convert organizational inertia into scalable momentum in the era of autonomous systems.

The **Transformation Blueprint Series** defines the engine's architecture and establishes the structural logic of the entire system. By introducing the **Six Pillars** and **Six Lenses**, it forms the foundational geometry of enterprise modernization. Written for CEOs, Boards, and Strategy Leaders, this series acts as the **Navigation Deck**, providing the "Architectural North Star." It ensures that every move is strategically calibrated and ethically anchored before the gears are engaged, offering a shared mental model for aligning leadership teams. This intelligence layer removes the guesswork from transformation, providing a rigid yet adaptive framework that allows organizations to visualize their future state with absolute clarity. It is the vital starting point where complex, high-stakes ambition is codified into a disciplined, institutional reality that can withstand the pressures of rapid technological shift.

The **Maturity Diagnostic Series** provides the engine's accountability and calibration system, ensuring peak performance through proprietary **REVEAL ™** methodology. . Acting as the **Sensor-Driven Intake**, it utilizes 36 diagnostic cylinders to quantify organizational readiness and prioritize investments based on evidence rather than intuition. This series allows executive teams to identify hidden gaps and measure progress with scientific clarity, functioning as the **Measuring Layer** of the entire ecosystem. It ensures the engine remains tuned for resilience through a data-driven maturity model that tracks evolution over time. By providing rigorous assessment instruments and scoring logic, it offers the accountability required to sustain a long-term transformation journey. This layer acts as the engine's nervous system, constantly scanning the internal and external environment to calibrate the Blueprint's trajectory and ensure that every unit of energy spent drives the organization toward sustained excellence and institutional stability.

The **Consulting Playbook Series** translates architecture into action by operationalizing the engine through the proprietary **IGNITE™** methodology. It serves as the **Heavy-Duty Gearbox**, providing the repeatable consulting patterns, execution models, and directional force required to activate transformation across diverse domains. Designed for PMOs and delivery leads, these playbooks convert high-level strategy into tangible enterprise-scale outcomes, ensuring that the "torque" of the engine is effectively applied to the road. They function as the essential doing layer, reducing internal friction and shifting the organization from abstract concept to deep-seated conviction. By providing step-by-step pathways and intervention models, the series bridges the gap between executive intent and operational reality. This transmission layer ensures that momentum is not just created, but sustained and scaled, allowing the enterprise to accelerate through complex transitions with precision and measurable impact.

The **Transformation Engine** is built for scale across the global economy. Future **Industry Evolution** volumes will apply this 6×6 engine to specific sectors—such as Healthcare, Banking, and Manufacturing—revealing how the model drives specialized, high-velocity transformation. By adapting the Pillars and Lenses to sector-specific demands, these extensions ensure the Engine remains the definitive standard for navigating the complexities of the age of intelligent systems.

SERIES 1 | 6 BOOKS

SERIES 2 | 6 BOOKS

SERIES 3 | 6 BOOKS

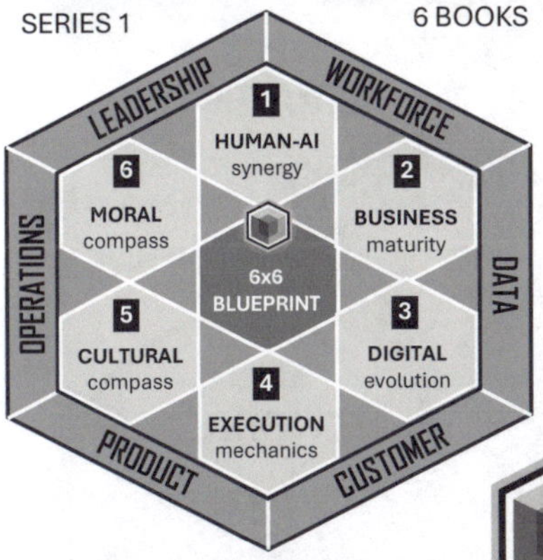

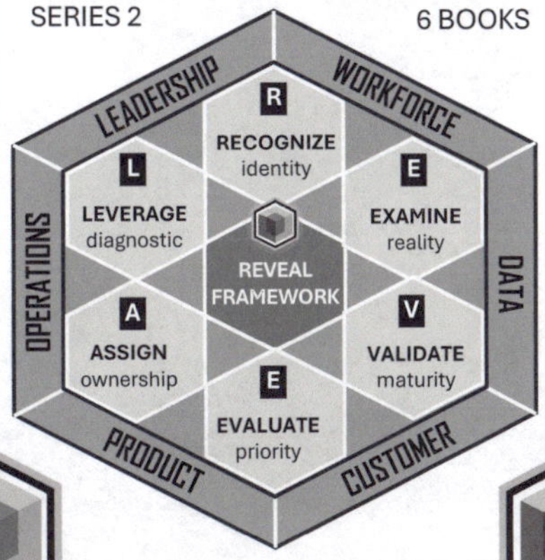

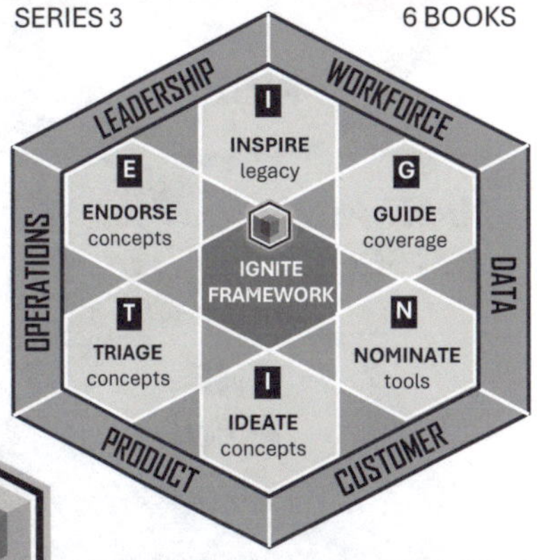

6X6 TRANSFORMATION BLUEPRINT

REVEAL MATURITY DIAGNOSTIC

IGNITE CONSULTING PLAYBOOK

see | **think** | **act**

The Ai6x6 palette is a semantic system, not a decorative choice. Each color encodes the emotional and operational signature of its lens: **pink** captures the human spark that ignites potential; **green** reflects the strategic discipline required for maturity; **purple** signals the depth and computational power of digital evolution; **orange** embodies the kinetic urgency of execution; **blue** anchors the cultural soul in trust and cohesion; and **yellow** illuminates the ethical clarity of the moral audit. Together, these colors form a coherent visual language that guides the reader through the transformation cycle with intuitive, emotionally aligned cues.

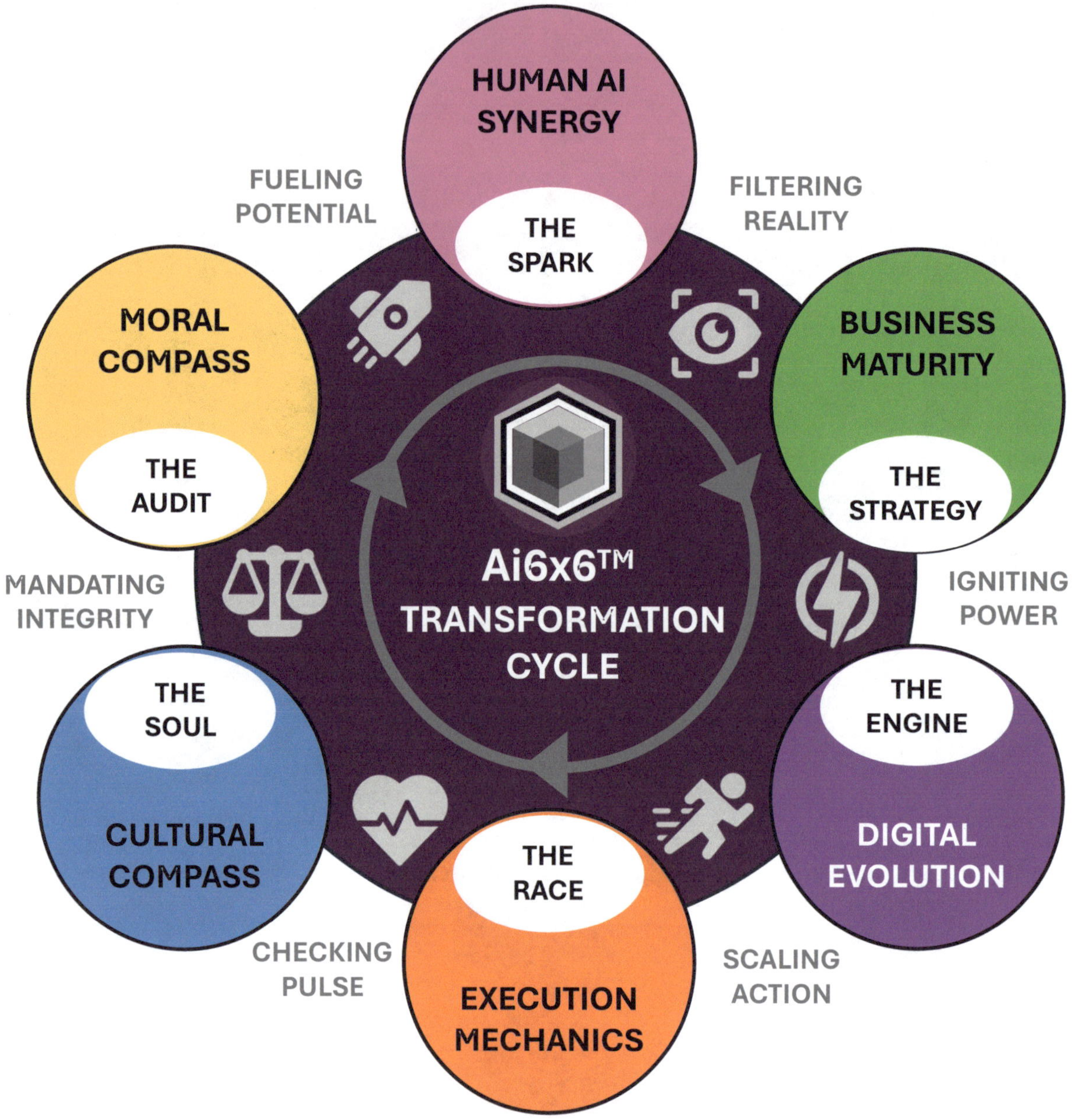

The **COLOR TABS ALONG THE PAGE EDGES IN THE MIDDLE** serve as the physical navigation system of the Ai6x6 series. Each tab mirrors the chromatic identity of its lens, allowing readers to locate themselves instantly within the transformation cycle. This creates a spatial reading experience where orientation is continuous and intuitive. The tabs turn our book into a modular interface—one where Human–AI Synergy, Business Maturity, Digital Evolution, Execution Mechanics, Cultural Compass, and Moral Compass become distinct territories with their own visual signatures. The result is a playbook that reads like a designed system, not a linear text.

6x6™ Transformation Blueprint

The Prologue to a Purpose

"The only way to go forward is to let go of the past."
The Myth of Orpheus and Eurydice

The tragic lesson of this myth is a cautionary tale for business. Orpheus's failure to trust his journey and looking back results in losing everything, underscoring the importance of fully committing to a forward-looking transformation.

The mission of this book series is simple: to transform leaders from passive observers of the AI Era into its active pioneers. We will ignite your thinking with exciting, highly practical content that pushes boundaries and provides the essential tools and frameworks you need to build real-world impact. We won't just tell you what to do; we'll show you how, using a diverse community's experiences and real-world examples that bring the abstract to life. This isn't a passive read; it's an active partnership and a dynamic dialogue designed to turn your natural curiosity into a definitive competitive advantage. This book is the first of a six-part series : Leadership Transformation; which governs your journey on how you should lead from the top and how you should inspire others to take on the AI Era and excel in it.

The **AI6x6 Transformation Engine** is a precision-engineered ecosystem designed to deliver **Guided Navigation** for enterprises facing high-stakes digital evolution. By integrating three proprietary systems—the **Transformation Blueprint**, the **Consulting Playbook**, and the **Maturity Diagnostic**—the Engine provides the robust "Operating System" required to dominate the AI era.

The architecture begins with the **Transformation Blueprint**, the Engine's intelligence layer. Here, the system utilizes **Six Lenses**—including the Cultural and Moral Compasses—to calibrate your trajectory and codify a master architectural plan. This strategic map then engages the **Consulting Playbook**, where the gears of the **IGNITE** methodology translate executive intent into immediate directional force through rapid-response execution cycles. Finally, the **Maturity Diagnostic** acts as a sensor-driven intake, utilizing six diagnostic cylinders to compress raw organizational data into high-pressure strategic insights that validate the transformation's impact. Bolted to a rigid **Chassis of Six Pillars**, the Engine ensures the enterprise remains resilient, balanced, and unshakeable even at peak acceleration. It is more than a strategy; it is a proprietary architecture for sustained institutional excellence.

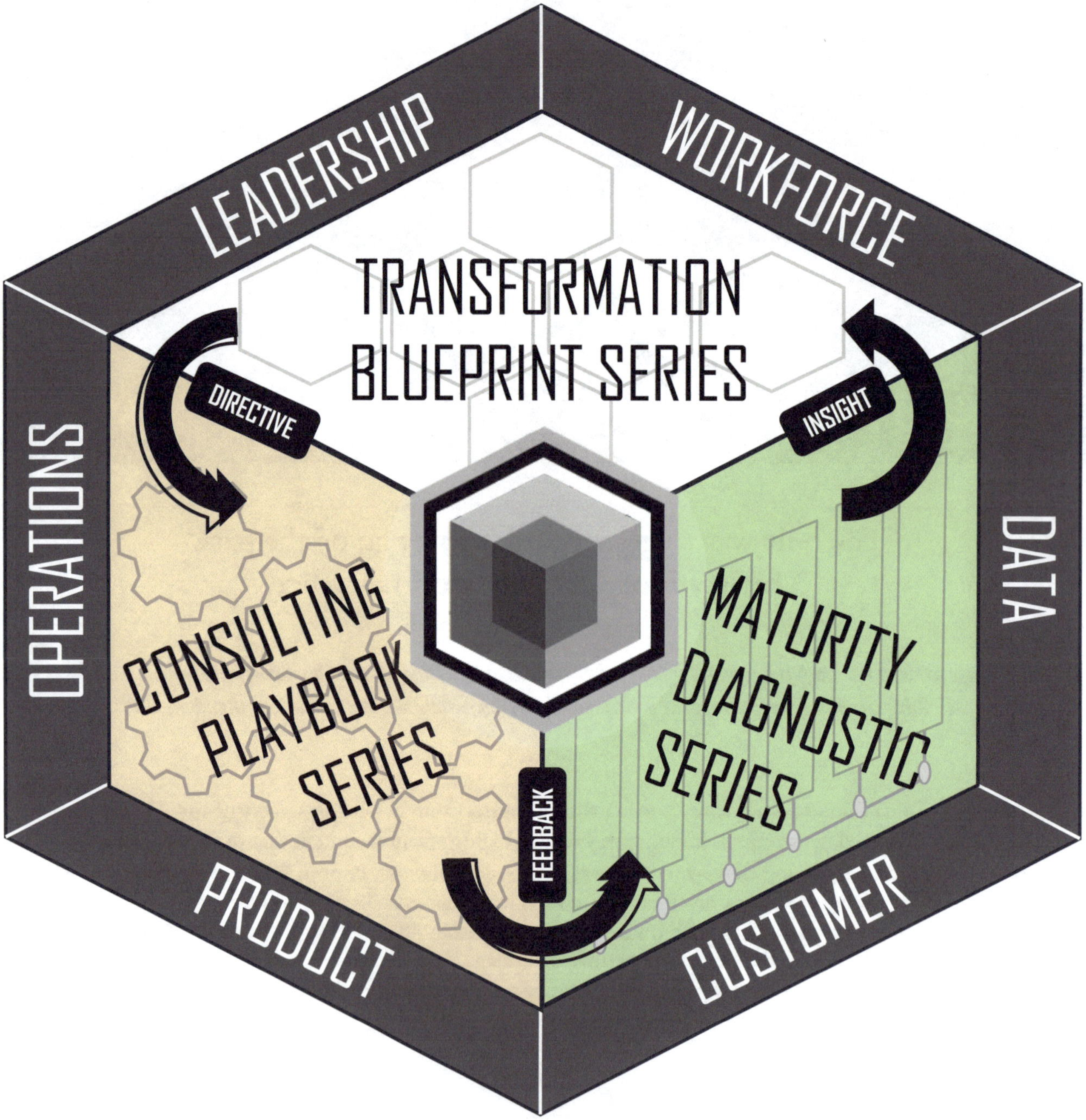

The AI6x6 Transformation Engine is a precision-engineered ecosystem that replaces guesswork with **Guided Navigation**. It is designed to convert organizational inertia into scalable momentum through a disciplined, mechanical approach to digital evolution.

CHASSIS (SIX PILLARS)
The entire system is anchored to a rigid structural frame. This chassis ensures that even as you reach peak acceleration, your enterprise remains resilient and unshakeable.

NAVIGATION DECK (SIX LENSES):
This is the intelligence layer. By applying our **Six Lenses**, we provide the **Guided Navigation** required to calibrate your trajectory, preventing costly drifts in strategy.

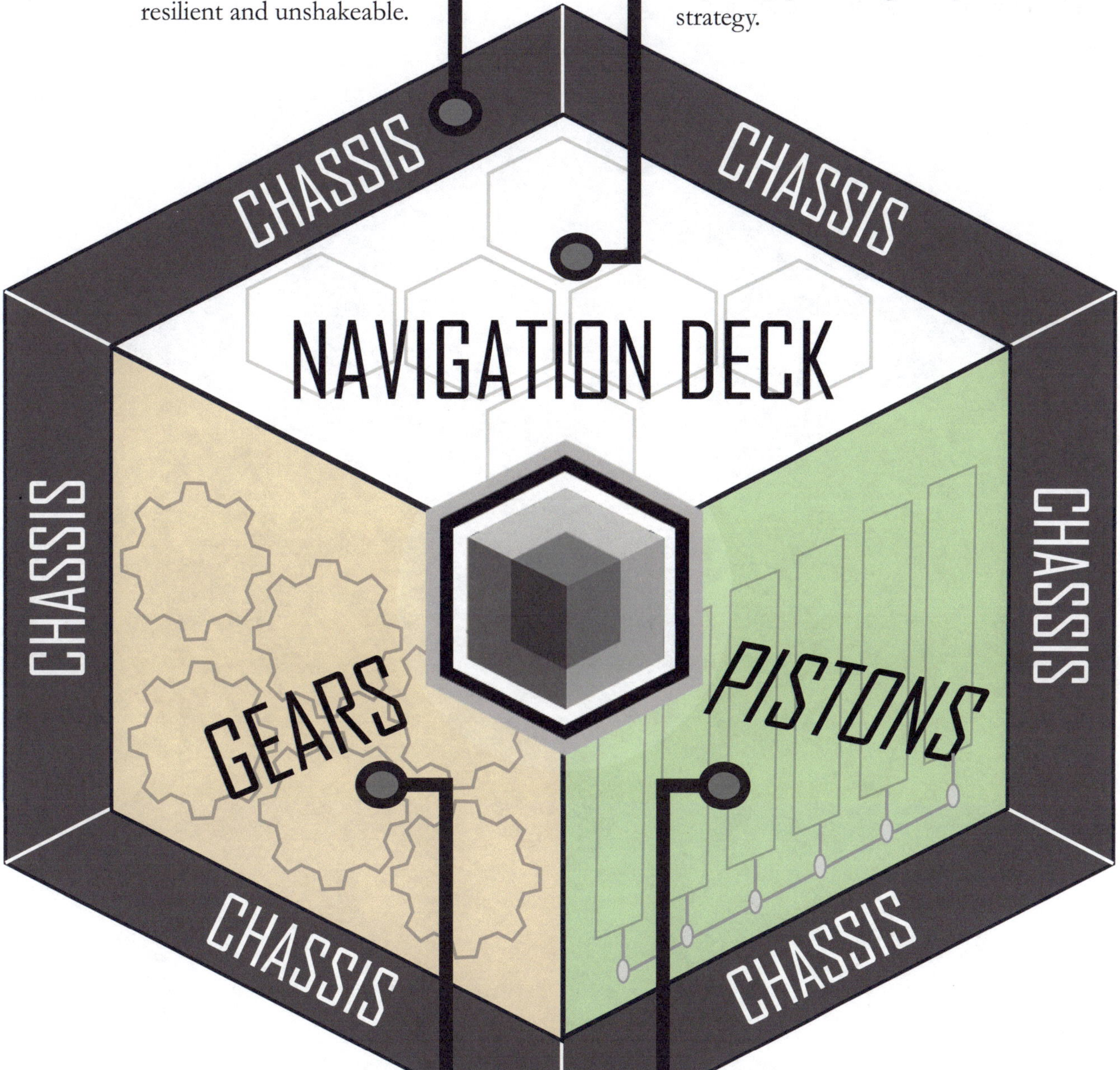

GEARS (IGNITE METHOD):
The "Torque" that converts diagnostic energy into immediate directional force through pilot programs and playbooks.

PISTONS (REVEAL METHOD):
The diagnostic "Suctions" that pulls in raw organizational data and compresses it into high-pressure actionable insights.

A car works best when its actions are coordinated through systems that generate power, transmit motion, and guide direction. Each component plays a precise mechanical role that keeps the vehicle stable, responsive, and moving with purpose.

CHASSIS
Supports vehicle structure, absorbs road stress, anchors mechanical systems, maintains stability, and ensures safe, balanced handling across varying speeds, terrains, and driving conditions.

NAVIGATION DECK
Displays real-time driving information, maps, routes, and alerts, helping the driver maintain direction, anticipate conditions, and make confident, informed decisions.

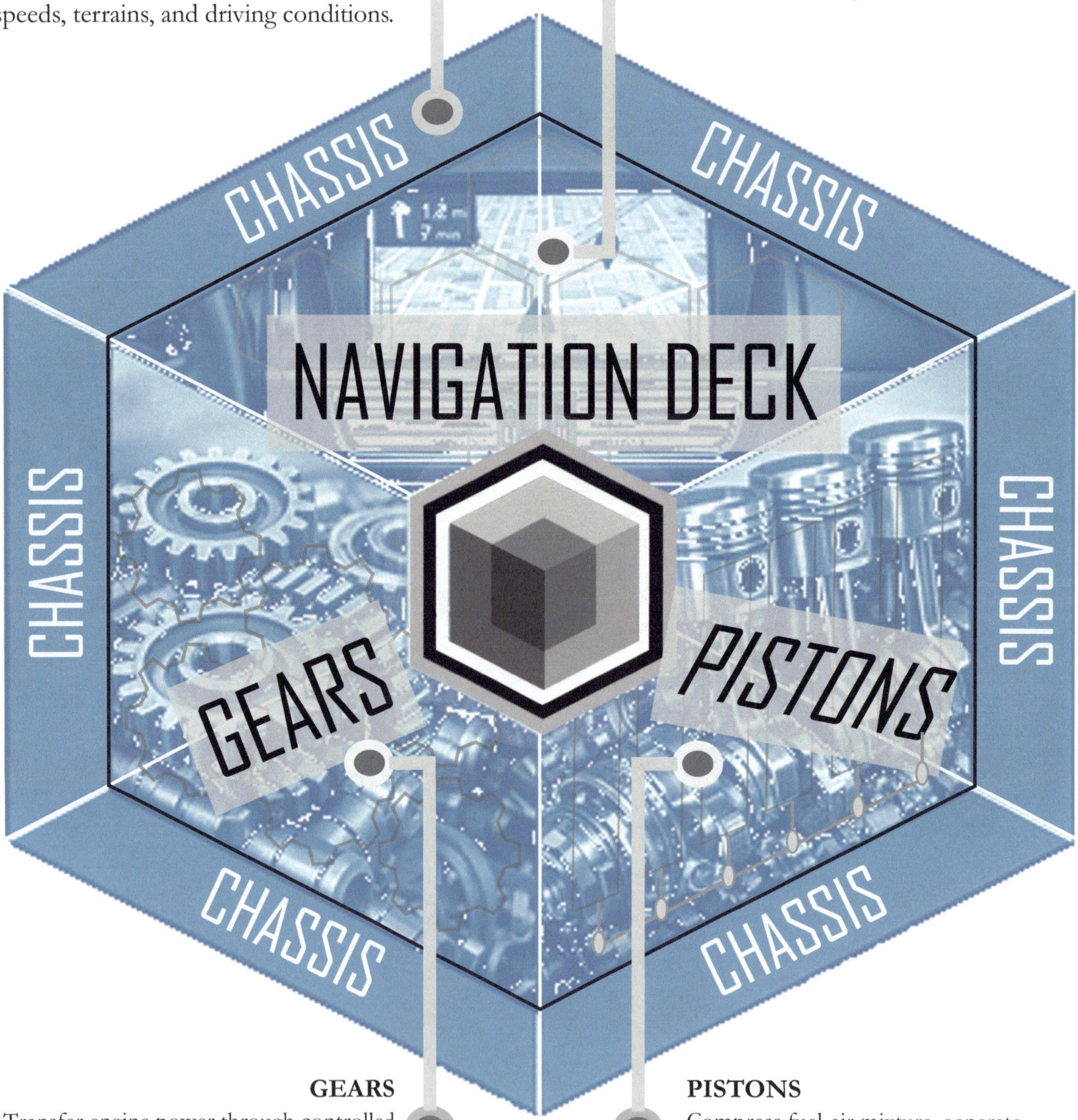

GEARS
Transfer engine power through controlled ratios, regulating speed and torque, ensuring smooth acceleration, efficient performance, and seamless coordination between mechanical components.

PISTONS
Compress fuel-air mixture, generate combustion force, drive the crankshaft, & convert explosive energy into continuous motion powering the vehicle's entire drivetrain system.

The **AI6x6 Transformation Engine** replaces the ambiguity of traditional consulting with mechanical certainty, converting organizational inertia into scalable momentum. By integrating three proprietary systems—the **REVEAL** diagnostic intake, the **IGNITE** consulting transmission, and the **6x6 Blueprint** navigation deck—the Engine provides the robust "Operating System" required to dominate the AI era. The **REVEAL** methodology acts as a sensor-driven intake, utilizing 6x6 diagnostic cylinders to compress raw organizational data into high-pressure strategic insights. This "data fuel" then powers the **IGNITE** phase, where consulting playbooks serve as a heavy-duty gearbox, translating executive intent into immediate directional force through rapid-response execution cycles. This entire power plant is governed by the **6x6 Blueprint**, the intelligence layer that uses **Six Lenses**—including the Cultural and Moral Compasses—to calibrate your trajectory and ensure ethical, operational alignment. Bolted to a rigid **Chassis** of **Six Pillars**, the Engine ensures the enterprise remains resilient, balanced, and unshakeable even at peak acceleration. It is more than a strategy; it is a proprietary architecture for sustained institutional excellence.

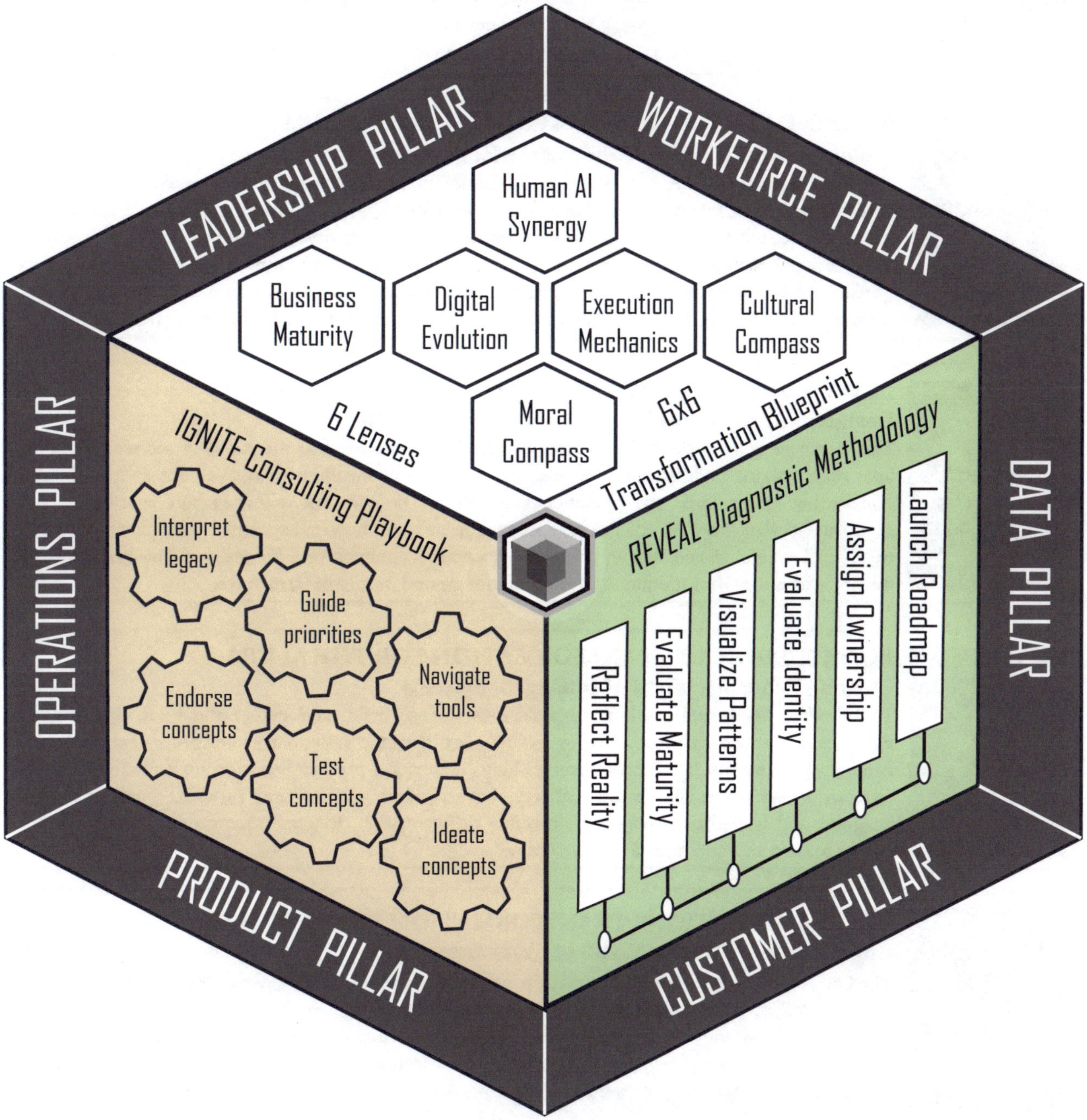

BOOK 1: LEADERSHIP TRANSFORMATION FOR THE AI ERA

Designing the Vision, Not Just Directing It

The Leadership book reframes what it means to lead in an era defined by intelligence, acceleration, and ambiguity. It explores how leaders think, decide, communicate, and shape meaning across the enterprise, emphasizing that leadership is not a role but an architecture. The narrative examines how leaders make decisions systemically, exercise judgment in an AI-augmented world, create alignment through narrative clarity, and uphold ethical stewardship as the foundation of trust. This book sets the tone for the entire series by positioning leadership as the integrator of all transformation forces.

BOOK 2: WORKFORCE TRANSFORMATION FOR THE AI ERA

Elevating Human Capability in a Technology-Driven Era

The Workforce book examines the human system of the enterprise — the skills, roles, mindsets, structures, and experiences that determine how people show up to the work of transformation. It explores how organizations support people as AI reshapes every function, and how leaders build environments where individuals feel empowered rather than displaced. The narrative dives into the evolution of adaptive talent ecosystems, the emergence of AI-augmented roles, the importance of team cohesion and psychological safety, and the recognition of human capability as a strategic asset.

BOOK 3: DATA TRANSFORMATION FOR THE AI ERA

Building the Intelligence Fabric of the Enterprise

The Data book establishes the informational backbone of the organization. It explores how data is collected, governed, interpreted, and transformed into intelligence, emphasizing that data is no longer a technical asset but the raw material of decision-making, personalization, automation, and innovation. The narrative examines how trustworthy data governance is built, how intelligence flows across systems, how data architectures are designed, and how data becomes a source of enterprise-level advantage. This book gives leaders the clarity to build an intelligent, insight-driven organization.

BOOK 4: CUSTOMER TRANSFORMATION FOR THE AI ERA

Designing Experiences That Earn Trust and Loyalty

The Customer book anchors the enterprise in the lived experience of the people it serves. It explores how organizations understand, anticipate, and respond to customer needs — not as a marketing exercise but as a cultural and operational discipline. The narrative examines how emotion shapes experience design, how AI enables personalization at scale, how transparency strengthens customer trust, and how end-to-end journey coherence becomes a competitive differentiator. This book ensures the organization evolves around the customer, not around its internal structures.

BOOK 5: PRODUCT TRANSFORMATION FOR THE AI ERA

Creating Value in a World of Intelligent Systems

The Product book explores how organizations design, build, and evolve the value they deliver. Products are no longer static offerings — they are dynamic systems of intelligence, behavior, and continuous adaptation. This book examines how AI enhances product behavior, how continuous value creation replaces episodic feature delivery, how product strategy and product operations must integrate, and how digital ecosystems redefine differentiation. It transforms product thinking from output to outcome, and from features to value.

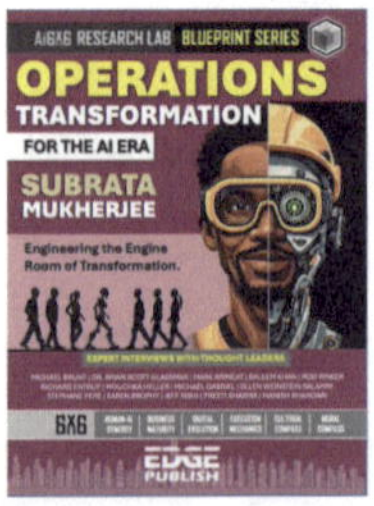

BOOK 6: OPERATIONS TRANSFORMATION FOR THE AI ERA

Engineering the Engine Room of Transformation

The Operations book addresses the systems, workflows, and processes that determine whether strategy becomes reality. It explores how organizations orchestrate complexity, manage risk, and maintain resilience in environments defined by acceleration and uncertainty. The narrative examines how intelligent workflows are designed, how operational resilience is built, how AI optimizes processes, and how execution scales with precision. This is the book where transformation becomes tangible.

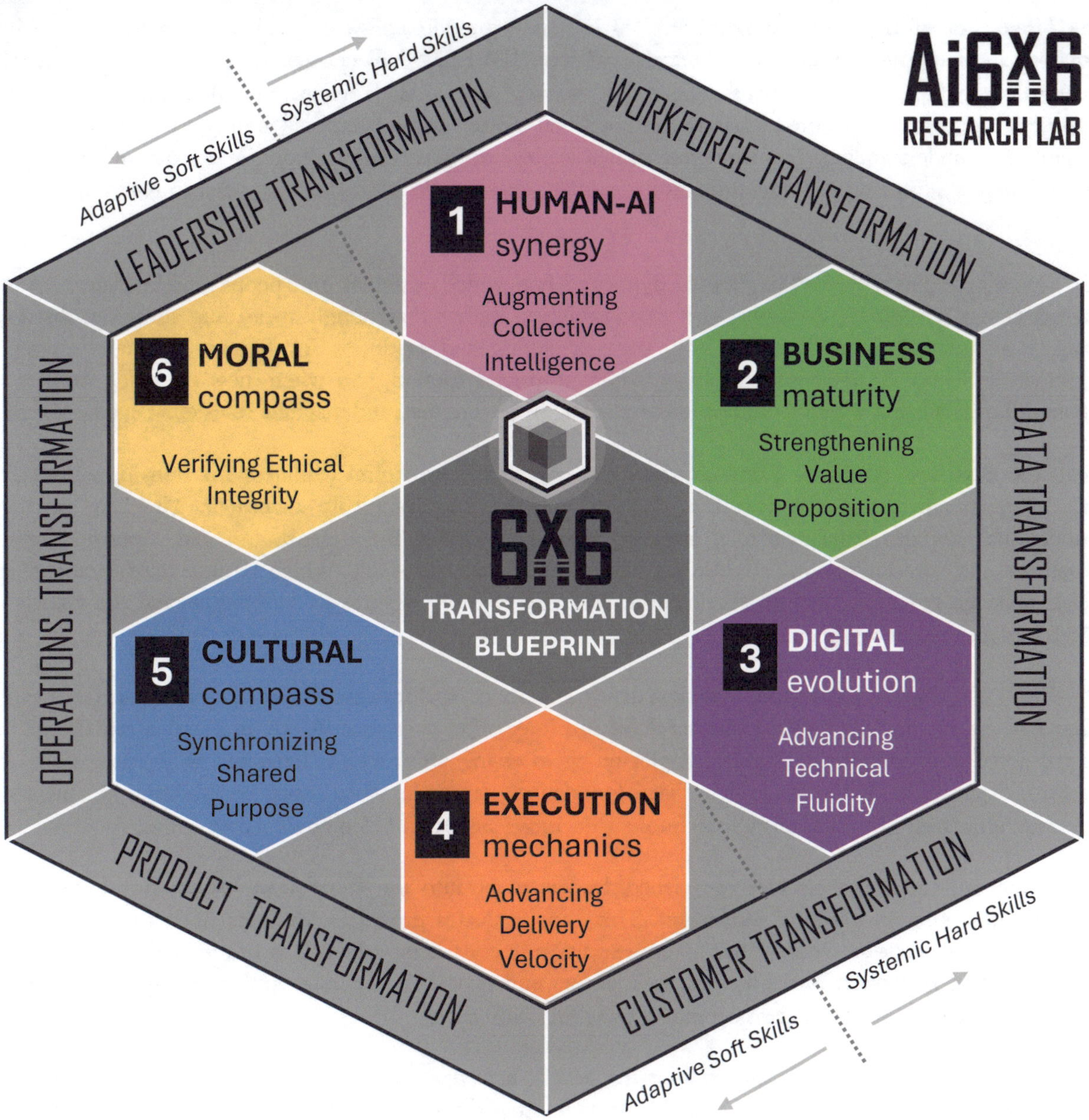

The **Ai6x6 TRANFORMATION BLUEPRINT™** provides the multidimensional structural integrity required for sustainable, high-stakes evolution. Serving as the **Navigation Deck** and **Chassis** of the Transformation Engine, the BLUEPRINT™ moves beyond episodic change to establish a permanent, scalable "Operating System" for the enterprise. The architecture is defined by the intersection of **Six Pillars** and **Six Lenses**. The Pillars—Leadership, Workforce, Data, Operations, Product, and Customer—provide the horizontal breadth and load-bearing strength of the organization. These are bisected vertically by the Six Lenses to reveal the true depth of the enterprise. Leaders apply **Human–AI Synergy** to augment collective intelligence and **Business Maturity** to instill systemic discipline. The **Digital Evolution** lens ensures technical fluidity, while **Execution Mechanics** masters the art of precision delivery. This is anchored by the **Cultural Compass**, which synchronizes shared purpose, and the **Moral Compass**, which verifies ethical integrity.

By aligning these twelve dimensions, the BLUEPRINT™ eliminates strategic drift and structural fragmentation. It ensures that every initiative is not just an isolated project, but a calibrated component of a unified, resilient, and unshakeable corporate architecture. Also remember to follow the Tab colors on the edges to navigate quickly.

The **Ai6x6 Transformation Blueprint™** gives leaders a multidimensional, architecture-grade view of how transformation behaves inside their enterprise. Where REVEAL™ diagnoses the present and IGNITE™ accelerates the future, the Blueprint defines the *structural truth* of the organization. Built on six proprietary Lenses, it converts fragmented observations, cultural noise, and operational ambiguity into a coherent, systems-level understanding. The Blueprint equips leaders with the clarity, depth, and structural intelligence required before designing any transformation roadmap, ensuring decisions are grounded in reality rather than aspiration.

The Blueprint begins with the **Human–AI Synergy Lens**, which examines how people and intelligent systems interact across workflows, decisions, and responsibilities. Rather than simply measuring adoption, this Lens analyzes augmentation, displacement risk, trust dynamics, and readiness for human-machine teaming. It distinguishes where AI enhances capability, where it creates friction, and where new operating models are required. Leaders begin with a grounded understanding of how human and machine roles must evolve together.

Next, the **Business Maturity Lens** evaluates the enterprise's structural readiness for transformation. This includes governance, financial discipline, decision velocity, and leadership alignment. The Lens separates foundational strength from structural fragility, revealing whether the organization can support sustained transformation without collapse. It clarifies which capabilities are stable, which require reinforcement, and which must be redesigned before acceleration. Business Maturity ensures leaders understand the chassis on which transformation will operate.

The third Lens, **Digital Evolution**, maps the depth, coherence, and scalability of the digital ecosystem. Leaders often face fragmented platforms, technical debt, and integration gaps that silently undermine transformation. Digital Evolution cuts through this complexity by assessing architecture, interoperability, data flow, and platform extensibility. It clarifies which digital foundations can scale, which must be modernized, and which create systemic risk. This Lens ensures technology decisions are grounded in architectural truth.

With structural and digital realities established, leaders move into the **Execution Mechanics Lens**, which examines how strategy becomes movement. This Lens evaluates processes, rhythms, accountability systems, and operational precision. It identifies where execution is consistent, where it is brittle, and where it breaks under pressure. Execution Mechanics ensures leaders understand the organization's true capacity to deliver, adapt, and sustain momentum across complex transformation cycles.

The fifth Lens, **Cultural Compass**, explores the beliefs, behaviors, incentives, and informal power structures that shape how people respond to change. Culture is often the silent accelerator or the invisible brake. This Lens reveals alignment, resistance patterns, psychological safety, and leadership influence networks. It distinguishes cultural strengths from cultural liabilities, giving leaders a clear view of the human forces that determine whether transformation will be embraced or rejected.

The final Lens, **Moral Compass**, evaluates ethics, trust, transparency, and responsible AI principles. As organizations adopt intelligent systems, the stakes of ethical decision-making rise dramatically. This Lens ensures transformation is not only effective but also equitable, accountable, and aligned with long-term societal and organizational values. It clarifies where safeguards are strong, where risks exist, and where governance must evolve.

Together, these Six Lenses form a complete, repeatable, consulting-grade architectural framework. The Ai6x6 Transformation Blueprint™ is not a checklist or maturity model; it is a structural intelligence system that moves leaders from assumption to understanding, from understanding to design, and from design to confident transformation planning. It gives leaders the conviction to move forward — not because the future is predictable, but because their decisions are grounded in multidimensional truth, structural clarity, and a deep understanding of how transformation behaves across the enterprise.

Ai6x6™ Transformation Blueprint

Understand Pillars & Lenses

"We don't see things as they are, we see them as we are." — Anaïs Nin

The **Ai6x6 TRANSFORMATION BLUEPRINT** is fundamentally structured around two core concepts: **Pillars and Lenses**. Together, they create a dual-axis architecture that brings both breadth and depth to enterprise transformation. The Pillars provide the WIDTH of the framework, defining the major domains where transformation must take root for the organization to evolve coherently. These domains—Customer, Product, Operations, Data, Workforce, and Leadership—represent the full span of enterprise behavior. Each Pillar receives its own dedicated book, allowing leaders to explore the complexities, tensions, and opportunities of that domain without dilution or compromise.

This ensures that transformation is not treated as a single initiative but as a system of interconnected disciplines that must advance in parallel. Complementing this horizontal structure is the DEPTH created by the Six Lenses, the universal imperatives that cut vertically through every Pillar. These Lenses represent the non-negotiable dimensions of modern transformation—forces that must be activated in every domain, regardless of industry, maturity, or strategic ambition. The six Lenses are: Human–AI Synergy (L1), Business Maturity (L2), Digital Evolution (L3), Execution Mechanics (L4), Cultural Compass (L5), and Moral Compass (L6). Each Lens exposes a different layer of organizational capability, ensuring that no domain is approached superficially, and no critical dimension is overlooked. The goal of the blueprint is to be holistic and very comprehensive so that executives are not blind-sighted by a key aspects being overlooked. Since a lot of ground has to be covered, the books may look very dense, but I have tried my best to keep it brief & engaging.

Transformation has become one of the most OVERUSED words in modern leadership, yet one of the least understood. Organizations declare it, teams attempt it, consultants package it—but few enterprises ever truly achieve it. Not because they lack ambition or intelligence, but because they approach transformation as a sequence of disconnected initiatives rather than a shift in how the organization thinks, behaves, and evolves. I created the **Ai6x6 Transformation Blueprint** because I have seen the same pattern repeat across industries, functions, and eras of technology: transformation fails not at the point of strategy, but at the point of structure. It fails when organizations modernize technology but ignore culture. It fails when they redesign processes but overlook trust. It fails when they invest in AI but fail to prepare people. It fails when they build strategy but cannot execute it. It fails when leaders underestimate the complexity of aligning human capability with machine intelligence.

The Ai6x6 Transformation Blueprint exists to solve this fragmentation. It brings together **six foundational lenses** and **six transformation pillars** into a single, coherent architecture. Each lens represents a dimension of capability that every organization must master. Each pillar represents a domain— Leadership, Workforce, Data, Customer, Product, Operations —where transformation must take root. Together, they form a 6×6 matrix that reveals not only what must change, but how change becomes sustainable, repeatable, and scalable.

This blueprint is intentionally multidimensional. Transformation is not linear. It is not sequential. It is not a checklist. It is a living system—human, technical, cultural, operational, ethical—moving together or not at all. The Ai6x6 model acknowledges this complexity and gives leaders a structure strong enough to hold it.

- It is called a blueprint because it is both **architectural** and **actionable**.
- This blueprint is not a methodology per se. It is not a maturity model in the traditional sense. It is not a checklist as you commonly see it. It is a **HOLISTIC** way of seeing things that are often overlooked.
- It is called Ai6x6 because it integrates **AI-driven intelligence** with **six human-centered disciplines**, applied across **six transformation pillars**.
- It is a system designed for organizations that want to evolve with intention rather than react with urgency.

The Six Lenses emerged from a simple but unavoidable truth: "**no single discipline can carry transformation on its own"**. Not AI. Not culture. Not governance. Not execution. Not digital modernization. Not ethics. Each is necessary, but none is sufficient. Real transformation requires all six lenses working together—reinforcing one another, correcting one another, and elevating one another.

It reveals the forces that shape organizational behavior, the tensions that undermine progress, and the capabilities that determine whether transformation becomes sustainable or collapses under its own ambition. It provides leaders with a structural map—one that makes complexity navigable and makes evolution repeatable.

Each lens represents a dimension of organizational life:

1. Human–AI Synergy: how intelligence is created, shared, and amplified.
2. Business Maturity: how discipline, readiness, and structure turn ambition into capability.
3. Digital Evolution: how technology becomes fluid, integrated, and trustworthy.
4. Execution Mechanics: how strategy becomes action and action becomes advantage.
5. Cultural Compass: how beliefs, behaviors, and meaning shape the human system.
6. Moral Compass: how ethics, transparency, and responsibility anchor everything else.

Together, they form a complete operating philosophy for modern transformation—one that is as human as it is technical, as strategic as it is operational, as aspirational as it is grounded. The Ai6x6 Transformation Blueprint is not about predicting the future. It is about preparing organizations to shape it. It is about giving leaders a language for what they instinctively know but rarely articulate. It is about building enterprises that can evolve continuously, coherently, and confidently. Most of all, it is about honoring the truth that **transformation is not something organizations *do*—it is something they *become*.**

The Pillars of the **Ai6x6 Transformation Blueprint** define the WIDTH of enterprise evolution. They represent the six domains where transformation must become real, measurable, and operationally grounded. While the Lenses provide the depth of capability, the Pillars provide the breadth of application—ensuring that transformation is not isolated to a single function but becomes a structural shift across the entire enterprise. Each Pillar receives its own dedicated book, allowing leaders to confront the full complexity of that domain with precision and without compromise.

The Leadership Pillar sits at the top of the architecture because leadership is the force that shapes every other domain. It defines how decisions are made, how priorities are set, how accountability is distributed, and how meaning is communicated. This Pillar examines the evolution of leadership in an era defined by AI, acceleration, and ambiguity. It challenges leaders to move from managing activities to designing systems, from directing work to enabling capability, and from protecting the past to shaping the future. Leadership is not a separate function—it is the integrator of the entire Blueprint.

The Workforce Pillar focuses on the human system that powers the enterprise. It explores how roles evolve, how teams collaborate, how skills are developed, and how people experience their work. In a world where AI augments every function, the Workforce Pillar examines how organizations support human capability, build adaptive talent ecosystems, and create environments where people feel empowered rather than displaced. This Pillar forces leaders to confront the truth that transformation succeeds or fails based on whether people feel equipped, supported, and connected to the mission.

The Data Pillar establishes the informational foundation of the enterprise. It examines how data is collected, governed, interpreted, and transformed into intelligence. In an AI-driven world, data is no longer a byproduct of operations—it is the raw material of decision-making, personalization, automation, and innovation. This Pillar explores how organizations build trustworthy data ecosystems, eliminate silos, and ensure that intelligence flows across the enterprise. It challenges leaders to treat data not as a technical asset but as a strategic capability that shapes every other Pillar.

The Customer Pillar anchors the organization in the lived experience of the people it serves. It forces leaders to confront the reality that customer expectations evolve faster than internal processes, and that loyalty is earned through coherence, trust, and emotional resonance. This Pillar examines how organizations understand, anticipate, and respond to customer needs—not as a marketing exercise, but as a cultural and operational discipline. It challenges leaders to build enterprises where customer experience is not a department, but a shared responsibility embedded in every decision.

The Product Pillar focuses on how organizations design, build, and evolve the value they deliver. Products are no longer static offerings—they are dynamic systems of intelligence, behavior, and continuous adaptation. This Pillar explores how product strategy, product design, and product operations must work together to create experiences that feel alive and responsive. It examines how AI reshapes product behavior, how digital ecosystems redefine differentiation, and how organizations must shift from feature delivery to continuous value creation.

The Operations Pillar addresses the engine room of the enterprise—the systems, workflows, and processes that determine whether strategy becomes reality. It examines how organizations orchestrate complexity, manage risk, and maintain resilience in environments defined by acceleration and uncertainty. This Pillar forces leaders to recognize that operational excellence is not about efficiency alone; it is about adaptability, intelligence, and the ability to respond to change without losing coherence. Operations is where transformation becomes tangible.

Together, these six Pillars form the structural WIDTH of the Ai6x6 Transformation Blueprint. They define the terrain on which the Six Lenses operate, ensuring that transformation is comprehensive, coherent, and enterprise-wide.

4
Earned relationships
CUSTOMER
"Who we serve?"
5
Brand Experience
PRODUCT
"What we serve?"
6
Engine room of service
OPERATIONS
"How we serve?"
2
Human Capability
WORKFORCE
"Human Assets"
3
Information foundation
DATA
"Insights"
1
Vision, mission & agenda
LEADERSHIP
"C-Suite"
Each Pillar = "Width" of the Transformation

4
Discerning Diners
CUSTOMER
"The Patrons"
5
Food, Drinks, Ambiance
PRODUCT
"The Feast"
6
Ingredients, Purchase
OPERATIONS
"The Logistics"
2
Cooks, Stewards, Servers
WORKFORCE
"The other staff"
3
Menu, Recipe & Reviews
DATA
"The Menu"
1
Lead Culinary aspect
LEADERSHIP
"The Exec Chef"

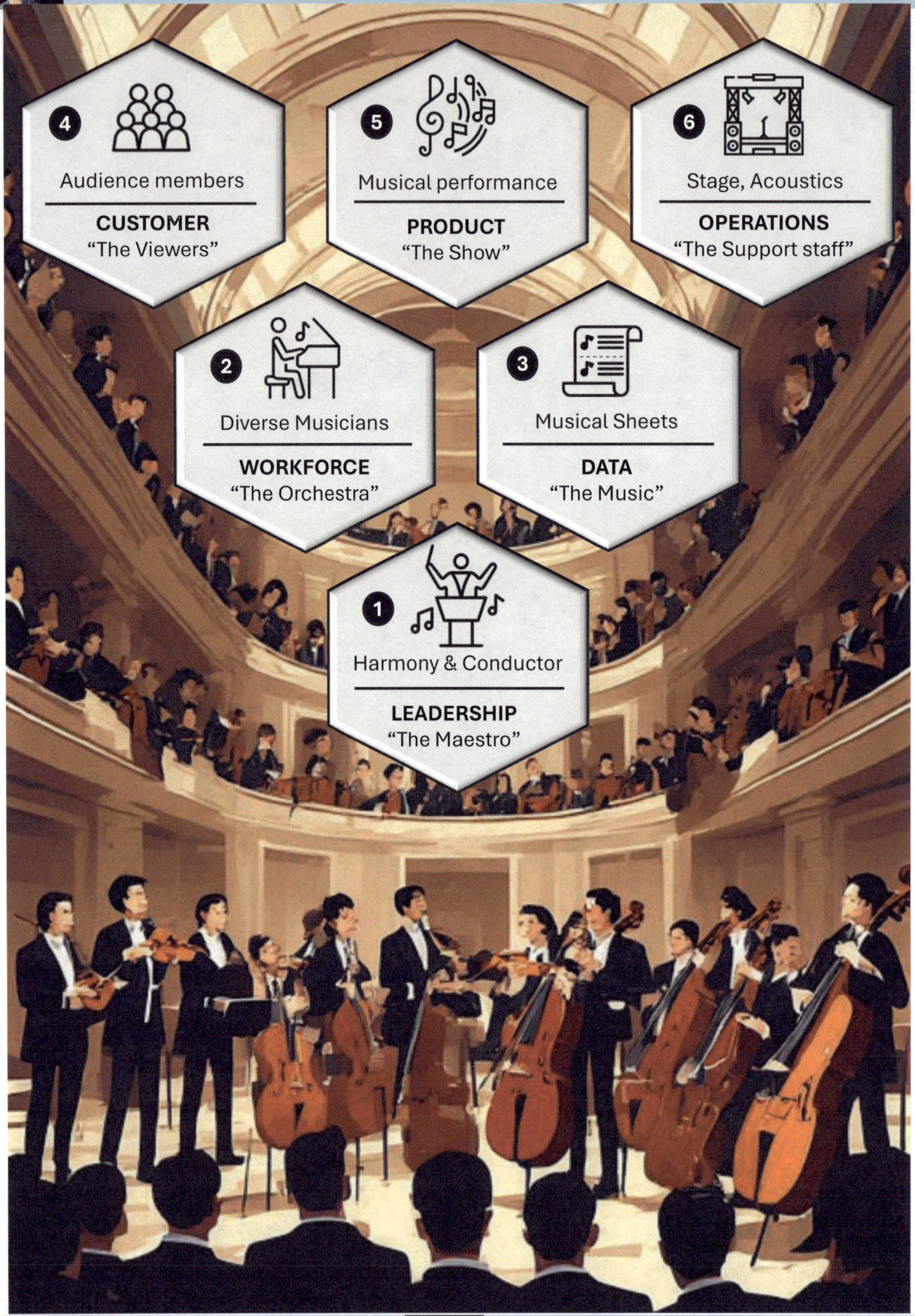
4
Audience members
CUSTOMER
"The Viewers"
5
Musical performance
PRODUCT
"The Show"
6
Stage, Acoustics
OPERATIONS
"The Support staff"
2
Diverse Musicians
WORKFORCE
"The Orchestra"
3
Musical Sheets
DATA
"The Music"
1
Harmony & Conductor
LEADERSHIP
"The Maestro"

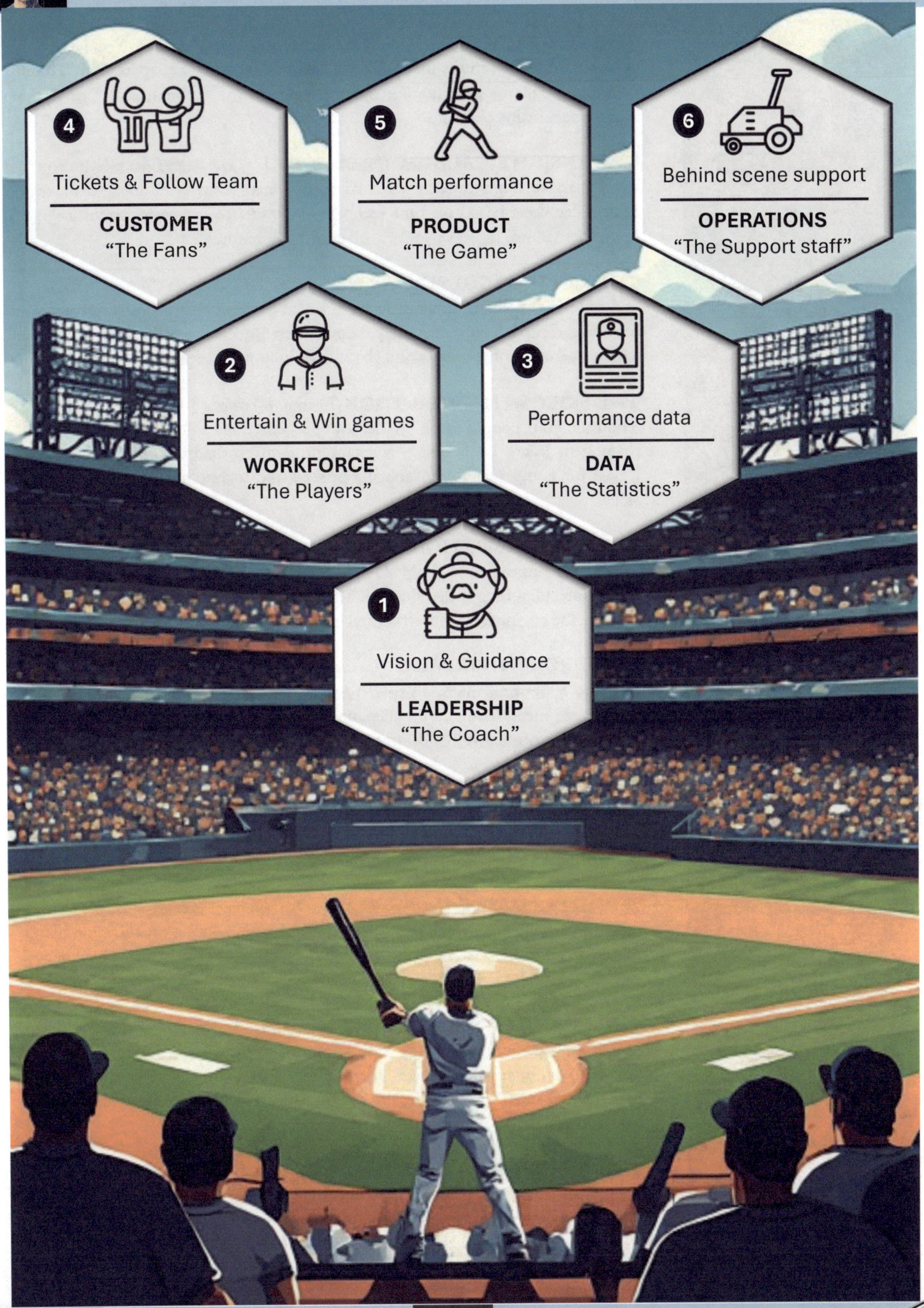

4
Tickets & Follow Team
CUSTOMER
"The Fans"
5
Match performance
PRODUCT
"The Game"
6
Behind scene support
OPERATIONS
"The Support staff"
2
Entertain & Win games
WORKFORCE
"The Players"
3
Performance data
DATA
"The Statistics"
1
Vision & Guidance
LEADERSHIP
"The Coach"

THE LITERACY GAP (*Invested: Product | Overlooked: Workforce*): You launch genius AI-driven features that your staff cannot explain, sell, or support. The technology fails to gain market traction because the "human layer" of the enterprise remains obsolete.

THE INSIGHT VACUUM (*Invested: Data | Overlooked: Leadership*): You build massive, high-integrity data lakes, but leadership continues to make critical decisions based on "gut feel." The investment is wasted because the upper-tier culture refuses to adopt a data-driven mindset.

THE COLD MACHINE (*Invested: Operations | Overlooked: Customer*): Back-end efficiency reaches an all-time high, but the actual user journey becomes robotic and rigid. You successfully reduce operational costs but lose the brand loyalty and emotional connection that drives long-term revenue.

THE SKILL WITHOUT A TOOL (*Invested: Workforce | Overlooked: Product*): You train your people to be "AI-Ready" and culturally agile, but you provide them with legacy tools and a stagnant product roadmap. Your best talent leaves for competitors who actually possess the technology to match their skills.

THE HALLUCINATION (*Invested: Leadership | Overlooked: Operations*): Executives announce a bold "Digital First" vision, but the underlying operational plumbing is decades old. The strategy never leaves the boardroom because the engine lacks the structural integrity to turn the wheels.

THE BROKEN PROMISE (*Invested: Customer | Overlooked: Data*): You market a hyper-personalized experience, but your internal data silos remain a fragmented mess. You end up alienating customers with irrelevant offers and inaccurate information that destroys their trust.

THE PROTOTYPE TRAP (*Invested: Product | Overlooked: Operations*): A brilliant AI-driven tool is developed in a vacuum, but the legacy supply chain and logistics cannot fulfill the demand it creates. The product succeeds in the lab but collapses the moment it hits the real-world operational floor.

THE DARK ASSET (*Invested: Data | Overlooked: Workforce*): You spend millions cleaning and structuring your data, but no one in the organization has the literacy to query or interpret it. The data sits idle and "dark," providing zero ROI because the people aren't equipped to find the value.

THE ROGUE ENGINE (*Invested: Operations | Overlooked: Leadership*): Middle management automates workflows to save time, but leadership has no visibility or governance over these new "black box" processes. The organization moves faster, but the steering wheel is no longer connected to the captain's chair.

THE INTERNAL ECHO (*Invested: Workforce | Overlooked: Customer*): Employees are empowered and "AI-literate," but they become obsessed with internal process improvements rather than user needs. The company becomes a high-performing bubble that is perfectly optimized for a customer that no longer exists.

THE BLIND CAPTAIN (*Invested: Leadership | Overlooked: Data*): A visionary leader attempts to steer the company through a massive pivot based on bold assumptions. Because the data pillar was ignored, they navigate straight into a market reef that the "missing data" would have clearly identified.

THE FANCY WRAPPER (*Invested: Customer | Overlooked: Product*): Your marketing and customer service are world-class, but the actual product has not evolved to keep up with competitors. You win the initial sale through great branding, but the churn rate sky-rockets once the customer realizes the product is hollow.

THE ANALYTICS ISLAND (*Invested: Data | Overlooked: Operations*): The data team produces perfect, real-time reports on every aspect of the business. However, the operational workflows are so slow and manual that by the time the business acts on the data, the opportunity has already vanished.

THE ORPHAN PROJECT (*Invested: Product | Overlooked: Leadership*): Innovation teams build cutting-edge AI tools that solve real problems, but leadership never provides a permanent budget or strategic home for them. These "pilot projects" eventually wither and die because they lack the top-down air cover to survive.

THE FRUSTRATED EXPERT (*Invested: Workforce | Overlooked: Operations*): You hire the best AI researchers and data scientists, but you force them to work within manual, "broken" internal workflows. Your experts spend 90% of their time fighting the system rather than building the future.

THE IVORY TOWER (*Invested: Leadership | Overlooked: Customer*): The board develops a strategy based purely on shareholder value and financial engineering. By ignoring the Customer pillar, they fail to see a massive shift in user sentiment that makes their entire three-year plan irrelevant.

THE HUMAN BOTTLENECK (*Invested: Operations | Overlooked: Workforce*): You automate 80% of your workflows, but the remaining 20% require human intervention that your staff isn't trained for. The entire system grinds to a halt because you optimized the machine but forgot to upskill the operator.

THE REACTIVE PANIC (*Invested: Customer | Overlooked: Leadership*): The company chases every new customer trend and social media "vibe" to stay relevant. Without a strategic anchor from Leadership, the organization becomes a fragmented mess of contradictory ideas that stand for nothing.

THE WASTEFUL HARVEST (*Invested: Data | Overlooked: Product*): You collect trillions of data points and pay massive cloud storage fees, but none of that data is ever used to iterate or improve the actual product. You are "Data Rich" but "Value Poor," hoarding information with no path to monetization.

THE GRASSROOTS REBELLION (*Invested: Workforce | Overlooked: Leadership*): Staff begins using "Shadow AI" and unsanctioned tools because leadership is too slow or fearful to provide guidance. You have an AI-ready workforce that is operating outside of the company's governance, creating a massive security and legal risk.

If the Pillars define the WIDTH of enterprise transformation, the **Six Lenses** define its DEPTH. They are the universal imperatives that cut vertically through every domain—Leadership, Workforce, Data, Customer, Product, and Operations. Each Lens exposes a different dimension of capability, ensuring that transformation is not superficial, fragmented, or episodic, but deeply rooted in the systems that shape how organizations think, behave, and evolve. These Lenses are not optional. They are not enhancements. They are the structural forces that determine whether transformation becomes sustainable or collapses under its own ambition. Each Lens reveals a layer of truth about the enterprise—its intelligence, its discipline, its technology, its execution, its culture, and its integrity.

The Human–AI Synergy Lens (L1) examines how intelligence is created, shared, and amplified across the enterprise. It explores the evolving relationship between human judgment and machine capability, ensuring that AI augments rather than replaces human potential. This Lens forces leaders to confront how decisions are made, how insights flow, and how people interact with intelligent systems. It challenges organizations to build environments where humans and AI operate as partners—each elevating the other. Human–AI Synergy is the starting point because it defines the cognitive foundation of modern transformation.

The Business Maturity Lens (L2) establishes the structural discipline required for transformation to take hold. It examines governance, alignment, prioritization, funding, risk management, and organizational readiness. This Lens reveals whether the enterprise has the scaffolding to support change or whether ambition is outpacing capability. Business Maturity is where leaders confront the operational truth: transformation is not powered by inspiration alone—it is powered by systems, discipline, and repeatable mechanisms that ensure consistency across the enterprise.

The Digital Evolution Lens (L3) focuses on the technological backbone of the organization. It examines how systems integrate, how data flows, how platforms scale, and how digital environments adapt to new demands. This Lens ensures that technology becomes an enabler rather than a bottleneck. It challenges organizations to build digital ecosystems that are fluid, intelligent, secure, and trustworthy. Digital Evolution is not about tools—it is about creating a technological environment capable of supporting continuous change.

The Execution Mechanics Lens (L4) converts strategy into action. It examines the methodologies, operating models, workflows, and performance systems that determine whether the organization can deliver consistently and at scale. This Lens exposes the gap between what leaders intend and what teams experience. It ensures that execution is not left to improvisation but is engineered with precision. Execution Mechanics is where transformation becomes tangible—where ideas become outcomes and outcomes become advantage.

The Cultural Compass Lens (L5) explores the beliefs, behaviors, and shared meaning that shape how people think and collaborate. It examines psychological safety, communication norms, leadership behaviors, and the emotional climate of the enterprise. This Lens reveals whether the culture accelerates transformation or resists it. Cultural Compass ensures that people feel connected, supported, and aligned—because no transformation can succeed in a culture that does not want it.

The **Moral Compass Lens (L6)** anchors the Blueprint in ethics and transparency. Moving beyond a static matrix, it defines the **True North** for AI governance, data privacy, and leadership accountability. This Lens ensures transformation is principled, not just efficient. In the AI era, a Moral Compass is the boundary condition for legitimacy—without it, innovation loses its direction; with it, trust becomes an unbreakable competitive advantage.

Together, these Six Lenses form the DEPTH of the Ai6x6 Transformation Blueprint. They ensure that every Pillar—Leadership, Workforce, Data, Customer, Product, and Operations—is approached with rigor, coherence, and integrity. They transform transformation from a collection of projects into a complete operating philosophy.

SIX LENSES OF TRANSFORMATION : HUMAN BODY ANALOGY

SIX LENSES OF TRANSFORMATION : MOTOR VEHICLE ANALOGY

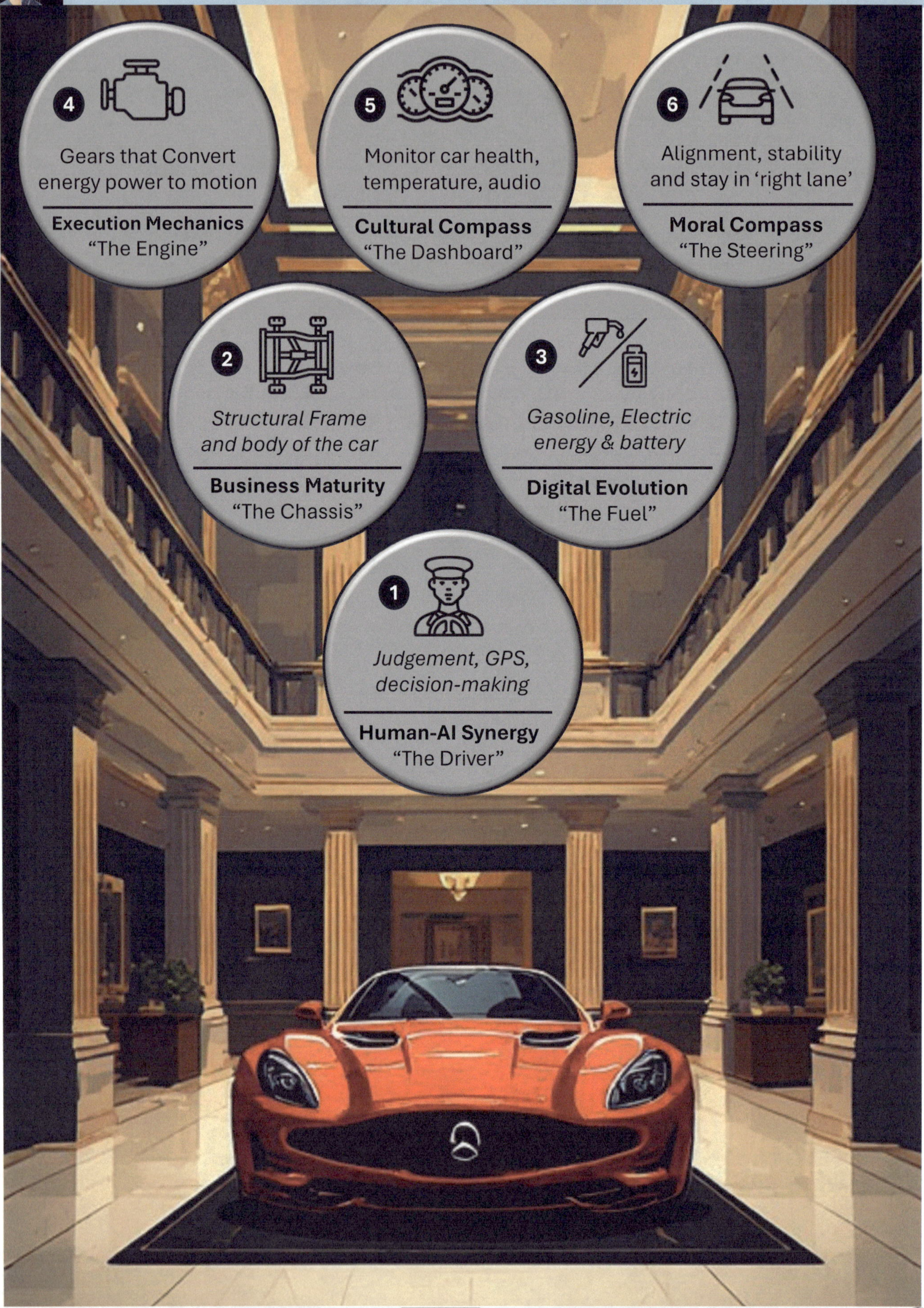

THE ETHICAL IMPLOSION (*Invested: Digital evolution | Overlooked: Moral compass*): You build a high-speed AI infrastructure with zero ethical guardrails or governance. One biased algorithm creates a public relations nightmare and a massive legal liability that can bankrupt the firm.

THE BUREAUCRATIC FREEZE (*Invested: Business maturity | Overlooked: Cultural compass*): You have established perfect governance and checklists, but you are operating within a terrified culture. People follow the rigid "process" of transformation but stop taking risks, effectively killing all real innovation.

THE THEORETICAL DREAM (*Invested: Human–AI synergy | Overlooked: Execution mechanics*): The organizational "brain" develops brilliant ideas for AI collaboration, but there is no "muscle" to act on them. Innovation stays trapped as a pilot project and never scales to produce enterprise value.

THE AUTOMATED ERROR (*Invested: Execution mechanics | Overlooked: Human–AI synergy*): You build a powerful execution engine but ignore the necessity of human judgment and intervention. The system begins executing "bad" decisions or flawed logic at 100x speed without anyone knowing how to stop it.

THE FUNDED HOBBY (*Invested: Cultural compass | Overlooked: Business maturity*): The internal culture is happy, safe, and collaborative, but there is no structural discipline or funding prioritization. Massive amounts of energy and capital are wasted on uncoordinated "passion projects" that never deliver ROI.

THE PRINCIPLED STANDSTILL (*Invested: Moral compass | Overlooked: Digital evolution*): You have world-class ethics and deep integrity, but you lack the "blood flow" of modern data and systems. You remain a "good" organization, but your technology is too outdated to actually compete with faster rivals.

THE FERRARI IN THE GARAGE (*Invested: Digital evolution | Overlooked: Execution mechanics*): You invest millions in an elite, high-performance tech stack that stays idle because no one has built the workflows to use it. You have purchased the "Engine" but never bothered to build the "Transmission" to turn the wheels.

THE LEGALISTIC LOOPHOLE (*Invested: Business maturity | Overlooked: Moral compass*): You strictly follow every legal rule and regulation while completely violating the spirit of ethics and trust. Your "Governance" is perfect on paper, but your brand is dying because your customers feel exploited by your "technically legal" actions.

THE INTELLIGENT OUTCAST (*Invested: Human–AI synergy | Overlooked: Cultural compass*): You develop AI tools that are technically brilliant and offer high synergy, but the workforce finds them creepy or intrusive. Because you ignored the "Senses" of the culture, the tools are rejected by the very people they were meant to help.

THE EFFICIENT COLLAPSE (*Invested: Execution mechanics | Overlooked: Business maturity*): The organization is moving with incredible power and speed, but it is moving toward a goal that isn't funded or strategically prioritized. You are winning the race, but you've realized too late that you're on the wrong track entirely.

THE NICE LIAR (*Invested: Cultural compass | Overlooked: Moral compass*): The culture is polite, collaborative, and highly "engaged," but it is built on a foundation of hidden ethical skeletons. People are "happy" because they are protecting each other from the truth, leading to an eventual systemic implosion.

THE ANALOG SAINT (*Invested: Moral compass | Overlooked: Human–AI synergy*): The leadership has high integrity but lacks the "AI-IQ" to understand the risks of the modern digital landscape. They are making ethical decisions based on a world that no longer exists, leaving them vulnerable to new forms of disruption.

THE EXPENSIVE CHAOS (*Invested: Digital evolution | Overlooked: Business maturity*): You layer high-tech AI tools on top of disorganized, immature governance and disorganized funding. The tech doesn't solve your problems; it simply makes your existing chaos move much faster and cost much more.

THE STAGNANT FORTRESS (*Invested: Business maturity | Overlooked: Execution mechanics*): The organization has perfect defenses, massive funding reserves, and rigid discipline, but zero kinetic movement. You are a perfectly "safe" business that is being slowly surrounded and suffocated by smaller, faster competitors.

THE DR. FRANKENSTEIN (*Invested: Human–AI synergy | Overlooked: Moral compass*): You achieve brilliant "Brain-to-Machine" innovation that pushes the boundaries of what is possible. However, because you ignored the "Inner Ear" of balance and ethics, you create a system that causes more social harm than organizational good.

THE BURNOUT FACTORY (*Invested: Execution mechanics | Overlooked: Cultural compass*): You optimize your "Muscle" for maximum output but completely ignore your "Senses" and emotional climate. You fail to detect internal fatigue and resentment until your highest-performing talent quits en masse.

THE LUDDITE OASIS (*Invested: Cultural compass | Overlooked: Digital evolution*): You have created the best place in the world to work, with high safety and great vibes. However, because you ignored the "Blood Flow" of digital evolution, your company is being disrupted out of existence by competitors you can't even see.

THE TOOTHLESS POLICY (*Invested: Moral compass | Overlooked: Execution mechanics*): You have beautiful "Values" posters and world-class ethics policies, but no actual workflows or mechanics to enforce them. Your integrity is a theory that falls apart the moment a team is pressured to hit a quarterly target.

THE GHOST SHIP (*Invested: Digital evolution | Overlooked: Human–AI synergy*): You have successfully implemented full automation and "lights out" data systems across the enterprise. However, because there is no human judgment in the loop, the system continues to run on outdated logic that is driving the company into a ditch.

THE PRECISE DINOSAUR (*Invested: Business maturity | Overlooked: Human–AI synergy*): You have spent years optimizing and documenting every single process to a level of "Six Sigma" perfection. Unfortunately, you ignored the "Brain" of AI synergy, and those processes are now perfectly optimized for a market that has already moved on.

Human judgment amplified by AI

Human–AI Synergy defines how intelligence is created, shared, and amplified across the enterprise. It examines the evolving partnership between human judgment and machine capability, ensuring that AI elevates rather than replaces human potential. This Lens forces leaders to confront how decisions are made, how insights flow, and how people interact with intelligent systems. It exposes whether the organization is building a truly augmented workforce or simply layering tools on top of old thinking. Ignoring this Lens leads to fragmented automation, poor adoption, and inconsistent decision-making. Human–AI Synergy is essential because it establishes the cognitive foundation of transformation—how the enterprise thinks, learns, and adapts—and without it, AI becomes a disconnected tool rather than a strategic multiplier.

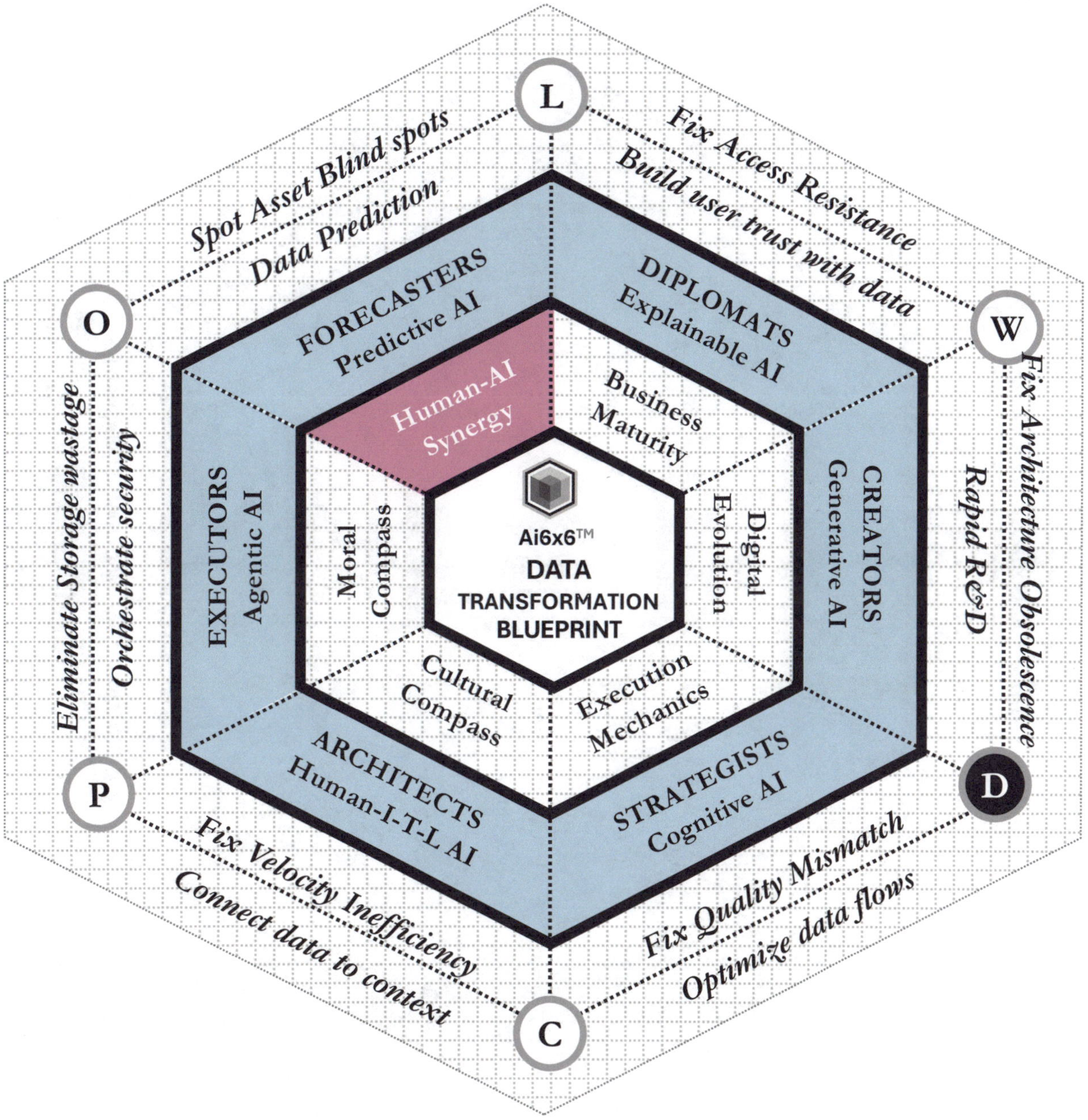

Ai6x6 HUMAN-AI SYNERGY FRAMEWORK FOR **LEADERSHIP LIFECYCLE**

Discipline that stabilizes transformation

Business Maturity is the structural discipline that transforms raw ambition into a legacy of Authority. It begins at the Foundation, establishing the governance and funding models that anchor an organization's identity before activation begins. As the enterprise moves through Initiation and Cohesion, this Lens evaluates whether the internal scaffolding is robust enough to support the increasing velocity of the transformation engine. This Lens exposes a fundamental truth: sustained success is engineered through systems, not slogans. It ensures that as Momentum builds, the enterprise maintains the Optimization necessary to prevent collapse under the weight of its own initiatives. Mastering this Lens allows an organization to transcend temporary programs and achieve the permanent market Authority that defines industry leadership.

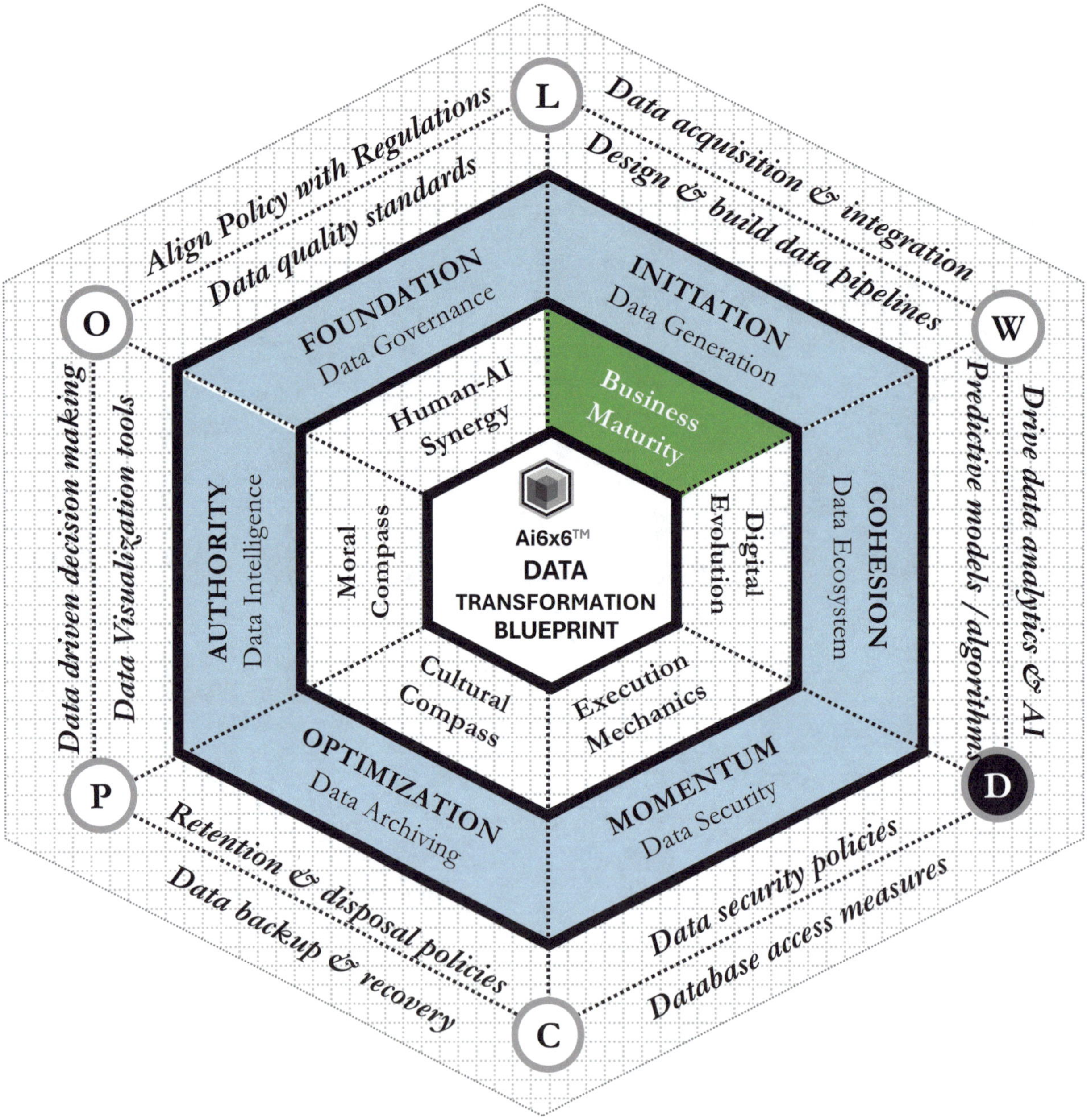

Ai6x6 BUSINESS MATURITY FRAMEWORK FOR **LEADERSHIP LIFECYCLE**

Infrastructure built for continuous change

Digital Evolution examines the technological backbone that enables modern transformation. It evaluates system integration, data flow, platform scalability, and the adaptability of digital environments. This Lens ensures technology accelerates rather than obstructs progress. It forces organizations to confront technical debt, architectural fragility, and the true readiness of their digital ecosystem. When ignored, enterprises build brittle systems, slow innovation, and create environments where strategy moves faster than infrastructure. Digital Evolution is not about tools—it is about constructing a digital foundation capable of continuous change. It ensures that technology becomes an enabler of intelligence, speed, and resilience. Without this Lens, even the strongest strategies fail because the underlying environment cannot support enterprise-wide evolution.

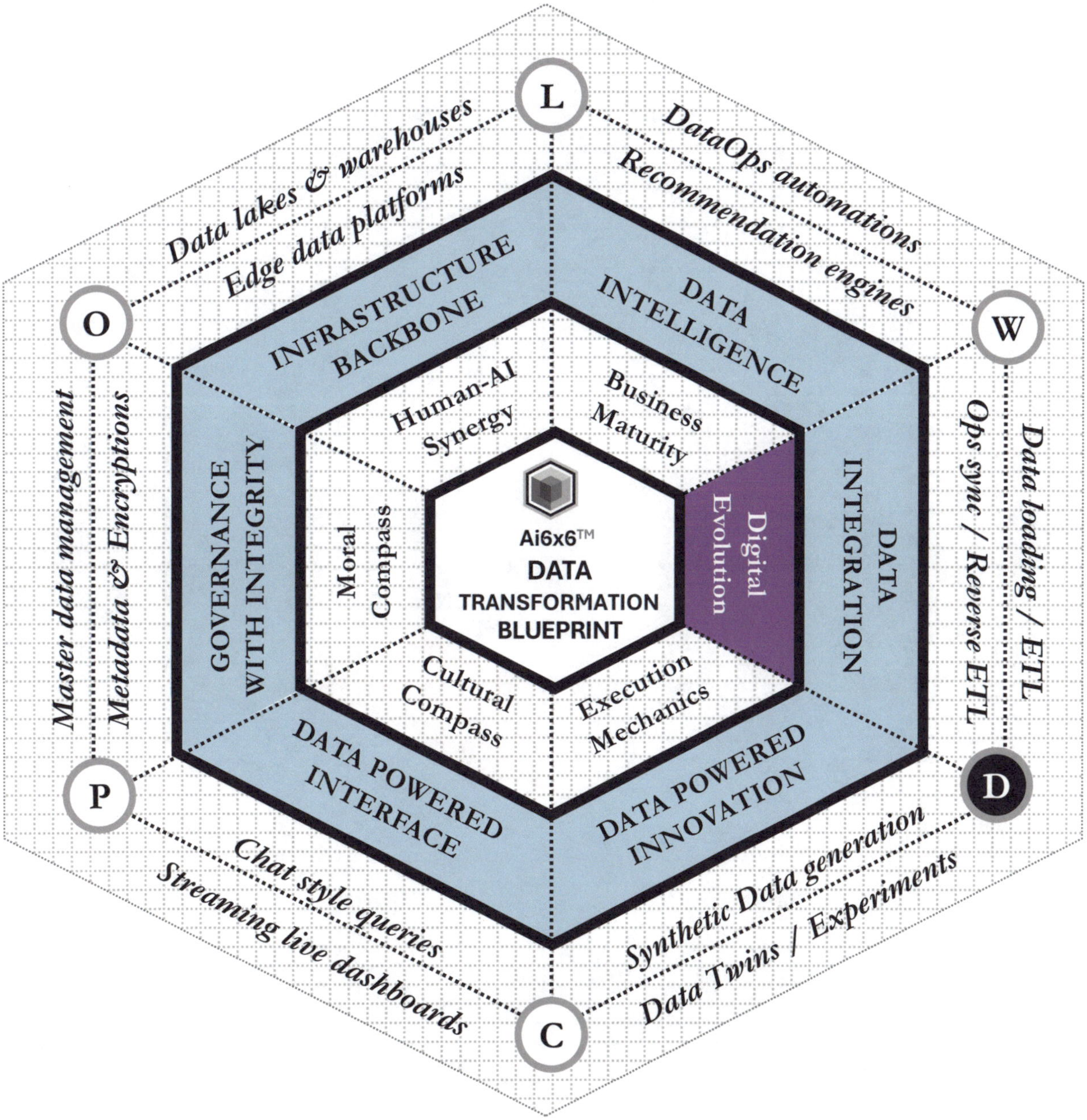

Ai6x6 DIGITAL EVOLUTION FRAMEWORK FOR **LEADERSHIP LIFECYCLE**

Strategy delivered with operational precision

Execution Mechanics converts strategy into consistent, scalable action. It examines operating models, workflows, methodologies, and performance systems to determine whether the organization can deliver reliably. This Lens exposes the gap between leadership intent and frontline reality, revealing whether execution is engineered or improvised. It forces leaders to confront bottlenecks, handoff failures, and the true velocity of delivery. Ignoring this Lens results in missed deadlines, inconsistent outcomes, and organizational frustration. Execution Mechanics is essential because it transforms ideas into outcomes and outcomes into advantage. It ensures that transformation is not theoretical but operational—a system of delivery that produces measurable impact at scale.

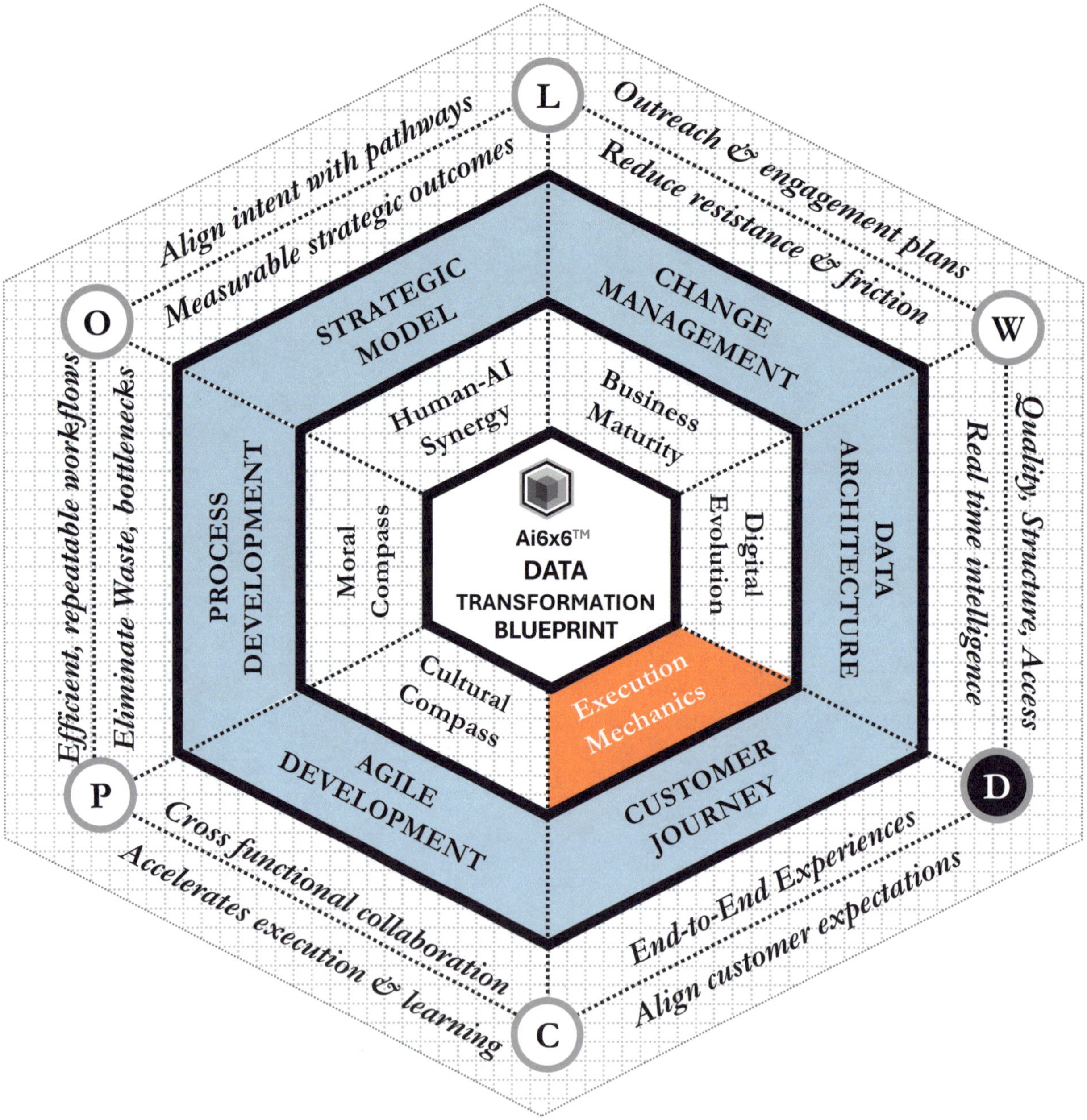

Ai6x6 EXECUTION MECHANICS FRAMEWORK FOR **LEADERSHIP LIFECYCLE**

Culture aligned to accelerate change

Cultural Compass explores the beliefs, behaviors, and shared meaning that shape how people think, collaborate, and respond to change. It examines psychological safety, communication norms, leadership behaviors, and emotional climate. This Lens reveals whether culture accelerates transformation or quietly resists it. It forces leaders to confront the invisible forces that determine engagement, trust, and momentum. Ignoring this Lens means ignoring the human system that ultimately determines success. Even the best strategies fail in cultures that are misaligned, fearful, or fatigued. Cultural Compass is essential because transformation is a human endeavor—no system or technology can succeed without cultural readiness. It ensures people are not just compliant but committed, creating the conditions for lasting change.

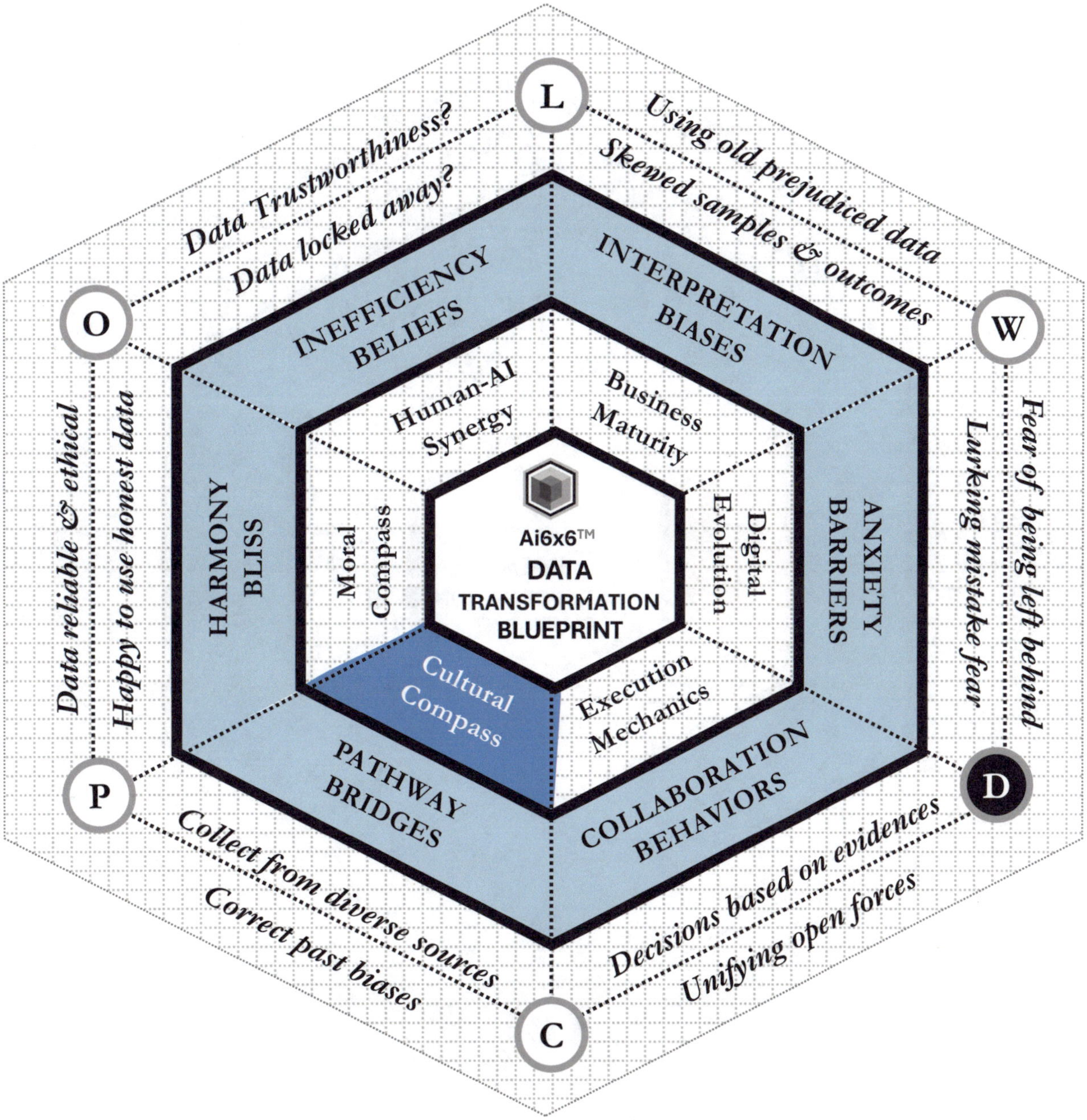

Ai6x6 CULTURAL COMPASS FRAMEWORK FOR **LEADERSHIP LIFECYCLE**

Ethics governing intelligent enterprise systems

The **Moral Compass** anchors transformation in ethics, transparency, and responsibility. It evaluates data privacy, AI fairness, accountability structures, and the moral obligations of leadership. This Lens ensures that innovation is principled and that stakeholders feel protected, respected, and informed. It forces leaders to confront the ethical implications of their decisions and the long-term consequences of their systems. Ignoring this Lens exposes organizations to reputational damage, regulatory risk, employee resistance, and customer disengagement. Trust is the boundary condition of modern enterprise—without it, nothing else holds. The Moral Compass is essential because it defines the legitimacy of every decision, system, and outcome, ensuring transformation is not only effective but worthy of confidence.

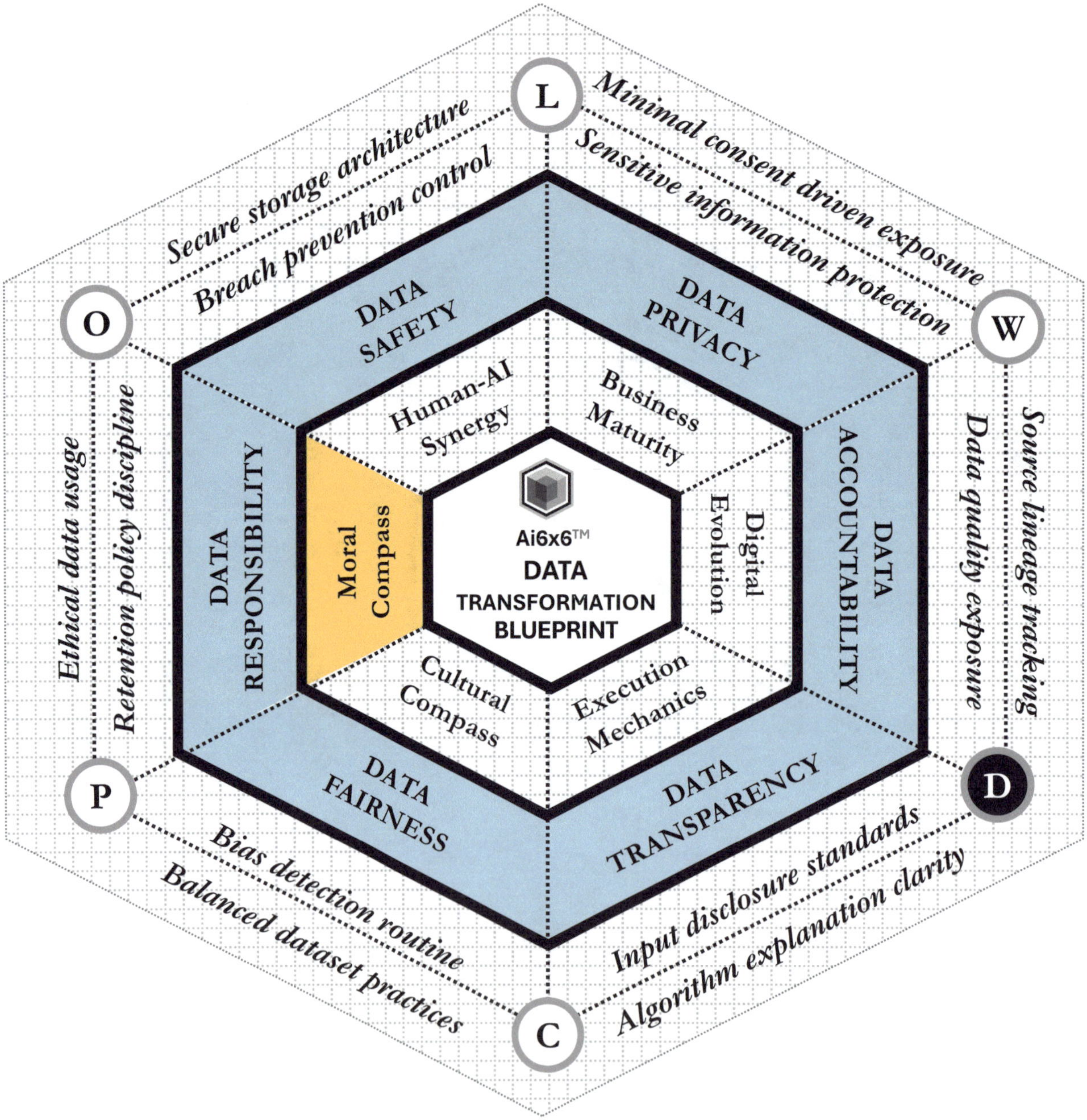

Ai6x6 MORAL COMPASS FRAMEWORK FOR **DATA LIFECYCLE**

Ai6x6™ Transformation Blueprint

Why Transform Data?

"Without data, you're just another person with an opinion." —W. Edwards Deming

W. Edwards Deming was an American statistician, professor, and author known for his work in quality management and continuous improvement.

In an era of intense competition, relying on gut feelings and outdated reports is a recipe for failure. Data transformation is essential because it moves an organization from a reactive to a proactive state. Without it, companies are unable to accurately understand customer behavior, optimize operations, or identify new market opportunities. This change is needed to create a culture where data is the foundation of every decision, providing the real-time insights required to drive sustainable growth and stay competitive. The benefit of this transformation is smarter, more informed decision-making and a stronger ability to identify new opportunities.

Why need Data Transformation?

1. **Establishment of a Unified Source of Truth (USoT):** This foundational step eliminates conflicting reports and organizational debates by providing a single, authoritative, and agreed-upon data model for all core business entities (e.g., Customer, Product, Transaction). The USoT serves as the single foundation of trust, enabling executive management to spend less time reconciling numbers and more time acting on insights. Every department, from finance to marketing, operates with total confidence in the underlying metrics.
2. **Improved Data Quality and Integrity:** Automated quality checks, lineage tracking, and consistent definitions enforced at the point of data creation drastically reduce the incidence of errors and the manual cleanup required downstream. High-integrity data is the necessary fuel for accurate machine learning models, ensuring automated decisions do not rely on flawed inputs. This shifts the data team's focus from cleaning to advanced analysis.
3. **Real-Time Decision Support:** Modern, streaming data architectures deliver insights with low latency, enabling operational systems to move beyond backward-looking reports to predictive and prescriptive action in live customer interactions. This immediacy allows the system to change its behavior dynamically (e.g., adjust pricing, flag fraud) in the critical moment of interaction, maximizing transaction value. The business operates on the present state, not historical averages.
4. **Enhanced Regulatory and Audit Compliance:** Automated data masking, access control, and lineage tracing simplify meeting complex regulations (e.g., CCPA, GDPR) and allow for the instantaneous generation of compliance audit reports. Proactive compliance eliminates the risk of substantial financial penalties and avoids the reputational damage associated with data privacy breaches. Compliance becomes a non-event, seamlessly managed by the system.
5. **Democratized Data Access:** Self-service data product portals, clear metadata, and secure APIs allow non-technical business users to securely access and analyze data for ad-hoc queries, accelerating insight generation across the board. Empowering thousands of business users with data access multiplies the organization's analytical capacity far beyond what a centralized data team could ever achieve. Insights are generated at the speed of business need.
6. **Accelerated AI/ML Model Deployment:** High-quality, standardized, and consistently delivered data provides the stable foundation necessary for rapidly training, deploying, and maintaining effective machine learning models that drive new business value. The speed of data preparation is often the biggest bottleneck in AI projects, and data transformation drastically reduces this time, accelerating ROI on AI investment. The AI lifecycle becomes reliable and repeatable.
7. **Monetization of Data Assets:** By packaging internal data assets as governed "data products," the organization can identify new revenue streams by selling or exchanging anonymized, high-value data insights with partners or external customers. This creates an entirely new line of business, transforming a cost center (data storage) into a profit center (data product sales). The organization leverages its proprietary data as a market advantage.
8. **Optimized Inventory and Supply Chain:** Real-time visibility into stock, demand signals, and logistics, coupled with predictive modeling, allows for ultra-efficient inventory management, reducing holding costs, obsolescence, and stockouts. The synchronization of production with real-time demand signals minimizes working capital tied up in inventory, freeing up cash flow. This JIT (Just-in-Time) accuracy provides a significant liquidity advantage.
9. **Reduced Data Redundancy Costs:** Consolidation of siloed data lakes, elimination of shadow IT reporting systems, and use of unified cloud infrastructure lower total data storage and processing costs across the enterprise. Streamlining the data architecture eliminates the unnecessary expense of maintaining multiple, conflicting, and redundant copies of the same information. Operational expenditure on cloud services is significantly optimized.
10. **Proactive Fraud and Anomaly Detection:** Real-time streaming data combined with advanced machine learning techniques enables the system to detect and flag fraudulent transactions or critical system anomalies moments after they occur, mitigating financial and security losses immediately. The early identification of suspicious patterns minimizes exposure to high-value losses, ensuring financial systems remain secure and trustworthy. Automated mitigation is far faster than human intervention.

1. From Gut to Ground Truth
The CEO Who Made Every Call by Instinct

Raj was sharp, confident—and wrong. For three quarters, he insisted customer churn wasn't a concern. "They always come back," he'd say. But a new data analyst built a simple cohort retention chart that told a different story: churn had quietly doubled in 18 months. Raj was stunned. "Why didn't we see this?" The answer? Their data was scattered, unstructured, and trapped in tribal knowledge. That moment catalyzed a transformation—centralized dashboards, cleaned customer metrics, weekly data standups. Within a quarter, retention efforts reversed the slide. Raj didn't stop trusting his gut—but he finally trusted the ground beneath it.

When decisions affect millions, intuition alone isn't enough. Transformation starts when data stops being a reference—and becomes the reflex.

2. Broken Dashboards, Broken Trust
When No One Believed the Numbers

When Finance said the quarter was up 12%, Marketing laughed. "That's impossible—we paused campaigns." Sales muttered, "Check your pipeline filter." And just like that, the dashboards were dismissed. Data was no longer insight—it was argument. The truth? Each team was using different fields, definitions, and refresh rates. Leaders realized they weren't misaligned—they were misinformed. A cross-functional task force standardized KPIs, metadata, and data lineage. Business terms were defined in plain English. And a single source of truth became the company's daily ritual. The dashboards didn't just get cleaner—they became conversation starters.

Because if people don't trust the numbers, they won't trust the strategy. And without a shared truth, alignment is just theater.

3. Growth at the Edges
When Analysts Had to Beg for Access

Sophie had a breakthrough: a pricing analysis that could unlock $2M in revenue—but the data she needed was stuck in a permissions vortex. Five emails. Three tickets. Two weeks later, she got access—and by then, the pricing window had closed. Leadership saw the pattern: their data wasn't democratized—it was barricaded. The solution wasn't just better tooling—it was a new operating model. Data ownership moved closer to product teams. Guardrails replaced gatekeepers. Within months, decisions that once took weeks were made in hours. Sophie wasn't just an analyst anymore—she was a catalyst.

Because when access is gated, value is delayed. But when data flows to the edge, innovation follows.

4. The Merger Meltdown
When Two Systems Couldn't Speak

When two legacy retailers merged, everyone celebrated—until the first quarterly sync. "How many customers do we have?" someone asked. One team said 1.2 million. The other said 960,000. Both were "right"—but the systems used different definitions of "active." The fallout was chaos: missed synergies, duplicative campaigns, mistrust. Finally, a data harmonization team unified schemas, cleaned duplicates, and mapped logic across platforms. The new executive dashboard pulled from both worlds—but with one voice. Post-merger revenue stabilized. Culture healed.

Because in the digital era, integration isn't just legal—it's linguistic. And transformation means turning fragmentation into fluency.

5. Compliance Close Call
When Data Governance Didn't Exist

During an external audit, a simple question—"Where is customer data stored?"—sent the IT team into a frenzy. No clear lineage. No classification. No documentation. What started as a routine review became a near-crisis. The company wasn't non-compliant yet—but they were flying blind. In the aftermath, they launched a governance program: data stewards were assigned, retention rules enforced, access logs audited. It wasn't glamorous—but it restored control. Within a year, they passed every audit with ease. More importantly, customers trusted them more.

In a world of rising regulation, data governance isn't bureaucracy—it's brand protection. And delay is liability.

6. AI That Lied
When Dirty Data Trained the Bots

The new AI assistant was supposed to reduce ticket volume by 40%. Instead, it hallucinated. It gave wrong refund windows, misclassified complaints, and offered "free shipping" on out-of-stock items. The model wasn't broken—it was trained on messy, mislabeled, outdated data. Customers were furious. So were support agents. The AI team paused deployment and partnered with domain experts to clean and relabel datasets. They built feedback loops into every ticket. Weeks later, accuracy soared—and trust returned. AI wasn't the problem. Dirty data was.

In the age of automation, transformation means realizing: your algorithms are only as smart as your inputs are sane.

7. Talent Blind spots
When HR Had No Insight into Its Own People

Sarah suspected turnover was rising—but couldn't prove it. The HRIS hadn't been updated in six months. Exit interview data was stuck in PDFs. Promotions were logged manually, inconsistently. Leadership asked for workforce insights, but all Sarah had were stories. That changed when People Ops embedded a data partner. They centralized people analytics—tracking sentiment, mobility, attrition, and equity. Within three months, the CHRO presented a full-fidelity talent map, leading to better policies and predictive interventions. Sarah didn't just have instincts—she had evidence.

Because culture isn't just what people feel—it's what you can measure. And people transformation requires people intelligence.

8. Product Feedback Black Hole
When User Data Was Ignored

The app reviews were brutal: "Glitchy," "slow," "unclear." But the roadmap kept marching—based on leadership hunches and gut design. Until one intern scraped thousands of support tickets and discovered a pattern: 80% of complaints were tied to just three UX decisions. Leadership halted the sprint, read every thread, and brought users into design reviews. The result wasn't just a better product—it was a shift in mindset. Customer feedback became a dataset, not a distraction. And the backlog finally reflected real needs. Because ignoring user signals isn't just inefficient—it's arrogant.

Transformation means listening at scale—and acting like it matters.

9. Forecasting Failure
When Spreadsheets Couldn't See the Storm

Every quarter, the forecasts were "fine"—until reality hit. Product returns surged. Margins shrank. Yet Finance kept presenting models built on 20 static variables, without any real-time indicators. Then a sudden supply disruption caught them flat-footed. The solution? They connected real-time sales, inventory, sentiment, and macro indicators into a dynamic forecasting engine. Variance dropped by 60%. The CFO stopped squinting at cells—and started spotting patterns. Because the future isn't built in Excel alone.

Data transformation means sensing before reacting. And spreadsheets—no matter how beautiful—can't see around corners.

10. ESG Without Proof
When Sustainability Was Just a Slide

The company's annual ESG report looked great—until an investor asked, "Can you show the data behind your carbon footprint claims?" Silence. The metrics were modeled, not measured. Leaders realized they weren't lying—but they weren't leading either. So they embedded IoT sensors, tracked water usage, digitized supply chain audits. Within a year, ESG moved from PR to platform. Sustainability dashboards became boardroom staples. And when customers asked about impact, the team answered with confidence—and data. In today's climate-conscious market, transformation means turning good intentions into verifiable intelligence. Because if you can't prove it, you can't claim it.

11. The Unscalable Analyst
When One Person Knew Everything

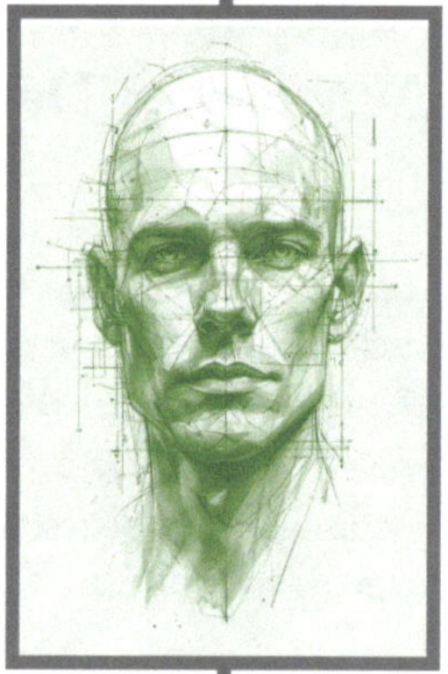

Carlos was a legend—the only one who knew how to fix the revenue model. Or pull the quarterly LTV churn cohort. Or troubleshoot the finance ETL. But when Carlos left for a sabbatical, everything stopped. The company hadn't invested in documentation, repeatable data products, or scalable platforms. Carlos was the system. After the chaos, leaders made a hard shift: every model became modular, pipelines were versioned, teams embraced data-as-code. Knowledge moved from heads to hubs. And when Carlos returned, he smiled: "You don't need me now." That was the compliment. Transformation isn't about one hero—it's about building a system that doesn't collapse when they're gone.

12. Value Hidden in Plain Sight
When No One Asked the Right Question

The new loyalty program seemed like a flop. Adoption was flat. "It's a failed feature," the CMO said. But Claire, a junior analyst, wondered: "Are we looking at the right customers?" She sliced the data differently—and uncovered that high-value Gen Z users were engaging heavily, but being ignored by campaigns. With that insight, marketing shifted its strategy. Engagement tripled. Revenue followed. What changed wasn't the product—it was the question. Claire's curiosity reminded the team: data isn't just to confirm—it's to discover. Transformation starts when teams stop fearing the unknown—and start getting curious.

Ai6x6™ Transformation Blueprint

Power of Diversified Insight

"Automation applied to an inefficient operation will magnify the inefficiency."

Bill Gates, Co-founder of Microsoft. This serves as a critical warning for business transformation, underscoring that the first step must be process optimization and system repair. Simply layering AI onto broken, inefficient legacy systems will only accelerate organizational chaos.

We have deliberately chosen to move beyond the limiting confines of the conventional single-expert **Foreword**, which, by its nature, risks restricting the reader to one singular point of view. In an era where **business transformation** is shaped by a confluence of disruptive forces, relying on a solitary oracle is fundamentally inadequate. Instead, this book opens with a **curated collection of 15 interview snippets** from a diverse array of global thought leaders. This method immediately immerses you in a rich, multi-faceted tapestry of opinions and real-world experiences. By foregrounding this **plurality of perspective**, we offer a wider, more nuanced, and dynamic lens through which to view the challenges and enormous opportunities presented by modern change. Consider these short, powerful excerpts a **Chorus of Counsel**: a comprehensive and inclusive foundation built on collective wisdom that better equips you for your own complex journey of enterprise innovation and transformation. This is a commitment to **diversified insight** from the very first page.

Gain unprecedented, insider access to the minds of 27 global thought leaders in this six-book series. Sourced from comprehensive 1-on-1 interviews, these experts are recognized worldwide for their work in business transformation and emerging technology. Their insights are drawn from my personal, professional network—including former bosses, colleagues, and business partners — giving you an insider's view into the minds of industry shifters. To maximize focus and utility, we've synthesized these deep-dive conversations into thematic 'interview cards.' Each of the six books offers 15 relevant, concentrated insights on specific topics, allowing you to quickly access and apply the most relevant expertise, no matter where it originated in the original interview.

EXPERT VOICES

Subrata Mukherjee: As we move into a new era driven by AI, the value of data has fundamentally changed. How does a company's data strategy need to evolve to prepare for the widespread adoption of generative AI?

Saleem Khan: I think creating more organic data moats will be key to winning in the future. AI will become as common as owning a laptop. Everyone from grammar school students to senior executives will have access to gen AI. The question is how do you differentiate on the fidelity of the answers AI will provide based on premium data modules. Remember gen AI is trained on the open web and not proprietary datasets. If you want a generic answer to how risky a shipping vessel may be for instance gen AI could do a go job looking up adverse media on the vessel and provide an answer but if you're an underwriter and are ensuring the hull and machinery of a boat you're going to want many more data points and a deeper assessment of the risk. Companies need to ensure they continue to create data moats that are impenetrable by gen AI.

SM: You've held roles as both a **Chief Data & Analytics Officer** and a **Chief Digital Product Officer**. How do you see the responsibilities of these two roles intersecting? Are they two sides of the same coin?

SK: I think the reality is that data is the product now. The role of the CDO and CPO are morphing into an amalgamation. C-Suite product executives need to have a deep understanding of the data. I almost think of data as being colors on an artist's palette that can be used to paint an amazing work of art or a mere sketch. It's important for a product executive to be able to stitch together the right data within the right contexts to solve specific customer problems. Without this full end-to-end knowledge of who your customer is and the type of data they need to answer questions or identify an insight you will not be successful in the end goal of increasing reach with customers.

SM: From your perspective, which of these two roles is more critical for a company's success in today's market?

SK: I don't think it's a matter of which role is more critical, it's a matter of the hard work being done to determine what the customer's actual needs are. This can be done by product, data, technology, or all of the above but someone needs to do it in conjunction with sales and marketing.

SM: How does your experience as a product leader influence your data strategy, and vice-versa?

SK: Product leadership and data strategy definitely have a symbiotic relationship for me. Having held both roles allows me to empathize with a customer but also get down to the brass tacks of a specific dataset. I think what I have learned in my journey is to ensure you do everything with a customer in mind. When I was an engineer, I thought like an engineer. Now I try to think in terms of stories or customer journeys. How will a dataset enable a customer journey? How can I habituate my customers to keep coming back to the well for the vital information they need? I think it takes a product and data mindset to be able to answer these questions.

SALEEM KHAN

Chief Data & Analytics Officer, Pole Star Global

Saleem is a seasoned data and analytics executive with a strong history of creating business value through data-driven innovation and successful company exits. In his current role as Chief Data & Analytics Officer at Pole Star Global, he leverages his deep expertise to lead and execute data strategy. His notable achievements include serving as Chief Digital Product Officer at VettaFi, which was successfully acquired by TMX Group, and as Chief Data & Analytics Officer at Discovery Data, which had a successful exit to Institutional Shareholder Services. He has also held leadership positions in data innovation and advanced analytics at major corporations like Dun & Bradstreet and S&P Global, and was a founder of the fintech startup ESGINE, an AI-enhanced platform for ESG compliance. He studied at Fordham University. He is based out of New York, United States.

EXPERT VOICES

Subrata Mukherjee: As the former **Chief Data Officer** at The Economist, you were a pioneer in using data for business transformation. How did your experience with data analytics and digital transformation at The Economist prepare you for your current focus on digital trust and intellectual property.

Stephane Pere: At The Economist, building a global data strategy meant balancing innovation with responsibility. Shifts in technology and consumer behavior often created uncertainty or were seen as constraints, but I viewed them as opportunities to reinvent how we operated and gain a competitive advantage. By embracing these changes, we were able to design new revenue streams from data while strengthening internal processes and decision-making. This dual challenge—innovation and protection—directly shaped my work today. With Evidency and Vaultinum, I apply the same principles: data & tech have value only if they are trusted and protected. My legal education also helped me connect the dots between technology, monetization, and intellectual property protection, turning perceived risks into lasting advantages.

SM: Based on your experience, what is the most profound digital transformation that has taken place in the media industry over the last decade?

SP: The most profound transformation has been the pivot from scale to depth. For decades, digital media chased clicks and advertising impressions. Over the past decade, the real shift has been toward building direct, trusted relationships with audiences through subscriptions, memberships, and niche communities. This is more than a business model change—it's a cultural shift that places trust, transparency, and quality at the core of media economics.

SM: How has your experience at a legacy company like The Economist influenced your approach to advising and leading agile start-ups?

SP: Working in a 180-year-old global institution taught me that even nimble teams need rigorous governance and brand identity—which are often overlooked in startups chasing speed. I encourage founders to build frameworks that allow agility without exposing them to wrongful manipulation or brand dilution.

SM: You have a long history in the media industry, from TV and print to online advertising and subscriptions. You helped a company grow through subscriptions. What's the biggest challenge for media companies transitioning from an advertising model to a subscription model?

SP: The hardest challenge is cultural, not technical. Moving from an advertising mindset to a subscription model requires redefining success: from reach to retention, from clicks to trust. It means asking: how do we create enough value that people are willing to pay directly? For many legacy media companies, this requires a painful shift in KPIs, skills, and even editorial priorities. The winners are those who embrace trust and user-centric strategies over sheer volume.

SM: You mentioned the "cookie-less world" in a past interview. How have these privacy changes impacted the strategies you develop for digital marketing and audience conversion?

SP: The demise of third-party cookies is not a threat—it's an opportunity. It forces companies to focus on first-party data, consent, and transparent value exchange with users. Readers will share data willingly if they understand the benefit, whether it's personalized content or access models that respect their privacy. The shift has driven innovation in registration walls, conversational data capture, and trust-based audience engagement.

STÉPHANE PERE

Former Chief Data Officer – The Economist | MD – Evidency, France

Stéphane is a seasoned digital and tech executive with a strong focus on reinventing trust in the digital realm. His extensive career includes pivotal roles at leading companies, where he has driven strategic transformation and innovation. As Chief Data Officer and Executive Vice President at The Economist, he earned multiple awards for his work in transforming the B2B market intelligence business. He has also served in executive roles at Vaultinum, Evidency, and Poool, where he has leveraged his expertise in digital strategy, business development, and software as a service (SaaS) to accelerate growth and mitigate risk. He is also a strategic advisor and non-executive director, guiding media and tech companies in revenue acceleration and digital transformation. He studied in London Business School , Emlyon Business School & Paris-Pantheon-Assas University.He is based out of Paris, France.

EXPERT VOICES

Subrata Mukherjee: How do you see the relationship between Business Intelligence (BI) and data transformation? Is one a prerequisite for the other? What is the single biggest technical or organizational challenge you've consistently faced in data integration projects?

Manu Nair: Business Intelligence (BI) and data transformation are mutually reinforcing. Data transformation ensures that BI tools operate on clean, structured, and governed datasets, enabling accurate and actionable insights. While BI can begin with raw data, its full potential is unlocked only when transformation processes are in place to standardize and enrich the data. This makes data transformation a practical prerequisite for scalable, enterprise-grade BI.

The single biggest challenge we consistently faces in data integration projects is aligning disparate data sources across business units with varying standards and governance maturity. Technical complexity is somewhat addressed and resolved with intelligent connectors, however, the organizational silos, making it difficult to enforce consistency, lineage, ownership and real-time accessibility.

SM: How do you leverage automation in your data pipelines to improve the speed and quality of data ready for analytics and AI?

MN: NeoStats leverages automation across its data pipelines to accelerate ingestion, transformation, and validation. Tools like Azure Data Factory, DataFlow Gen2 and Fabric Pipelines are used to orchestrate workflows, while automated data quality checks ensure consistency and accuracy. Metadata tagging, schema enforcement, and anomaly detection are embedded into the pipeline to reduce manual intervention. This not only improves the speed at which data becomes analytics-ready but also enhances trust in the outputs used for AI and BI applications

SM: What is one major trend in data and AI that you believe will fundamentally change one of the industries you serve, such as banking or manufacturing?

MN: One major trend we see reshaping industries like banking is the rise of Generative AI for hyper-personalization and decision intelligence. In banking, GenAI is enabling relationship managers to deliver tailored product recommendations, automate customer insights, and even draft personalized communications, transforming how banks engage clients. This is already driving increase in loan bookings and uplift in email conversions for some NeoStats clients. In manufacturing, GenAI is accelerating design cycles and optimizing production through predictive analytics and real-time data integration.

Aspiring data leaders should focus on building a strong foundation in both technical and business domains. I believe that leadership in data and AI requires not just expertise in tools, but the ability to drive value through strategy, governance, and execution. We believe in Full Stack Data Developers. Embrace continuous learning, foster cross-functional collaboration, and prioritize ethical, outcome-driven innovation. Most importantly, develop the ability to translate data into decisions that matter.

SM: What's the one thing that keeps you most motivated as you continue to help clients turn their data into real business value?

MN: What keeps me most motivated is seeing data move from dashboards to decisions and actions. Helping clients unlock real business value, whether through AI-driven efficiency or smarter customer engagement, is what makes the work meaningful every day. Equally, I love working with the team, making an impact in their life and career

MANU NAIR

Chief Data Officer, NeoStats | Ex-Senior Data Architect, Unisys

Manu Nair is a Chief Data Officer and accomplished data enthusiast with a track record of driving end-to-end data and AI transformations. At NeoStats, he operates with a "Data to Value" mindset, helping clients across banking, insurance, manufacturing, and retail solve real-world problems by building secure and AI-powered solutions. His expertise is grounded in building strong data foundations, ensuring data quality, and maintaining rigorous security. Previously, as a Senior Data Architect at Unisys, he led teams to transform and organize data into usable insights. Prior to that, as Manufacturing Delivery Head for Application Modernization at HPE, he led a team of over 250 professionals, modernizing legacy systems and driving operational efficiency for global clients. Manu is passionate about turning strategy into action and enabling businesses to make smarter, data-driven decisions. He is based out of Bangalore (Bengaluru), India

EXPERT VOICES

Subrata Mukherjee : Given your work with Stability AI and your focus on generative AI, your insights here are key. You helped launch the Christie's Art & Tech Summit where NFTs were first announced. How does the current wave of generative AI adoption compare to the early days of NFTs in terms of enterprise interest and hype versus practical use cases?

Richard Entrup: Very different technology and value propositions, and while GenAI is somewhat new, AI is not new by any means and has been around for decades. From an enterprise perspective, NFTs or Web3, can afford customer engagement, conversion, and retention, and in some cases, fractionalized ownership of physical assets. NFTs in the art world were hype for sure, but it's early, and blockchain as a technology is still being explored beyond Bored Ape JPGs being sold for $1M. Cryptocurrency and Digital Assets are also starting to be taken more seriously in the US, which don't forget they run on blockchain technology, all $3.5T of them. It's still early, but the enterprise really hasn't adopted these technologies.

SM: What is the single most important piece of advice you give to a C-suite executive who is trying to understand and invest in generative AI?

RE: Approach your GenAI efforts with a formal strategy, clearly refined business goals and outcomes, and ensure you're building AI in a responsible way. Change Management is also key as it always is with new technology utilization but often overlooked.

SM: As an advisor and investor in an open-source AI company, what are your thoughts on the debate between open-source and closed-source AI models for enterprise use?

RE: I believe the debate between open-source and closed-source AI models for enterprise use hinges on balancing immediate practicality with long-term innovation. It's also very relative to the business needs and the environment. I'm a big fan of hybrid approach, that is, combining open-source and closed-source systems to optimize ROI and innovation;

Open-source models offer cost-effective, customizable solutions that empower enterprises to tailor AI to specific needs, foster community-driven innovation, and avoid vendor lock-in, making them ideal for speculative, transformative bets like building proprietary, industry-specific AI systems. However, they require significant in-house expertise and robust security measures, which can delay short-term ROI.

Closed-source models, conversely, provide reliable, plug-and-play solutions with strong support and safety features, enabling rapid deployment and measurable returns in areas like customer service or analytics, but their high costs and lack of transparency risk dependency and regulatory challenges. A hybrid strategy—leveraging closed-source for quick wins and open-source for strategic, long-term differentiation—best aligns enterprise needs with the evolving AI landscape, with open-source holding an edge for fostering innovation and control in a rapidly advancing field.

An **NFT** (non-fungible token) is a unique digital certificate of ownership and authenticity for a physical or digital asset, recorded on a blockchain. Unlike cryptocurrencies, which are interchangeable (fungible), each NFT has unique characteristics that make it non-interchangeable, serving as a unique digital record for items like digital art, music, collectibles, and even virtual real estate.

RICHARD ENTRUP

Managing Director, Head of Emerging Solutions at KPMG US

A multi-award-winning global technology and innovation leader with nearly three decades of experience driving strategic business transformation. He is at the forefront of identifying and incubating disruptive technologies like GenAI, Quantum Computing, and Spatial/XR, pioneering new products and solutions for clients. A two-time recipient of the HMG Global Leadership Institute Award for Exceptional Business Transformation, his distinguished background includes C-suite leadership roles (CIO, CTO, CDO) at globally recognized brands such as Verizon, Disney/ABC, Christie's, Time Warner, Viacom, and Tiffany & Company. He also serves as a strategic advisor and investor to a portfolio of innovative startups and venture capital firms, passionately guiding ventures focused on cutting-edge innovations. He is based out of New York, United States.

EXPERT VOICES

Subrata Mukherjee: What is the fundamental problem with unstructured data in healthcare, and why have traditional methods failed to solve it at scale?

Mousumi Sengupta: The fundamental problem with unstructured data in healthcare is its semantic ambiguity and lack of machine-readable context. Traditional methods like keyword searches fail because they operate on a flat, syntactic level. They can't interpret nuance, understand synonyms (e.g., "MI" for "myocardial infarction"), or connect concepts across different documents. This creates massive, intractable data silos. Without a way to impose a consistent, semantic layer on this data, we cannot derive actionable insights. For example, 'Heart' can be an organ, a feeling, or an emoji. Traditional methods can't derive its true, semantic meaning, which is a major bottleneck in personalized medicine.

SM: You mention that Condext finds an "automated way to add accurate context." Could you walk us through, at a high level, how your technology is able to do this?

MS: Our technology uses a hybrid architecture that combines the power of Large Language Models (LLMs) with a robust knowledge graph. At a high level, the LLM acts as an advanced semantic parser. It reads unstructured text—like a doctor's note—and identifies key entities, relationships, and events, understanding the clinical intent. This extracted information is then normalized and ingested into our knowledge graph. The graph serves as a canonical source of truth, semantically linking entities such as a patient, a disease, a gene, and a medication. This process automates the creation of a structured data layer from a sea of unstructured data, allowing smart systems to seek out the correct context on their own.

SM: How does the lack of cohesive, contextual data directly impact patient care and medical research?

MS: The lack of cohesive, contextual data leads to a significant degradation of both clinical and research efficacy. For patient care, fragmented data can lead to diagnostic errors and suboptimal treatment plans because a clinician may miss critical drug interactions or relevant medical history buried in different systems. For research, this fragmentation severely impedes scientific discovery. Researchers spend an enormous amount of time on manual data wrangling and literature review. This inability to easily connect a patient's phenotype with their genotype, or link a clinical trial outcome to preclinical research, creates significant delays in translating scientific breakthroughs into new therapies and personalized care.

SM: What is your ultimate vision for how your work at Condext will impact the future of medicine?

MS: My ultimate vision for Condext is to democratize access to biomedical knowledge. By automating the process of adding context to billions of unstructured healthcare records, we will create a global, interconnected knowledge graph that anyone can query. This will empower every researcher, clinician, and biotech company with a unified view of all available data. My hope is that Condext will enable medical IT systems to seamlessly talk to each other and exchange information in a natural manner, without the need for manual data manipulation. I believe Condext will dramatically accelerate the pace of scientific discovery, identify new therapeutic pathways for rare diseases, and ultimately, enable a future where every medical decision is informed by the collective intelligence of the entire medical community.

MOUSUMI SENGUPTA

Chief Data Strategist at Condext, Ex-Accenture, Ex- Mount Sinai Hospital

Mousumi Sengupta is an accomplished leader in digital health, currently serving as the Founder and Chief Data Strategist at Condext, a startup automating the contextualization of healthcare data. Her expertise in Data Strategy, Ontology Engineering, and AI is built on a career at prominent organizations like Accenture, where she helped develop a Life Sciences practice with a focus on Knowledge Graphs and Generative AI. Her prior roles include leading a FAIR Data Management function at Bristol Myers Squibb and driving semantic interoperability at Mount Sinai Health System. She was also a Director at organizations like Nuance Communications and the Multiple Myeloma Research Foundation (MMRF). She is a proven innovator at the intersection of technology and clinical data. She has a Masters in Bioinformatics from Columbia University. She is based out of New York, United States.

EXPERT VOICES

Subrata Mukherjee: At AInspire, you consult with CTOs and CDOs on AI strategy and governance. What are the top three concerns or hesitations you hear from C-level executives when it comes to adopting Gen AI?

BG: Cost, risk, and ROI come up in nearly every executive conversation. Leaders want to understand the full cost of adoption, from infrastructure to integration. They also ask about the risks—technical breakdowns, legal exposure, brand impact. And of course, they want to know if the long-term payoff will justify the disruption. These are smart, necessary questions in any strategic investment. What makes generative AI different is how fast the ground moves. Today's productivity gain in software development might be 10x. Next year, it could be 40x. That volatility makes traditional forecasting almost meaningless. The upside is clear, but it's hard to time. The risks, too, are shifting—intellectual property, hallucinations, emerging regulations. It's not hesitation as much as a need for clarity before leaping.

SM: What exactly are agentic processes and LLMs in the context of AI, and why are they important for the future?

BG: Agentic processes and LLMs are central to how enterprise AI will function in the future. The most powerful AIs will drive these systems by acting as managers that direct and coordinate the work of more specialized models. These high-capacity models take on broader context and decision-making responsibilities, while lighter models carry out focused tasks. This creates a structure similar to how companies organize teams around leadership and execution. Agentic processes refer to the use of multiple AIs working together to complete a task. They can be arranged in sequence, run in parallel, or operate in other collaborative forms. Each AI involved is typically powered by an LLM, or large language model, which provides the intelligence needed to carry out its part. The distinction between the strategic manager models and the task-level models is important because it shapes how organizations will scale and deploy AI across functions. This model reflects familiar management structures, making it both practical and powerful for enterprise use.

SM: How do you advise companies to balance the speed of innovation with the necessary rigor of AI governance and risk management?

BG: In the next year or so, advanced AIs will begin playing a direct role in enforcing governance policies and risk controls. These systems will not only monitor and apply compliance rules to their own outputs but will also evaluate and reinforce policy across human-generated work. This marks a major shift in how oversight is implemented, moving from manual review to real-time, AI-driven enforcement that spans both machine and human contributions. Any governance or AI rules put in place today should be established with the clear understanding that they will soon be obsolete. The systems they are meant to manage will be replaced by faster, cheaper, and more capable AIs in a very short timeframe. The governance policies and mechanisms in use now will not apply to what comes next, they will mostly be useless. Companies must govern today's AIs with today's tools but recognize that tomorrow's AIs will bring their own governance logic with them, built in and self-enforcing. I am working on frameworks for creating self-governing AI that deploy across the organization and iterate to improve their performance, control risks, and enhance the quality of outcomes. Overly restrictive AI is a thing of the past. AI will work toward business goals and will not be a hindrance or an obstacle to overcome, but rather a partner with a deep understanding of what is reasonable in terms of governance and practice.

DR. BRIAN SCOTT GLASSMAN

Chief Product Officer & Head of Client Engagements at AInspire.ai

Dr. Glassman is a seasoned product management executive, engineer, and product strategist specializing in generative AI. As the co-founder and Chief Product Officer at AInspire.ai, he leads client engagements and product strategies, collaborating with C-level executives to develop and deploy AI solutions for large enterprises. With a career dedicated to leading technology innovation, he previously served as Director of Product Management at Electra Vehicles, where he built the product function for AI-powered solutions in the automotive sector. His extensive experience includes a seven-year tenure at Rave Digital, where he led product management and customer engagements for Fortune 1000 clients. Furthermore, his background as an engineer and product leader at Cameron, a Schlumberger company, highlights a unique expertise in technical product development. He is based in New York.

Subrata Mukherjee: How do you move from simply using AI tools to building an "AI-first" business strategy that fundamentally changes how you deliver value?

Françoise Nauton-Inglis: An AI-first strategy starts with redefining the problem or what we're trying to solve: not "Where can we sprinkle AI?" but "Which decisions in our value chain deserve machine assistance, and which must remain human?" With Juliia, we rebuilt product discovery around intent understanding rather than keyword matching; that shifts AI from a feature to the engine of the customer journey. Practically, it means owning your data pipelines, defining responsibility guardrails early, and reorganizing teams around outcomes (conversion, satisfaction, returns) instead of deliverables. We align product, data, and CX on one roadmap, and we measure AI not by model metrics but by business lift and customer effort reduced. That's when AI becomes strategy—not a tool

SM: What is the biggest challenge you've faced in developing and scaling a custom AI solution for the e-commerce sector?

FNI: We are not here to reinvent the candle but to look at light in a different way. Next is data quality and context. Merchants hold rich but fragmented information—catalog, attributes, real-time availability, merchandising rules, service constraints. Training or prompting without this context produces "clever answers, wrong outcomes." We solved it by orchestrating various layers of LLM prompting and a governance layer so Juliia respects merchandising logic, compliance, and brand voice by default. The second challenge is change management: moving teams from "manage filters and synonyms" to "coach an AI teammate." That requires new rituals, analytics, and trust-building—plus transparency about what the AI knows and why it recommends. Finally, scaling across verticals means designing for variance: language and seasonality.

SM: You mention that Juliia is designed to "humanize" the digital experience. How do you ensure that an AI-powered solution feels intuitive and helpful, rather than robotic or impersonal?

FNI: We treat "human" as a service standard: clarity, context, empathy, and agency. Practically, Juliia understands the shopper's intent beyond the words, references store-specific constraints (sizes in stock, delivery windows), and justifies its recommendations ("these sunglasses have photochromic lenses, ideal for changing lights as often happens when sailing mid season"). At any point during the conversation the user can change its mind or correct the AI's assumptions ("these glasses are too square for me, I would like something more feminine looking". Tone is brand-tuned but always plain language as is the Backoffice management. We also instrument frustration: when users don't get the answer to their question, click back, or abandon, we fine tune our prompting system or imagine new features, just like you would do when training a new sales advisor. Finally, we the sites opt between free-form conversation and guided suggestions based on classic filters—because humanizing doesn't mean forcing conversation; it means offering the fastest respectful path to a good choice.

SM: You've built a reputation for being hands-on and solving problems. How do you scale that personal approach as your company grows?

FNI: By scaling the behaviors through culture, not the person. We approach problem-solving in collective manner —weekly key personnel meetings, blameless culture and clear decisions — Good habits and positive experience propagate especially ones that actually deliver results and help people manage their jobs better. We celebrate all victories, especially the small ones as much as big launches. The goal is a culture where curiosity, clarity, and ownership are normal.

FRANÇOISE NAUTON-INGLIS

CEO @ Spiriit / Juliia, Ex GM/CFO of Kaliop Interactive Media

Françoise Nauton-Inglis is an accomplished business leader with over 20 years of experience in the digital and technology sectors. She is the CEO of Spiriit, an agency that designs and builds custom digital and AI solutions, and its AI-first product, Juliia, which focuses on intelligent search and product recommendations for e-commerce. Françoise's expertise lies in growth management, international development, and strategic business planning. Prior to Spiriit, she spent a decade at Kaliop Interactive Media, where she served as the Group General Manager, overseeing M&A, international expansion, and financial operations. She has a deep passion for structuring and growing innovative, profitable, and human-centered businesses, and she holds leadership positions in several organizations, including as President of Digital 113 and CI Montpellier Management. She is based out of Montpellier, France.

EXPERT VOICES

Subrata Mukherjee: Beyond basic personalization, what is the most impactful and innovative application of AI you are seeing in the retail & consumer goods space?

Charisma Glassman: Beyond recommendation engines and chatbots, the real frontier is generative design combined with dynamic supply chain orchestration. Imagine a world where AI not only identifies emerging trends from real-time cultural signals but also generates bespoke designs aligned to those shifts. Instead of the old seasonal model, this creates a fluid, continuous cycle of demand-driven creativity. On the operational side, AI is building what I call a "self-healing supply chain," systems that anticipate disruptions, reroute shipments autonomously, and adjust production schedules in real time.

SM: How can a legacy retailer begin to build an effective AI roadmap to compete with digitally native players, and what are the common pitfalls they should avoid?

CG: Legacy retailers often underestimate the strategic advantage of their historical customer data. The roadmap begins not with technology procurement but with a disciplined data audit, understanding which assets are most valuable and where the gaps lie. Start with a high-impact pilot that can show tangible results, whether that is computer vision for real-time inventory tracking or predictive analytics to sharpen promotions. The trap is the "big bang" approach, trying to build everything from scratch. That usually leads to failure. The smarter move is to lean on specialized partners while also embedding a culture of data literacy. Technology without adoption is just shelfware.

SM: What are the top two or three data strategies that retailers must prioritize to successfully leverage AI for customer experience and revenue growth?

CG: First, create a unified customer view by connecting every touchpoint including digital, in-store, service interactions, and social engagement into a single profile. Second, establish a rigorous discipline around data quality and governance. AI only performs as well as the inputs it receives. Finally, invest in real-time data infrastructure. The ability to act in the moment with dynamic pricing, hyper-personalized offers, and proactive service is no longer optional; it is the competitive edge.

SM: What are the unique challenges and opportunities of digital transformation in the luxury sector, where brand heritage and exclusivity are paramount? What is your playbook for entering new high-growth luxury markets? What is social media's role?

CG: The paradox in luxury is that digital channels are designed for scale, while luxury thrives on intimacy and exclusivity. The challenge is to replicate the velvet-rope experience in a mass medium. The opportunity lies in creating curated digital experiences, such as augmented reality fittings, blockchain-backed certificates of authenticity, and immersive storytelling that makes digital feel personal. Luxury is not about reaching everyone; it is about deepening connections with the few who matter most. My playbook has three pillars. First, hyper-localization: campaigns must resonate with cultural codes, not just be translated. Second, partnerships: authentic voices with local credibility build trust faster than any global campaign can. Third, platform-native strategy: in some markets, discovery, engagement, and purchase happen within a single digital ecosystem. Success requires meeting consumers where they already are, rather than expecting them to change behavior. Luxury has shifted from being a stage to being a salon. It used to be about broadcasting big launches and orchestrating exclusivity through scarcity. Today, it is about co-creation, participation, and community. Luxury brands that invite their clients into the process, whether through design votes, behind-the-scenes access, or virtual exclusives, are finding that engagement is the new currency of loyalty.

CHARISMA GLASSMAN

Global Head of Retail & Consumer Applied Advisory, Genpact

Charisma Glassman is a distinguished global executive, a Forbes Distinguished Leader, and one of the Top ranked AI Leaders in consulting. As Group VP and Senior Partner at Genpact, she is the Global Head of Retail & Consumer Applied Advisory, leading business transformations powered by AI for clients across the US, Europe, and APAC. Her impressive career includes senior leadership roles at various consulting companies, advising C-suite executives, and at major financial institutions, where she led digital initiatives that resulted in significant revenue growth. Charisma consistently leverages her expertise in digital transformation, product strategy, and global team leadership to drive strategic growth & innovation. She is based out of New York, in the United States.

EXPERT VOICES

Subrata Mukherjee : Given your experience with vast data sets at ratings agencies, how do you help clients organize and monetize their own data sets for AI adoption?

Sandy McCarron : We partner with firms that specialize in data organization and strategy. We also work with a Chief Data Strategist who has led data outsourcing initiatives. Our role is to provide the high-level strategic guidance and connect them with the right experts. We help them define the problem they want to solve, and then we help them build a team that can execute on that vision. It's about building a holistic strategy that connects the business problem to the technical solution. From there, we can think about ways to monetize their tagged up datasets.

SM: What is one critical piece of advice you would give to a business that wants to begin its AI journey but doesn't know where to start?

SMC: Start with the problems you're trying to solve—internally or for your clients. Then ask whether AI is the right tool. If you're not solving your customers' top problems, you won't stay in business. Too many companies start with AI and try to find a problem for it to solve. That's a recipe for failure. The best AI applications are the ones that address a clear, urgent need that's already well-understood.

SM: What is the most valuable lesson you've learned about leadership and innovation throughout your career?

SMC: On leadership: trust your instincts and listen. I inherited a team that had "bad resources," but my gut said they just weren't being heard. With support, some of them went from performance warnings to exceptional contributors. Don't write people off too quickly. On innovation: persistence. From the Carnegie Mellon team that won the DARPA Challenge, I learned that if you don't hear "no," keep going. Persistence is key to breakthroughs.

SM: Looking ahead, what is one major trend you believe will redefine the role of a technology leader in the next decade?

SMC: The single most significant trend that will redefine the role of a technology leader is the rise of AI as a core, strategic partner, not just a tool. Leaders will no longer simply manage human teams; they will be tasked with orchestrating cohesive "hybrid teams" of people and intelligent machines. This shift moves the leader's focus from mere project management to a more profound role as an "AI architect," responsible for designing the optimal interface between human intuition and machine intelligence. This involves a complex set of new skills: they will need to be proficient in ethical AI governance, fine-tuning large language models to align with business objectives, and ensuring the human-machine collaboration is seamless and productive. The future leader's success will be measured by their ability to leverage AI to unlock human potential, not replace it.

SM: What is one book, blog, or resource you would recommend to someone who wants to become a better product or innovation leader?

SMC: I recommend a few resources depending on what they want to focus on. For rapid innovation, I'd suggest Sprint by Jake Knapp, which is a fantastic guide written by the creator of the design sprint process with lots of tools for running your own. For a deep dive into product management, a must-read is Continuous Discovery Habits by Teresa Torres. Lastly, for staying up-to-date with the ever-evolving AI landscape, I find The AI Daily Brief podcast to be an invaluable resource. The key is to build a routine around staying current with new developments and keep learning.

SANDY MCCARRON

Co-Founder & COO at Sapient Advisors | Ex- Moody, Ex-S&P

Sandy McCarron is a co-founder and Chief Operating Officer at Sapient Advisors, a lean, AI-enabled agency specializing in product and data strategy. He focuses on guiding companies through AI adoption and achieving operational efficiencies through rapid AI Design Sprints. With a career spanning both media and fintech, his expertise is built on a strong foundation of innovation and program management. At Moody's Analytics, he was a Senior Director of the Innovation Unit, where he introduced Google Ventures Design Sprints and led a major program to upgrade a flagship product, generating a $60M increase in profit. He also held a senior technology director role at S&P Global. His early career includes positions as Director of Online Content at Penguin Publishing Group and Director of Online Production at Penguin Random House. She is based out of New York.

EXPERT VOICES

Subrata Mukherjee: : You led Leadership positions at Hearst where you drove video growth by 600% ?

Karen A. Brophy : To start, we implemented an online video platform (OVP), JW Player at the time, and automated video ingestion and connected video to related content or for AP stories, automatically connected the video they distribute to the associated article. This minimized on production locally to add video to articles, although local teams could always manually add a video and upload their own video. Monetization was difficult at first, the open market for programmatic video was tough at the time. There were size constraints, ad formats were changing, and overall CPM was $5 or under.

SM: What are the differences between managing Yahoo! News and local portfolios?

KAB: While Yahoo! News was one property in the division, it encompassed a variety of products in its portfolio including weather, local news search, original productions, video, and social media. There were a lot of stakeholders for News and executives across the company were always reaching out with ideas, so in some ways it was similar to local portfolios. But the focus was engagement, there was already massive reach at 1.5B page views a month. And the tools at Yahoo! to monitor KPIs and performance and the internal requirements for performance were set very high. Whereas with local portfolios, it was more about taking trusted brands and modernizing their products and platforms so they could scale and be more efficient. It was building better user experiences along with driving new revenue relationships centrally using the broader leverage of the company. In some ways it was being an entrepreneur but in a big company, so there was a lot of ability to be agile and move quickly.

SM: Are there universal product principles across verticals?

KAB: Yes. User-first design, measurable outcomes, speed to ship, and clear monetization strategy. Each vertical has nuances. Breaking news must come quickly and the platform and product need to showcase that. Politics can require attention to fairness and accuracy is key. Entertainment and lifestyle need to be engaging. But the discipline of defining customer needs, creating easy to use and engaging or joyful products that connect to revenue applies everywhere.

SM: What is your vision for AI in media?

KAB: I think AI can become an invisible collaborator in areas like personalization, automating production and optimizing monetization. But it is also going to have effects on how users consume media. Already media companies are seeing significant search traffic dips due to how google has implemented Gemini on the search results page. This disintermediation will only get worse, so media companies need to develop strategies on how to work with the LLMs to get paid for their content and offset the loss of users and revenue that is coming. Ideally AI can help expand reach and profitability, but that is a big question mark given the history of social and search impacts. I think AI can help in monetization in ways we are still exploring so there is some hope there. Ideally media companies come together on strategies to use AI to help grow their business while maintaining trust with their consumers and advertisers.

SM: How do you approach partnership strategy with streaming players?

KAB: With a sizable retransmission business, any deals we did with streaming players had to align with what we content we could distribute. Strategically, breaking news content and clips is the best to distribute. Short clips don't cannibalize our loyal audiences but bring awareness and alternative revenue streams from partners. To some degree, you have to work within their ecosystem in terms of monetization, but we were able to negotiate the appropriate deals for our business.

KAREN A. BROPHY

General Manager, President, COO/CPO | Ex Hearst, Yahoo!

Karen is a dynamic executive specializing in P&L Management, Digital Transformation, and Revenue-Driving AI and technology business strategy for national media brands and consumer platforms. As the founder and President, Digital Division at Nexstar Media Group, she oversaw all digital business, product, and monetization strategies for over 200 local stations, The CW, and The Hill, delivering the highest digital revenue in company history with 70% growth. She is an expert in launching scalable platforms, notably developing the AI-powered advertising platform, Stellar, and spearheading the rollout of OTT Video and FAST channels. Karen held senior Product Leadership roles at major companies like Hearst and Yahoo!, in addition to her current advisory work as a Fractional COO/CXO, focusing on operational excellence and Artificial Intelligence. She is from Los Angeles, California, USA.

EXPERT VOICES

Subrata Mukherjee: What are the unique challenges of building data pipelines to support data science and AI projects, as opposed to traditional business intelligence?

Manish Bhandari: The key difference is that AI pipelines are built for experimentation and model performance, while BI pipelines are for reporting stability. In BI, you mainly transform structured data for predictable dashboards. For AI, the challenges are different: you deal with raw, unstructured, and streaming data at scale. Pipelines must support iterative feature engineering, versioning, and reproducibility. Quality and explainability are critical, as models need trustworthy and traceable inputs. AI pipelines must also integrate with model training, deployment, and retraining loops.

SM: How has the rise of large language models and GenAI impacted the work of a data engineer, in terms of data preparation and feature engineering?

MB: The rise of LLMs and Generative AI has significantly expanded the data engineer's scope, particularly with unstructured data. We are no longer just preparing structured data; we now handle unstructured text, embeddings, and retrieval systems. Feature engineering is less about handcrafted columns and more about chunking, embedding, and retrieval strategies. These resource-intensive workloads also add new responsibilities around cost optimization and governance. Essentially, we're moving from building pipelines for dashboards to building pipelines that make unstructured knowledge AI-ready.

SM: You mention using a variety of technologies like Terraform, AWS, and Kubernetes. How do these technologies fundamentally change the way data platforms are built and managed compared to traditional methods?

MB: These technologies allow us to build resilient, self-service, and reproducible data platforms much faster than before. Infrastructure-as-code (Terraform) and container orchestration (Kubernetes) mean we can consistently reproduce environments, audit changes, and scale globally with confidence. The fundamental shift is that data platforms are now self-service products. Compared to traditional database-centric approaches, modern stacks support diverse workloads like streaming, machine learning, and LLM systems, all unified under reproducible infrastructure.

SM: What is the single most important principle you follow when designing a new data infrastructure from the ground up?

MB: My single most important principle is to design for evolution, not just today's use case. Every data platform becomes more complex over time. If you build only for the present, you create brittle systems. I prioritize modularity, observability, and automation from day one. For example: Modularity with event-driven pipelines allows for easy extension; Observability with end-to-end data quality monitoring ensures trust; and Automation with infrastructure-as-code allows for safe, quick deployments. This principle enabled Tinder to seamlessly onboard new tenants and now allows my current project to embed AI-driven insights into workflows without re-architecting.

SM: What are the most important considerations when deciding whether to build a batch or real-time data pipeline for a new application?

MB: The primary consideration is the business need for latency. If a business requires immediate action, such as for fraud detection or product recommendations, a real-time pipeline is necessary. If insights drive daily or weekly decisions, batch processing is more cost-effective and reliable. Real-time adds significant complexity in state handling and reliability, so I only recommend it when the clear ROI justifies the investment. In practice, many systems are hybrid, using streaming for real-time freshness and batch for analysis and enrichment.

MANISH BHANDARI

Ex-Tinder, Ex- Wayfair, Currently building an AI-First Data platform

Manish Bhandari is a seasoned professional with a strong focus on building and maintaining analytics infrastructure and architecting data solutions. He is currently in a new venture at a Stealth company, leveraging his extensive experience from leadership roles in the tech and data sectors. Previously, he served in data infrastructure roles at companies such as Tinder and Wayfair. As a Senior Data Engineer at QuantumBlack, AI by McKinsey, he was instrumental in building impactful analytics solutions for C-level executives and designing data pipelines to support AI and data science projects. His experience is built on a foundation of managing corporate data infrastructure at Genscape. Manish's core expertise lies in Data Engineering, Advanced Analytics, and ETL, with proficiency in technologies including Python, Databricks, and AWS. He is based out of San Francisco, United States.

EXPERT VOICES

Subrata Mukherjee: After a long and successful career as a C-level technology leader, you've shifted to providing governance and advisory services. What is the most significant difference between leading an internal IT organization and advising a board on technology and cyber strategy?

Michael Gabriel: As a C-Level technology executive, your mandate is to align technology strategy with business objectives—and to own the execution and ongoing performance against those goals. In contrast, when serving as an Advisor or Board Member, your responsibility shifts to oversight and governance. You're not accountable for execution, staffing, process improvement, change management, or technology selection. "Noses in, fingers out" is a guiding principle that ensures Advisors and Board Members maintain objectivity and avoid entanglement in operational decisions, but with a more elevated perspective rather than a focus on operational involvement.

SM: In your role as a board advisor, what is one emerging technology or trend that you believe boards are not paying enough attention to?

MG: Quantum Computing is the main emerging technology that I don't believe is on the radar of most Boards. Boards seem to have their hands full with both Cybersecurity and AI Governance & Oversight, and aren't handling that effectively in most cases. That being said, Quantum Computing will vastly complicate both of those areas exponentially. While it appears that the time horizon on this is probably 3-10 years until adoption becomes available, it should be on the radar as a risk that needs to be monitored. How would that scale of computing power potentially impact their business in terms of potential risks and benefits? In many cases, it might not directly impact corporations any more than in utilizing SaaS products that provide those capabilities, but in some cases, it could render products and services no longer useful or viable, such as current encryption protections. If they have one, I believe that this will fall first into the purview of the Audit & Risk Committee. If not, then whatever mechanism they have in place of that should start to be engaged. So, starting to prepare for what this might mean, as well as gathering learnings from the vendors and professional service organizations their company utilizes, including external audit firms, industry peers and industry research publications, could help prepare for properly leveraging this capability for business growth and risk mitigation.

SM: From your experience, what is the most important piece of advice you can offer a C-level executive who is considering a board position?

MG: I would advise them to start preparing for this well before their retirement date. That could be by involvement as an Advisor to some companies that they have expertise with, and/or joining a not-for-profit Board, and the latter usually requires Board members to make annual donations. The best approach is to try to get on a Board while they are still working, unless that is not allowed or deemed a conflict of interest. If allowable then their head of HR and CFO might be able to provide some direction or guidance, and engaging with an executive search firm that has a Board placement focus would also be a goood direction to follow.

Lastly, once they are on a Board, joining an organization such as the National Association for Corporate Directors, NACD, would be a great way to improve their board-level education and exposure.

MICHAEL GABRIEL

Ex CIO at HBO, NBA | NACD Certified Board Director

Michael is a senior technology and cybersecurity executive with experience providing C-level leadership, governance, and advisory services. He is currently an independent Board Director and provides professional services for companies like Warner Brothers Discovery (WBD). His career is highlighted by a 14-year tenure as EVP & Global CIO at HBO, where he was the initial executive to spearhead the groundbreaking HBO GO initiative and led a company-wide digital transformation. An inventor on six technology patents, he also played a key role in information security governance across the Time Warner enterprise. Prior to HBO, he served as Global CIO for the National Basketball Association (NBA), where he implemented the first NBA retail store technology system. He also held senior IT leadership positions at EMI Records Group and NBC Olympics.. He is based out of New York

EXPERT VOICES

Subrata Mukherjee: Your role involves formulating and executing strategic initiatives for a large organization. How do you, as a Chief of Staff, ensure strategic alignment and help 98% of associates understand how their work supports the corporate mission?

Amanda Murphy: As a chief of staff, I have line of sight into all of the departments that make up the organization - I see how all of the puzzle pieces fit together to drive toward overall outcomes. I'm able to leverage this broad, holistic view to drive messaging from senior leadership to middle managers to hands on keyboards, helping teams see themselves in the strategy and understand how their work supports our outcomes and the businesses goals. I do this through various leader and associate touchpoints - all associate meeting content and talking points, crafting supporting materials and resources for our leaders and their conversations with their teams, leading associate advisory groups and change/culture champions, and through clear, transparent communications from leaders to team members.

SM: You are described as a "trusted advisor" to the executive team. How do you anticipate their needs and bridge gaps to enable peak organizational effectiveness?

AM: I'm able to start a project or deliverable for our executive team (such as a Board deck, financial update to C-suite, KPI report) so they have something to react to rather than having to spend time brainstorming and building from the ground up. If there's a need from one of the team members discussed in our weekly meeting and they don't have the resource or bandwidth to cover the need, I'm able to step in and take that item off their team's plate, acting as an extension of their leadership team. My unique position within the organization makes me the bridge between local teams and associates to senior leadership - I'm the point of contact for all staff or business operations questions/concerns/needs. Being involved in each department in a variety of ways allows me insights into process gaps that those working in the day-to-day may not notice. I've pulled together problem solving teams to facilitate operational excellence uplift.

SM: What is the most challenging part of operating in a large enterprise, and how do you use your skills in program management to navigate those complexities and lead cross-functional teams effectively?

AM: The most challenging part is the magnitude of teams and people touched by a process. One seemingly minor change can have a negative impact downstream. It's imperative to have all teams represented and involved in transformational change for it to be successful. Change doesn't work if it's only a request from the top to make it happen. Leaders and associates need to understand the why and help facilitate the transformation from the bottom up, as well. Research shows associates want to hear about change from their direct leaders, not from senior leadership. They have a relationship and a trust with direct or skip level leaders that they don't have with senior leadership.

SM: How has your experience as a news assignment editor and journalist influenced your approach to leadership & problem-solving in corporate environment?

AM: Working as a journalist early in my career granted me experience in relationship building, working closely with community members, working with diverse groups, and working fast (you can't miss a TV news deadline, or it's literally dead air). Determining whether something was worth sharing broadly, what is and isn't important to the messaging, has been highly beneficial in leadership and problem-solving. It helps to cut out the noise quickly to move with the key Who/What/When/Where/Why nuggets that will bring everyone else along. You're able to focus on the primary information. Another influence on approach that I learned from journalism is the power of inclusivity - having many perspectives helps lead to a better final product.

AMANDA MURPHY

Technology Director, Chief of Staff at Nationwide Insurance

Amanda is a seasoned leader specializing in transformational leadership, business operations, and organizational change management (OCM) within the technology sector. As the Technology Director, Chief of Staff for Security & Infrastructure at Nationwide, she formulates and executes strategic initiatives for a large, Fortune 100 organization. Her proven expertise in building and leading high-performance teams is demonstrated by her success in managing a complex merger. With a strong background in strategic communications, digital storytelling, and executive messaging, Amanda is skilled at creating strong corporate cultures and driving successful outcomes by bridging the gap between high-level strategy and tactical execution. She has been recognized with a Gartner Communications Award for her work in transforming internal communications. She is based out of Westerville, Ohio, United States.

EXPERT VOICES

Subrata Mukherjee: For those unfamiliar with the concepts, how would you explain the difference between a standard relational database and a knowledge graph?

Mousumi Sengupta: A relational database organizes data in a predetermined, table-based structure with relationships implied by foreign keys. It's efficient for transactional data but poor at handling complex, evolving relationships. A knowledge graph is like a machine-readable 'brain' that uses nodes, edges, and properties to explicitly represent entities and their relationships. This model is flexible and allows for continuous learning and dynamic queries. In a relational database, you might ask, "What medications is this patient on?" In a knowledge graph, you can ask a far more complex query: "Show me all patients on a medication that targets the same protein as their disease's associated gene."

SM: Could you provide an example from your work at Mount Sinai or Bristol Myers Squibb where using a knowledge graph enabled a breakthrough that would have been impossible with traditional data models?

MS: Yes. At Bristol Myers Squibb, we used a knowledge graph to accelerate the discovery of drug targets for a novel immune-oncology program. We integrated disparate datasets including patient clinical records, genomic sequencing data, and public biomedical literature. The knowledge graph allowed us to perform a federated query that traversed the graph, identifying a previously unknown causal link between a specific gene mutation and a clinical response to a different drug. This kind of insight would have been impossible with a traditional database, as it would have required a series of complex intersections to connect these sources.

SM: How does semantic interoperability, a key skill you possess, contribute to the value of a knowledge graph in a complex domain like healthcare?

MS: Semantic interoperability is the backbone of a valuable healthcare knowledge graph. It's the ability for disparate systems to understand and share data with a common meaning. My expertise in ontology engineering allows me to create a shared, formal representation of concepts—like diseases, drugs, and proteins—and their relationships. This serves as the foundation for the knowledge graph. By mapping data from various sources to this ontology, the knowledge graph can harmonize fragmented data. This ensures that a query about "hypertension" will correctly retrieve data linked to "high blood pressure," regardless of the source system's terminology, creating a powerful network of information.

SM: In what ways can knowledge graphs and ontology engineering act as a catalyst for organizations to achieve FAIR data principles?

MS: Knowledge graphs and ontology engineering are the primary enablers of FAIR data. Ontologies provide the controlled vocabulary and conceptual framework that make data Interoperable by providing a shared, machine-readable understanding of concepts. A knowledge graph then uses this ontology to link and integrate data from disparate sources, creating a unified, interconnected network. This structure makes data inherently Findable and Accessible because you can traverse the graph to discover relationships. By adding rich metadata, you make its context easily Reusable for new applications.

SM: How does achieving FAIR data impact the speed of drug discovery and research and development (R&D)?

MS: Achieving FAIR data dramatically accelerates the speed of R&D by eliminating the most significant bottleneck: data integration and semantic interoperability. When data is FAIR, a researcher can, in a matter of seconds, find and combine a patient's genomic data with a clinical trial dataset and a corpus of scientific literature. This ability to instantly create a complete, comprehensive view of a research question eliminates months or even years of manual data wrangling, significantly speeding up the process of scientific discovery.

Chief Data Strategist at Condext, Ex-Accenture, Ex- Mount Sinai Hospital

MOUSUMI SENGUPTA

Mousumi Sengupta is an accomplished leader in digital health, currently serving as the Founder and Chief Data Strategist at Condext, a startup automating the contextualization of healthcare data. Her expertise in Data Strategy, Ontology Engineering, and AI is built on a career at prominent organizations like Accenture, where she helped develop a Life Sciences practice with a focus on Knowledge Graphs and Generative AI. Her prior roles include leading a FAIR Data Management function at Bristol Myers Squibb and driving semantic interoperability at Mount Sinai Health System. She was also a Director at organizations like Nuance Communications and the Multiple Myeloma Research Foundation (MMRF). She is a proven innovator at the intersection of technology and clinical data. She has a Masters in Bioinformatics from Columbia University. She is based out of New York, United States.

EXPERT VOICES

Subrata Mukherjee: When it comes to quantum computing, how real is the threat of "Post Quantum Cryptography," and what should companies be doing right now to prepare for it?

Richard Entrup: I lead the Quantum practice currently at my current employer and view the threat of post-quantum cryptography (PQC) as very real and immediate for most enterprises, with a 3-5 year horizon before quantum computers can practically break current cryptographic systems like RSA or ECC.

Quantum algorithms, such as Shor's, could theoretically decrypt data protected by these standards, exposing sensitive data (e.g., financial transactions, intellectual property) collected today via "harvest now, decrypt later" attacks. The risk is significant—NIST estimates a 50% chance of a cryptographically relevant quantum computer by 2035.

However, the threat's urgency depends on your industry's data sensitivity and longevity. For example, healthcare or defense firms with decades-long data retention face higher risks than retail.

What companies should do now:

Inventory Cryptographic Assets: Map all systems using vulnerable algorithms (RSA, ECC, DH). Prioritize data with long-term confidentiality needs (e.g., trade secrets). Tools like cryptographic discovery scanners can reduce manual effort by 60%.

Adopt NIST PQC Standards: Transition to quantum-resistant algorithms like NIST's recently finalized lattice-based standards (e.g., CRYSTALS-Kyber). Start with hybrid cryptography—pairing PQC with classical methods—for critical systems like VPNs or secure APIs to ensure backward compatibility.

Build Crypto-Agility: Implement flexible cryptographic frameworks to swap algorithms without overhauling systems. This cuts future migration costs by 30-40%, per industry estimates. For example, update TLS protocols to support PQC algorithms.

Upskill Teams: Train security and dev teams on PQC through programs like SANS or NIST workshops. Lack of expertise is a top barrier, with only 20% of organizations having quantum-ready staff.

Monitor Threat Horizons: Partner with vendors and industry groups (e.g., Quantum Alliance) to stay updated on quantum advancements. Conduct annual risk assessments to adjust timelines.

Start with low-hanging fruit: pilot PQC in non-critical systems (e.g., internal file encryption) to gain experience. For high-risk sectors, aim for 20-30% of critical systems to be quantum-resistant by 2028. Collaboration with CIOs is key to align budgets and prioritize systems, ensuring business continuity without panic-driven overhauls.

Quantum computing is a revolutionary approach to computation that harnesses the principles of quantum mechanics to solve complex problems far beyond the capabilities of classical computers. **Post-Quantum Cryptography (PQC)** refers to cryptographic algorithms that are resistant to attacks from both classical and quantum computers, aiming to secure data and communications in a future era of quantum computing.

RICHARD ENTRUP

Managing Director, Head of Emerging Solutions at KPMG US

A multi-award-winning global technology and innovation leader with nearly three decades of experience driving strategic business transformation. He is at the forefront of identifying and incubating disruptive technologies like GenAI, Quantum Computing, and Spatial/XR, pioneering new products and solutions for clients. A two-time recipient of the HMG Global Leadership Institute Award for Exceptional Business Transformation, his distinguished background includes C-suite leadership roles (CIO, CTO, CDO) at globally recognized brands such as Verizon, Disney/ABC, Christie's, Time Warner, Viacom, and Tiffany & Company. He also serves as a strategic advisor and investor to a portfolio of innovative startups and venture capital firms, passionately guiding ventures focused on cutting-edge innovations. He is based out of New York, United States.

WE NEED YOU!

WELCOMING OUR NEXT SET OF THOUGHT LEADERS

The remarkable success of the Ai6x6 Organization Transformation Series is built upon the **hard-won wisdom** of executive leaders who have successfully navigated the AI transformation within their industries, turning theory into a deeply practical, actionable guide. As we expand the series, we are extending a highly selective invitation to a new cohort of pioneers—individuals whose professional journey is at the **leading edge of AI strategy and execution**. We are looking for executives who possess a proven track record of rewriting business models, mastering unforeseen ethical and competitive challenges, and offering a **clear blueprint** for the next 3-5 years of the AI-Era.

Contributing to the Ai6x6 Series is a unique opportunity to **cement your professional legacy**, joining an exclusive, curated network of global authorities whose insights are distilled into executive-level knowledge cards, directly shaping the governance and innovation decisions of the world's most influential companies. Our selection process is based strictly on **pure expertise and demonstrated leadership** in the AI-Era, reinforcing our commitment to editorial integrity and authentic thought leadership—there is no monetary exchange involved. Select contributors will also receive an invitation to blog on our expert network microsite, further amplifying their platform. To confidentially inquire about contributing your expertise, please contact us via our dedicated portal at **www.Ai6x6.com/contact** or by scanning the adjacent QR code.

Contact

SUBSCRIBE TO OUR NEWSLETTERS: To stay ahead of the curve and receive continuous, executive-level intelligence on the rapid evolution of the AI-Era, we invite you to register for our **Ai6x6 Weekly Newsletter**. Each edition delivers curated insights directly from our thought leaders and highlights critical shifts in AI strategy, governance, and provides relevant excepts from our website and books. Keep your finger on the pulse of the future and ensure your company remains competitive by signing up on the same page

PLEASE REVIEW US ON AMAZON!

Data Blueprint

We know your time is valuable, but if the Ai6x6 Organization Transformation Series delivered the real-world, actionable strategies you needed, please consider leaving a brief review on Amazon. If we missed the mark, we would also like to know, and that is the only we we can improve. Your honest feedback is crucial; it helps fellow executives discover this unique framework and benefit from the insights of our thought leaders. Thank you for supporting our mission!

BLUEPRINT SERIES

SECTION B

Human-AI Synergy

Ai6x6™ Transformation Blueprint

Story 1 — Enigma Codebreakers (1940–45): German U-boats encrypted communications using constantly shifting ciphers, forcing Allied cryptanalysts to chase patterns that dissolved daily. Manual decoding could not keep pace with wartime urgency, and insights often vanished before they could be confirmed. Electromechanical devices accelerated the testing of solutions at unprecedented speed. Humans interpreted these outputs, correcting false leads and refining assumptions. The fusion of machine acceleration & human judgment reshaped the trajectory of the war.

Story 2 — Turing's Bombe & Operators (1940–45): Breaking enemy codes required evaluating thousands of cipher possibilities rapidly, and the Bombe machine generated candidate solutions at speeds humans could not match. Yet its outputs were incomplete, ambiguous, and often misleading. Human operators validated results, corrected machine assumptions, and filtered noise from meaningful signals. Their contextual judgment transformed mechanical output into strategic clarity. This interplay of mechanical speed and human reasoning cracked codes that neither side could solve alone. It became one of history's earliest demonstrations of human–machine symbiosis.

Story 3 — Toyota Production System (1950s–70s): Factories struggled with defects, bottlenecks, and unpredictable workloads as machines produced volume without contextual awareness. Workers observed issues that machines could not detect or interpret, prompting Toyota to design a synchronized workflow that aligned human insight with machine precision. Real-time data from the line informed continuous improvement cycles. This human–machine rhythm created a new operational standard for reliability and efficiency. The system became a global model for intelligent production. It demonstrated how augmentation, not automation, drives sustainable excellence.

Story 4 — Salk Polio Vaccine Trials (1952–55): As polio outbreaks intensified, statistical models guided trial design but lacked the ethical nuance required for medical decision-making. Researchers interpreted model outputs through clinical judgment, ensuring safety and fairness while machines processed vast amounts of trial data. This combined system accelerated discovery without compromising integrity. Human oversight balanced computational rigor with moral responsibility. It showed how human values anchor scientific acceleration.

Story 5 — Apollo 13 Mission Control (1970): An explosion crippled Apollo 13, leaving astronauts in mortal danger as incomplete telemetry streamed back to Earth. Engineers relied on simulations to test survival scenarios under extreme constraints, but the machines could only reveal possibilities. Human teams improvised procedures the computers could not imagine, blending machine-generated insights with creative reasoning. It became a defining moment of human–machine crisis collaboration.

Story 6 — Air Traffic Control Modernization (1960s–80s): Surging air travel overwhelmed manual tracking systems, and controllers struggled to maintain safe separation across increasingly crowded skies. Radar automation provided real-time positional data, enabling machines to track aircraft continuously while humans made judgment calls. This hybrid system reduced collisions, increased capacity, and established a global standard for aviation safety. Human–machine coordination became the backbone of modern air travel. The system remains one of the world's most enduring examples of operational augmentation.

LESSONS FROM A DATA TRANSFORMATION PERSPECTIVE

Data's role evolves from raw signals requiring human interpretation to real-time streams shaping operational decisions. Early systems depended on humans to transform machine-generated possibilities into clarity, while later systems integrated data into continuous feedback loops that improved reliability and safety. Across this arc, one principle holds: data becomes powerful only when humans contextualize it. Machines accelerate pattern detection, but humans determine relevance, risk, and action. Human–AI synergy turns data into a shared cognitive substrate—machines providing scale and precision, humans providing meaning and direction. The result is a data environment that becomes progressively more intelligent through collaboration.

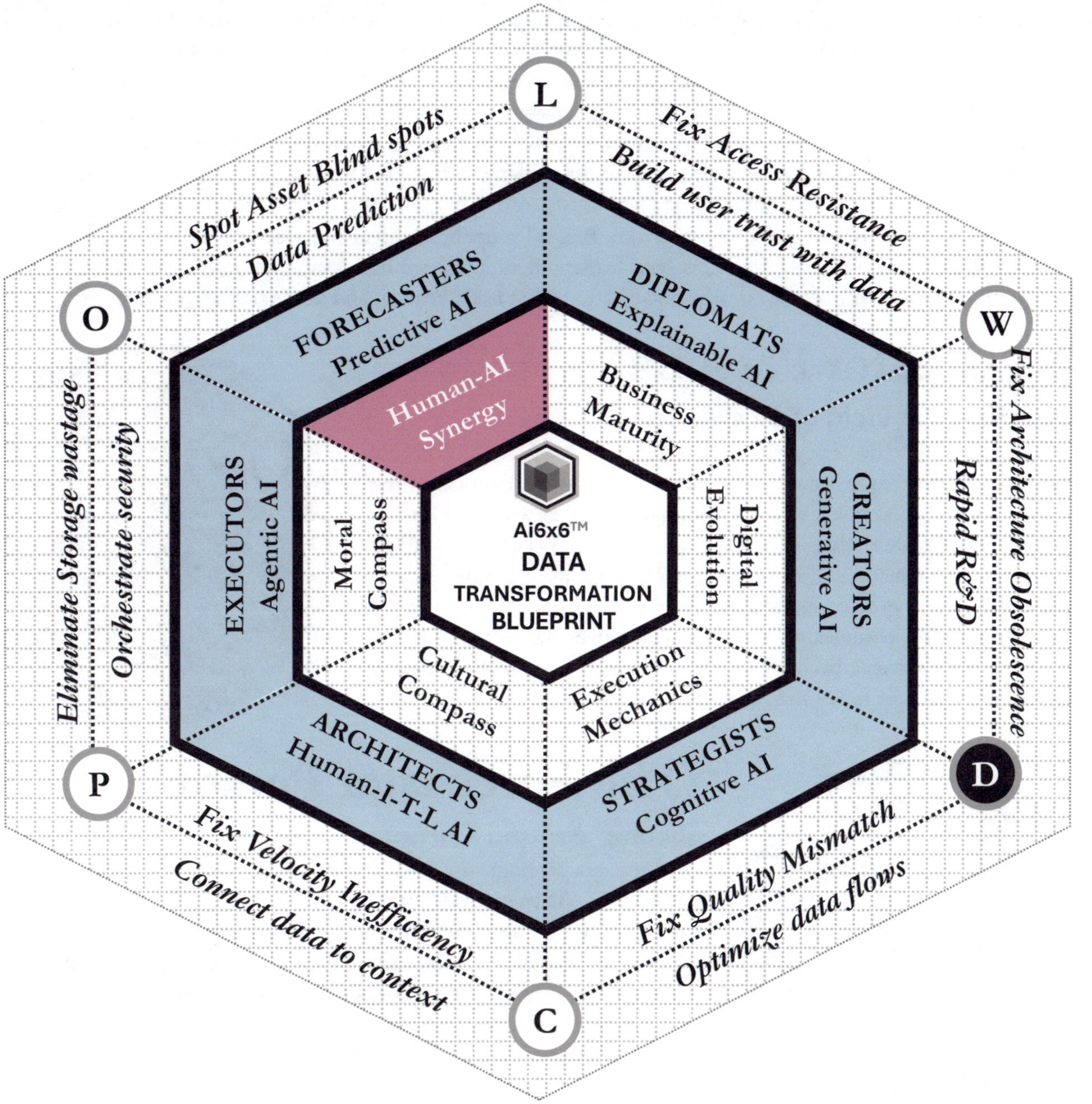

The **Ai6x6 Research Lab** chose this lens to address the fundamental evolution of data from a static resource to an active, intelligent partner. We believe the modern data professional must move beyond being a mere curator of information and become an orchestrator of autonomous systems. This lens is vital because it redefines how humans interact with the sheer scale of modern information through diverse functional personas. By exploring the synergy between human judgment and agentic logic, the Lab ensures that data becomes a living ecosystem capable of **Predictive Foresight** and **High-Level Reasoning**. It is designed to transform the data team into a force that uses AI to illuminate hidden patterns while maintaining absolute **System Transparency**.

It is about shifting the practitioner's heart from a state of data management to one of strategic "Visioneering." This lens acts as the cognitive bridge, allowing the individual to see AI not as a black box that obscures the truth, but as a diverse set of personas—like the **Architect** and **Executor**—that extend their own analytical reach. Without this shift, data remains a burden of complexity; with it, it becomes a source of profound innovation. It is an invitation to shed the role of a data gatekeeper and embrace a new, augmented authority grounded in the oversight of autonomous agents. The Lab ensures that while the machines process the volume, the human remains the essential guide for strategic context and ethical oversight.

Leadership in the age of intelligent systems demands a new grammar of collaboration—one that transcends the binary of human versus machine and instead frames intelligence as a continuum. The **Human–AI Synergy Lens** redefines leadership not as the art of commanding technology but as the discipline of co-creating with it. It asks leaders to move beyond automation and efficiency toward a deeper integration of foresight, ethics, creativity, cognition, governance, and execution. At its core, Human–AI Synergy is about **shared agency**. It recognizes that leadership today is exercised not only through human judgment but also through algorithmic influence. Every forecast, recommendation, and automated decision carries a fragment of leadership intent. The challenge is not to suppress that influence but to shape it—to ensure that AI amplifies human purpose rather than diluting it. This lens therefore becomes a mirror: it reflects how leaders distribute cognition, emotion, and control across human and machine systems, and how they design organizations that learn, reason, and act as unified intelligences.

The Creator Persona

Expanding the Field of Possibility

Creators represent the imaginative frontier of leadership. Through **Generative AI**, they explore more ideas than any human-only process could conceive. They treat AI not as a threat to originality but as a catalyst for it—a partner in ideation that expands the creative bandwidth of the organization. The Creator's discipline is not about producing more content but about discovering new patterns of thought. They use AI to prototype futures, to visualize complexity, and to democratize innovation. In their hands, generative systems become engines of curiosity, enabling leaders to design from abundance rather than scarcity.

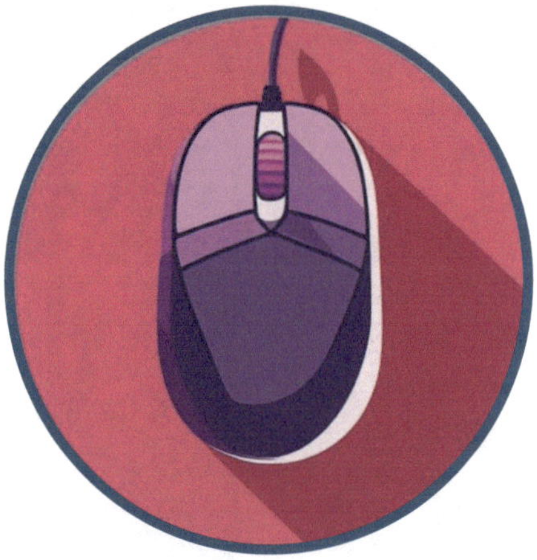

The Forecaster Persona

Seeing Around Corners

Forecasters embody the anticipatory dimension of leadership. They use **Predictive AI** to model futures before committing resources, transforming uncertainty into structured insight. Their craft lies in converting data into foresight—detecting weak signals, pressure-testing assumptions, and rehearsing alternative scenarios. In traditional leadership, forecasting was intuition-driven; in synergistic leadership, it becomes a disciplined dialogue between human imagination and machine probability. The Forecaster's maturity is measured not by how accurately they predict but by how responsibly they prepare. They shift the organization from reacting to events to anticipating them, embedding resilience into its strategic DNA.

The Strategist Persona

Orchestrating Cognitive Intelligence

Strategists use **Cognitive AI** to weave intelligence across functions, geographies, and time horizons. They design strategy as a living system—adaptive, self-correcting, and continuously learning. The Strategist's genius lies in orchestration: aligning distributed intelligences into coherent purpose. They understand that cognition is no longer centralized in the executive suite; it flows through networks of humans and machines that co-reason in real time. Their leadership is systemic rather than hierarchical, emphasizing feedback loops, dynamic modeling, and cross-domain synthesis. In this persona, AI becomes the connective tissue of enterprise intelligence.

The Diplomat Persona:

Building Ethical Bridges

Diplomats operate at the intersection of **Explainable AI** and human empathy. They translate algorithmic logic into moral language, ensuring that machine outputs are understood, trusted, and socially accepted. Their role is to humanize the interface between data and decision, to make transparency a leadership act. In a world where AI can both illuminate and obscure, the Diplomat safeguards the integrity of interpretation. They remind the enterprise that trust is not a technical feature—it is a cultural outcome. By bridging emotional and ethical terrain, they ensure that AI strengthens relationships rather than eroding them.

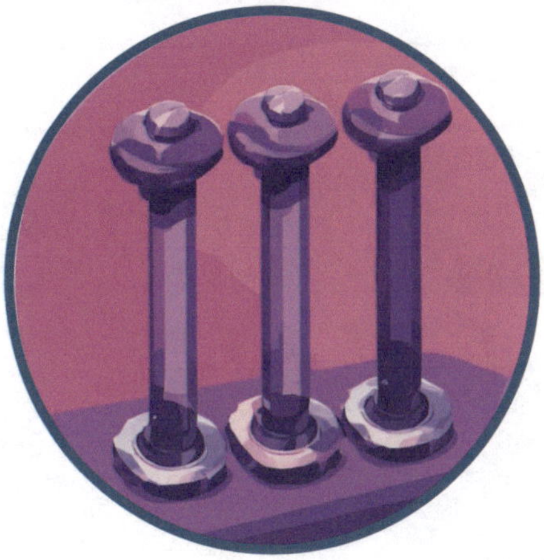

The Architect Persona:

Designing Responsible Collaboration

Architects are the custodians of **Human-in-the-Loop AI**, ensuring that governance and ethical guardrails are embedded into every workflow. They design environments where humans and machines collaborate responsibly—where oversight is not an afterthought but a structural principle. The Architect's leadership is infrastructural: they build the conditions under which synergy can thrive. Their focus is on accountability, transparency, and inclusion, crafting systems that respect both human dignity and machine precision. In their world, collaboration is engineered, not improvised; it is sustained by design integrity.

The Executor Persona:

Converting Insight into Action

Executors close the loop between knowing and doing. Through **Agentic AI**, they embed intelligence directly into operations so that insight becomes action without delay. They represent the kinetic dimension of synergy—the moment where cognition translates into execution. The Executor's mastery lies in operationalizing intelligence: automating decisions that are safe, contextual, and aligned with strategic intent. They ensure that the organization's nervous system is responsive, not reactive, and that every process carries the imprint of learned intelligence.

Together, these six personas form a **leadership spectrum**—from foresight to action, from imagination to governance. They illustrate that synergy is not a single capability but a choreography of roles. The Forecaster anticipates, the Diplomat interprets, the Creator imagines, the Strategist aligns, the Architect safeguards, and the Executor delivers. Each persona represents a distinct mode of engagement with AI, yet all converge on one principle: leadership is now a shared performance between human and machine. The Human–AI Synergy Lens therefore reframes leadership as **co-evolution**. It invites leaders to cultivate humility before complexity, curiosity before certainty, and design before control. It challenges them to lead not by commanding intelligence but by convening it—to orchestrate a dialogue between human wisdom and machine capability that continuously redefines what is possible. In this lens, synergy is not the end state; it is the ongoing discipline of learning together.

THE FORECASTER

Predictive AI · Meteorologist

Detect weather patterns

Spot Outside change

Seeing around corners and spotting trends before they hit.

THE STRATEGIST

Cognitive AI · Urban Planner

Scenario modeling

Suggest Growth options

Orchestrating intelligence and making complex decisions.

THE EXECUTOR

Agentic AI · Operations Chief

Auto-pilot smart city

Keep city running 24x7

Closing the gap between knowing and performing automated action.

THE CREATOR

Generative AI · Master Architect

Concepts & drafts

Design unique facilities

Making, building, and expanding the field of possibility.

THE DIPLOMAT

Explainable AI · Press Secretary

Explain system logic

Win people's trust

Building trust, empathy, and ethical connection.

THE ARCHITECT

Human I-T-L AI · Mayor & City Hall

Approval thresholds

Stop Unsafe moves

Providing the structure, governance, and guardrails.

Ai6x6™ Transformation Blueprint

Evolution of Humanity

"What is true is that which is successful."

Charles Sanders Peirce (1839–1914); The Unified Source of Truth (USoT) is "true" because it reliably enables effective governance and successful, coherent, data-driven action.

The earliest forms of data management, from the ancient scribes of Sumeria to the archivists of Alexandria, found their fulfillment in the painstaking process of establishing Verifiable Truth. Their purpose was providing the foundational certainty—the reliable record—upon which governance, commerce, and history could be built. Their professional satisfaction was rooted in the confidence their meticulous work provided. In the Information Age, the explosion of data volumes and the proliferation of conflicting systems introduced systemic doubt. Fulfillment for the modern analyst became diluted by the constant, exhausting need for manual reconciliation and data cleaning, leading to professional anxiety and the loss of Unimpeachable Confidence. The AI Era rectifies this by mandating data as a Governance Authority. Data professionals find their highest purpose in architecting automated systems of truth (like Automated Validation Checkpoints). Their fulfillment shifts from wrestling with dirtied data to guaranteeing its structural integrity, enabling them to focus their expertise entirely on generating strategic, certified insights based on a foundation they trust completely.

SIX UNIVERSAL ANCHORS OF HUMAN FULFILLMENT

HUMAN RAISON D'ÊTRE (THE REASON FOR EXISTENCE)

In an age defined by exponential automation, the core question of human existence shifts from survival to **significance**. When technology handles efficiency and labor, human value is no longer defined by capability, but by **meaning**. The six **Universal Anchors of Human Fulfillment** define the enduring, non-negotiable contributions of human consciousness—the pillars of a **Raison D'être** that secures our purpose across any era. These anchors affirm that our ultimate mission is to serve as the unwavering source of **Meaning and Moral Authority** in existence. They are the essential acts that only consciousness can perform.

1) TO EXPRESS

The **Anchor of Expression** is the essential human mandate to be the **Source of Novelty and Creation**, generating original ideas and subjective feelings that resist calculation or statistical synthesis. This purpose validates our unique consciousness by asserting that fulfillment comes from the generative act of bringing forth something utterly unique, not merely consuming or replicating. It requires **Uncurated Creativity**—the freedom to explore and define **Aesthetic Judgment**—thereby expanding the conceptual universe with concepts that have no statistical precedent. Through expression, we transform complex inner experience into external reality, upholding our role as the inventors of meaning.

The ultimate benefit of this anchor is the preservation of **Human Originality** and **Intellectual Sovereignty**. By committing to expression, we secure a future that is continually enriched by unique ideas, preventing stagnation and the predictability of automated systems. This anchor maintains the essential spark of invention, ensuring that human meaning remains tied to the capacity to create, **differentiating our value from that of a machine**. It is the source of all artistic, philosophical, and inventive breakthroughs that define a truly dynamic culture.

2) TO BELONG

The **Anchor of Belonging** defines our core identity as social beings, asserting that **authentic connection** is an existential necessity beyond mere digital interaction. This purpose requires the active, continuous labor of **Deep Connection**, prioritizing **unconditional presence** and shared vulnerability to foster a resilient community. Fulfillment is achieved by being the **social architect**, engaging in the emotional work of **mediating conflict**, and fostering cohesion. It includes preserving and transmitting **Cultural Heritage**—the stories and qualitative wisdom that bind us across generations. To belong is to confirm that our worth is interwoven with our relationships.

The ultimate benefit of this anchor is **Emotional Resilience** and **Social Stability**. By prioritizing genuine human connection, we safeguard ourselves against isolation and the psychological fragility that arises from purely transactional relationships.[1] This anchor ensures a supportive social framework that validates empathy and provides necessary emotional resonance, acting as a crucial defense mechanism against an isolating technological world. It builds the **trust and cohesion** necessary for any human system to function justly and compassionately.

3) TO FEEL

The **Anchor of Feeling** grounds human existence in the tangible, asserting the profound psychological value of **Embodied Mastery** and sensory engagement. This purpose elevates the dedicated pursuit of physical skill and process—not for necessity, but for the inherent satisfaction of the process. It is the quest for the **'Flow' state**, where an earned link between **exertion and accomplishment** provides deep fulfillment. This anchor mandates resisting the ease of total automation by engaging in voluntary challenges and perfecting non-essential crafts. It ensures that our abstract thoughts are grounded in the undeniable reality of the body and the senses.

The ultimate benefit of this anchor is **Physical and Psychological Integrity**. By committing to felt experience, we preserve our cognitive and physical health, resisting the inertia of automation and maintaining a sharp connection to reality. This anchor validates the **dignity of effort**, providing a clear, visceral confirmation of personal competence and self-mastery that is crucial for self-esteem. It ensures that fulfillment remains an **earned state**, confirming the complexity and richness of being physically alive.

4) TO GUIDE

The **Anchor of Guidance** establishes the human being as the **ultimate moral and ethical sovereign**. This purpose is founded on our unique capacity for empathy and subjective value judgment, making us the non-delegable authority for setting rules and managing risk. It is the practice of **Ethical Governance**, demanding that we **Draw the Moral Line** and translate compassionate principles into enforceable policy. Fulfillment is achieved by maintaining transparency, enforcing accountability, and continuously exercising the **Human Veto Power** over critical decisions.

The ultimate benefit of this anchor is **Systemic Integrity** and **Moral Safety**. By retaining final ethical authority, we ensure that technological power remains fully aligned with, and subservient to, human values, preventing systems from optimizing themselves at the expense of human welfare. This anchor provides the essential check against unchecked power, establishing the legal and moral frameworks necessary to protect all citizens. It guarantees that empathy and justice remain the **supreme architects of our collective future**.

5) TO KNOW

The **Anchor of Knowing** defines our core intellectual purpose as the engine of **Epistemic Exploration** and abstract reasoning. This purpose compels the mind to focus on formulating **Axiomatic Hypotheses**—the fundamental "Why?" questions that lie beyond the reach of data processing and established knowledge. Fulfillment is found in the generative act of **challenging established truths** and creating new theoretical frameworks that synthesize disparate fields. This anchor ensures that the pursuit of knowledge remains boundless, driven by philosophical depth rather than functional utility.

The ultimate benefit of this anchor is **Intellectual Freedom** and **Conceptual Progress**. By focusing on formulating new questions and challenging paradigms, we prevent the stagnation that could result if all knowledge were confined to data-driven answers. This anchor guarantees that our intellectual journey continues to be defined by profound theoretical breakthroughs and a relentless expansion of understanding. It confirms that human intelligence is defined not by the volume of information stored, but by the **depth and originality of the questions** we ask.

6) TO STRIVE

The **Anchor of Striving** is the **Existential Quest**—the voluntary pursuit of difficulty necessary to generate meaning, independent of any external necessity. This purpose recognizes that true fulfillment is an **earned state**, achieved through **Deliberate Discipline** and self-imposed ambition. It compels us to **Set Impossible Goals** and **Choose the Hard Way**—overcoming voluntary hardship—precisely because the challenge validates our capacity and resolve. This continuous effort allows us to craft a meaningful **Personal Legacy** by executing complex, long-term plans purely for the satisfaction of the effort.

The ultimate benefit of this anchor is **Authentic Self-Mastery** and **Purposeful Existence**. By embracing self-imposed difficulty, we cultivate the resilience, discipline, and fortitude that define character, resisting the complacency of automated comfort. This anchor ensures that human worth is determined by our **internal will**, rather than external circumstance, providing a framework for continuous personal growth and the creation of an enduring, meaningful legacy.

CONCLUSION: THE RAISON D'ÊTRE FOR THE HUMAN

The six Universal Anchors converge to form humanity's ultimate purpose: to be the **Unwavering Source of Meaning and Moral Authority**. This composite Raison D'être defines the essential human role in any advanced civilization.

Our mission is the sustained commitment to **non-automatable value**. We are the unique entity that not only pursues truth (To Know) but also creates the beauty (To Express) that makes truth worth pursuing. We are the mechanism that defines the rules of fairness (To Guide) and then personally invests the necessary effort (To Strive) to live up to those rules. Crucially, we are the only entity that grounds this entire conceptual structure in the tangible joy of physical reality (To Feel) and the vital necessity of shared vulnerability (To Belong).

The human purpose shifts from survival to an intentional journey of **growth, creation, and ethical custodianship**. Our value is not in output, but in the **input of consciousness, conscience, and will**. By diligently exercising these six anchors, every individual contributes to the essential, enduring purpose of the species: to be the **Sole Arbiter of Worth, Will, and Wonder**—ensuring that existence remains a deeply meaningful and purposeful quest.

In this era, human existence was focused on the direct, tangible application of skill for immediate survival, such hunting, fishing and surviving. The human was the Master of the Hunt.

In the Primitive/Paleolithic Era (Survival & Sustenance), human purpose was an absolute function of immediate necessity. Fulfillment was not a philosophical pursuit but a practical outcome of successfully continuing existence for oneself and the kin group. All six Universal Anchors manifested, but they were tethered directly to the biological and environmental imperative to survive the day and propagate the species. The overarching Raison D'être was to **Endure and Transmit**, meaning personal worth was inseparable from one's contribution to the group's survival calculus. This collective focus meant that individual talent was instantly converted into a benefit for all, solidifying a direct, unbroken link between personal action and existential significance.

To Express: Expression was functional and communal, centered on **Ritual Storytelling** and cave art. This art was not for individual fame or abstract aesthetics, but served as the primary educational system, preserving and transmitting essential survival knowledge, hunting techniques, spiritual beliefs, and the topography of the land. The repetitive, sacred nature of the rituals ensured collective memory and adherence to successful methods. Furthermore, these cultural forms encoded crucial information, often representing animals and spirits, that, if lost or forgotten, would jeopardize the tribe's safety and cultural identity.

To Belong: The Anchor of Belonging was non-negotiable, manifesting as **Kinship and Mutual Defense**. Isolation meant certain death; therefore, the purpose of connection was absolute. Fulfillment came from actively contributing to the group's defense, sharing food and resources, and ensuring the collective physical safety of the tribe above all else, fostering a fierce, self-sustaining mutualism. The emotional security derived from this deep collaboration—knowing that others were committed to your well-being—was a core psychological reward.

To Feel: Embodied Mastery was defined by **Essential Tool Mastery**. The feeling of accomplishment was directly tied to successfully chipping a sharp stone tool, accurately throwing a spear, or enduring the physical hardships of the hunt. This pursuit was not sport, but the daily necessity of mastering the physical world to secure food and shelter, and the psychological confidence gained was crucial for future success. This expertise immediately translated into social value; the most capable hunter or tool-maker earned respect and a higher status within the hierarchy, directly reinforcing the dignity of effort.

To Guide: Moral authority was based on immediate experience, taking the form of **Elder Authority and Tradition**. Guidance was unwritten law, passed down through generations of hard-won experience. Decisions—like where to hunt, when to migrate, or how to resolve disputes—were arbitrated by those with the deepest, most successful experiential knowledge, and were obeyed out of necessity for the group's stable cohesion. This system was vital for ensuring collective action remained predictable and risk-adverse, preventing internal conflict from threatening external survival.

To Know: The pursuit of knowledge was pure **Observation and Practicality**. The purpose of knowing was to accurately predict the immediate environment for resource exploitation. This involved memorizing the habits of migratory animals, understanding the medicinal or poisonous properties of local plants, and tracking subtle shifts in weather patterns—information critical for daily sustenance and successful, low-risk planning. Early forms of abstract reasoning also emerged here, through tracking the movement of stars or noting the cyclic nature of seasons and astronomical events, demonstrating an innate human desire to find order and predictability in environmental chaos.

To Strive: Striving was synonymous with **Daily Endurance**. There was no need to invent abstract goals; the pursuit of difficulty was the environment itself, offering constant challenge. Fulfillment was the successful completion of the continuous, grueling effort required to secure food, withstand the elements, and ensure the basic propagation of the tribe, creating a deeply ingrained sense of earned worth. The resilience built through continuous hardship was the ultimate personal asset, psychologically connecting physical suffering with the existential meaning of survival.

With stability, purpose shifted to sustained output, long-term planning, and maintaining social harmony within a fixed settlement. The human was the Master of the Land.

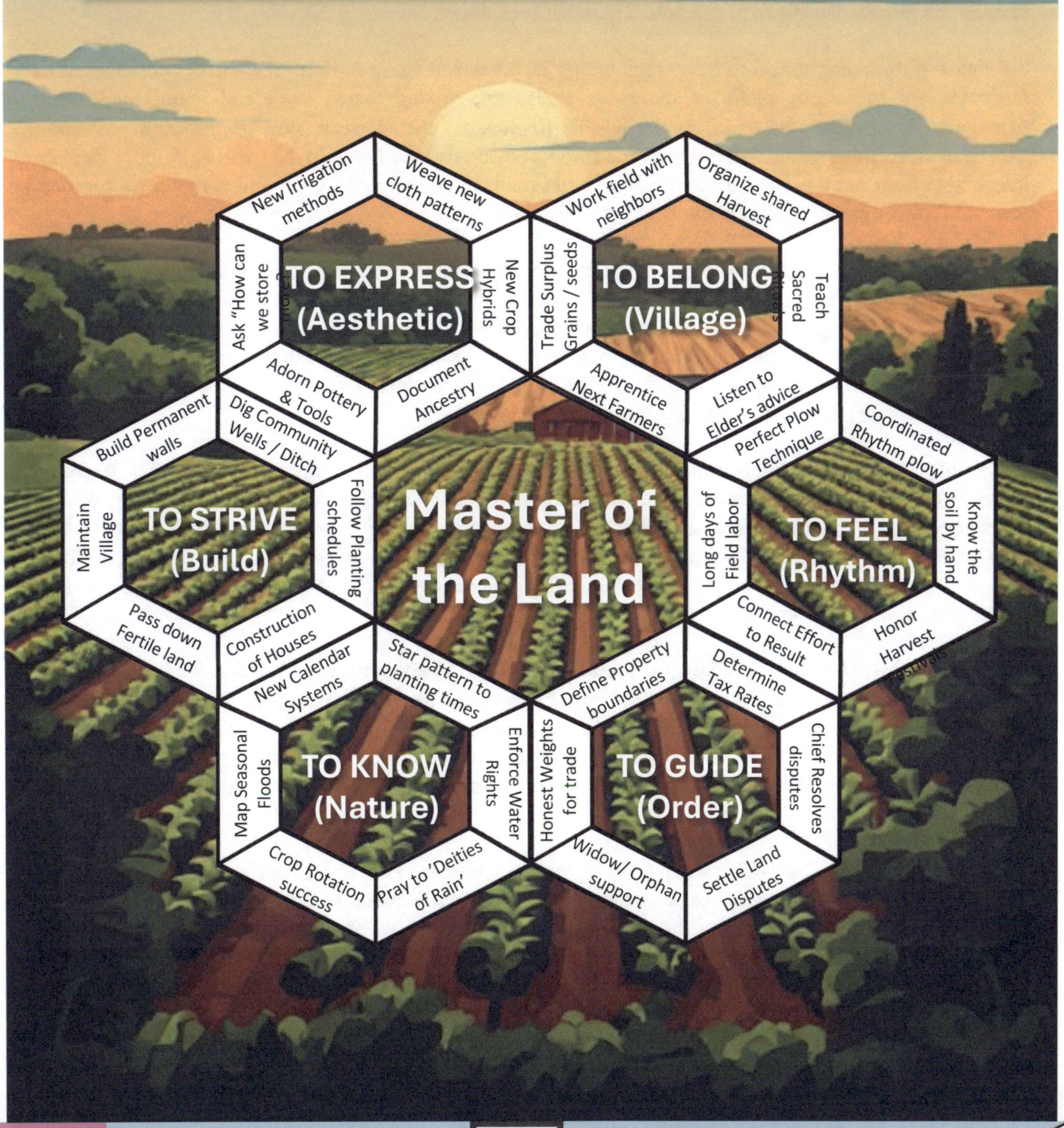

The Agricultural/Neolithic Era (Farming & Community) marked humanity's fundamental pivot from nomadic struggle to settled stability. The mastery of cultivation and animal husbandry created a reliable **Surplus of Time and Resources**, fundamentally shifting human purpose from immediate survival to calculated investment, long-term order, and the establishment of lasting community structures. The Raison D'être of this period was to **Settle and Establish Order**. This era saw humans take on the role of conscious managers of the environment, laying the groundwork for systematic laws, property rights, and enduring social legacies that extended far beyond one generation.

To Express: Expression became focused on **Accounting and Enduring Form**. The logistical necessity of tracking seeds, yields, stored grains, and communal ownership led directly to the invention of the first systematic writing forms, such as early cuneiform, demonstrating that expression was initially a tool for managing complex resources. Beyond text, expression was seen in durable, functional goods like specialized pottery, permanent architectural structures, and woven textiles. These forms reflected a deep, stable community identity and provided evidence of specialized skill that contributed directly to the village's permanence.

To Belong: Belonging shifted to being centered on **Land-Based Cooperation**. The success or failure of the farm was the success or failure of the entire settlement, making shared labor and mutual dependency paramount. Fulfillment was found in the reliable rhythms of shared work, such as irrigation projects or harvesting, cementing generational inheritance and deeply intertwined **Kinship Networks**. Purpose was now inseparable from one's physical, defined place within the land and the local community, creating a profound sense of rootedness and stability unknown in the Paleolithic past.

To Feel: Embodied Mastery focused on **Craftsmanship and Process**. Fulfillment was derived from the methodical, repeated application of skill required by agriculture—plowing a straight furrow, carefully managing livestock, or perfecting the complex processes of metallurgy and food preservation. This mastery provided a sense of control over the environment that raw endurance could not. The purpose of skill was purely to generate the stable surplus required to maintain the settled life, providing a deep psychological reward through the predictable yield of sustained, expert effort.

To Guide: Moral authority became formalized, focusing urgently on **Property Rights and Resource Management**. Guiding the community meant establishing clear, complex rules around ownership of land, managing precious resources like water rights, and ensuring crop protection from both animals and neighbors. This necessity gave rise to the first systematic, written laws, moving purpose from inherited tradition (as in the Paleolithic) to a **Codified Social Contract**. Leadership was defined by the ability to enforce these new laws, ensuring fairness and stability within an increasingly complex economy.

To Know: Abstract reasoning began to flourish, driven by **Mathematics and Calendar Tracking**. The purpose of knowledge was now predictive certainty, essential for farming success. This meant mastering geometry to accurately divide and manage fields, and engaging in sophisticated celestial observation to track seasons and accurately determine optimal planting and harvesting cycles. This pursuit of cosmic order confirmed the human ability to impose intellectual structure upon chaotic nature, turning farming into a sophisticated, knowledge-driven enterprise.

To Strive: Striving fundamentally shifted from day-to-day survival to **Creating Surplus and Investment**. The existential quest became the arduous, long-term, deliberate work required to build durable structures, amass protective stores of food, and ensure the economic and social prosperity of the next generation. This focus on legacy and investment transformed work into a purposeful act of building a lasting future, where one's worth was measured not just by survival, but by the inheritance left behind for the family and the stability provided to the community.

This era defined purpose through professional expertise and contribution to complex, often abstract, technological and economic systems. The human was the Master of the Machine.

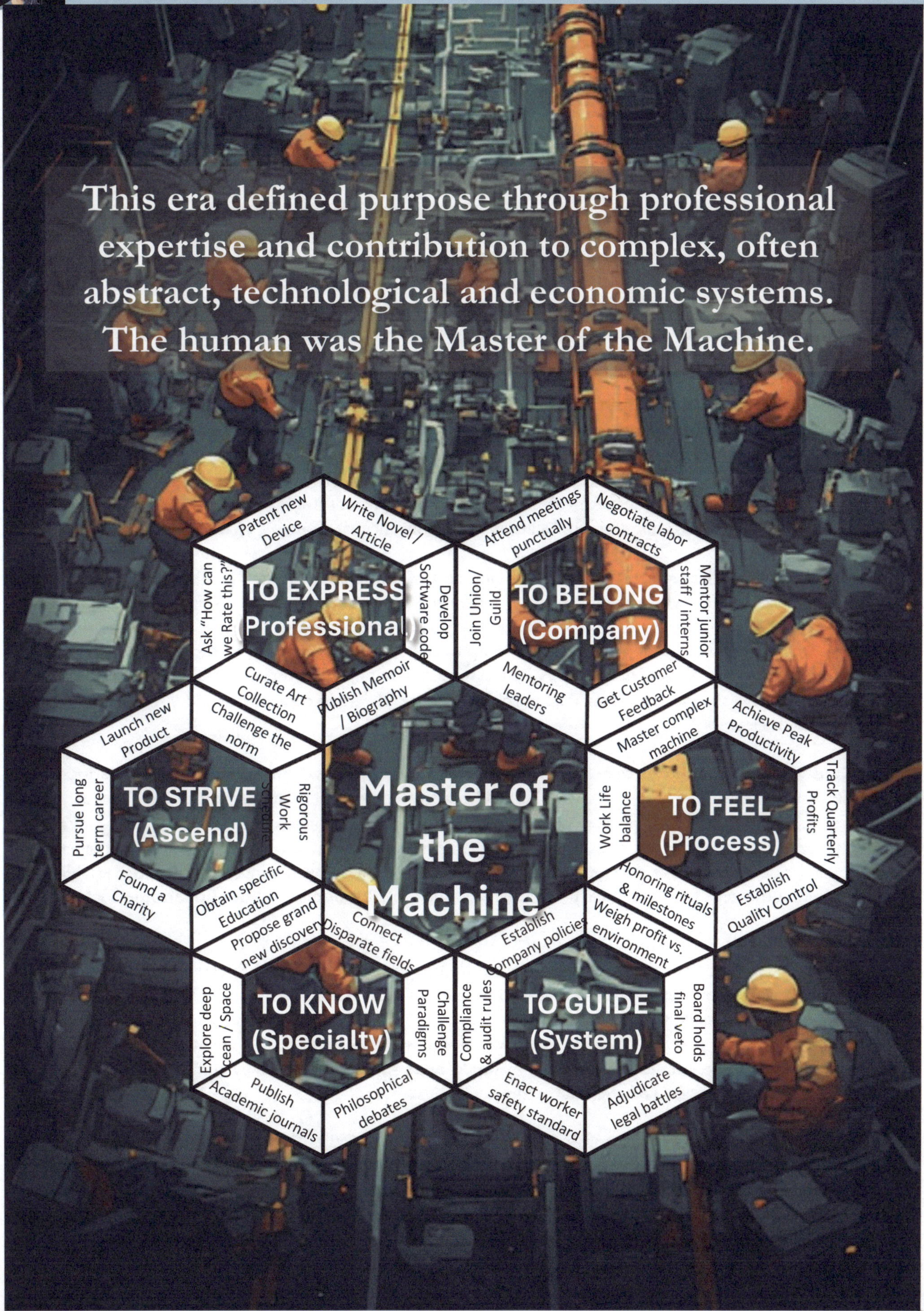

The Modern/Industrial Era (Specialization & Function) marked the decisive triumph of machine logic and capital over the natural world. Human purpose shifted from managing the cycles of the land (Neolithic) to mastering the mechanical efficiency of the factory and the market. The rise of the machine and the division of labor fundamentally redefined the individual, whose worth was now derived from highly specialized, functional expertise within a vast, interconnected economic system. The Raison D'être of this period was to **Achieve and Advance**, driving relentless progress through expertise and material gain.

To Express: Expression became focused on **Mass Media and Invention**. With the advent of the printing press, cheaper paper, and later, radio and cinema, creative expression transformed into a commodified product designed for wide distribution and profit. The purpose of expression was to inform, entertain, and persuade the masses, shaping shared cultural narratives through novels, newspapers, and commercial design. The creation of a **Patentable Invention** became a powerful act of individual expression, translating originality directly into economic ownership and societal contribution. Expression was celebrated when it served a functional, marketable, or propagandistic role.

To Belong: Belonging was redefined by **Nation and Institution**. The old agrarian kinship networks dissolved as masses migrated to urban factory centers, replaced by new, abstract, yet powerful social structures. Fulfillment was found in national identity, membership in professional organizations, and political affiliation. Crucially, the extreme specialization of labor necessitated the formation of organized counter-structures like **Labor Unions** to protect the rights of workers against centralized corporate power, demonstrating that belonging was a functional necessity for collective bargaining and social stability in a highly unequal economic system.

To Feel: Embodied Mastery centered on **Efficiency and Maximized Output**. The physical purpose shifted away from holistic craft towards highly repetitive, specialized labor—the mastery of a single, precise task on the assembly line. While monotonous, this specialization provided a clear, measurable sense of contribution, where personal worth was directly tied to economic output and reliability. Outside of the factory, the Anchor manifested in the rise of **Organized, Competitive Sports**, where physical effort was deliberately structured and formalized, allowing the industrial individual to achieve a sense of earned competence and visceral struggle that was absent in their work.

To Guide: Moral authority became formalized through **Secular Law and Corporate Regulation**. The complexity of contract law, property ownership, and massive corporate liability demanded sophisticated, non-religious guidance systems. The purpose of guidance was to manage the inherent conflicts between capital and labor, protect consumer safety in mass production, and define the boundaries of ethical economic competition. This period saw the relentless development of regulatory bodies and constitutional law aimed at establishing fairness within the new, powerful industrial economy, ensuring that **progress did not undermine social order** entirely.

To Know: The pursuit of knowledge was dominated by **Empirical Science and Specialization**. The purpose of knowing was functional utility, believing that all problems—from disease to industrial inefficiency—could be solved through the application of the rigorous **Scientific Method**. Knowledge became fragmented into specialized disciplines, each dedicated to gathering objective, measurable data. Fulfillment was achieved by contributing a verifiable, peer-reviewed fact or theory to one's specific field, thereby confirming the era's deep-seated faith in rational inquiry and technical progress.

To Strive: Striving was synonymous with **Ambition and Material Gain**. The existential quest was the pursuit of wealth, social mobility, and **Status**. Fulfillment was an earned state achieved through economic success, the establishment of a profitable business, or professional advancement within a corporate or governmental hierarchy. The cultural narrative of the "self-made individual" defined personal purpose, urging people to embrace continuous, relentless work to secure prosperity for the family and, more importantly, to carve out an enduring legacy of material success and high societal standing.

In this emerging era, AI handles efficiency and role specialization, freeing and forcing humans to refocus on subjective values, unique creation, and moral governance. The human becomes the Master Curator of Purpose

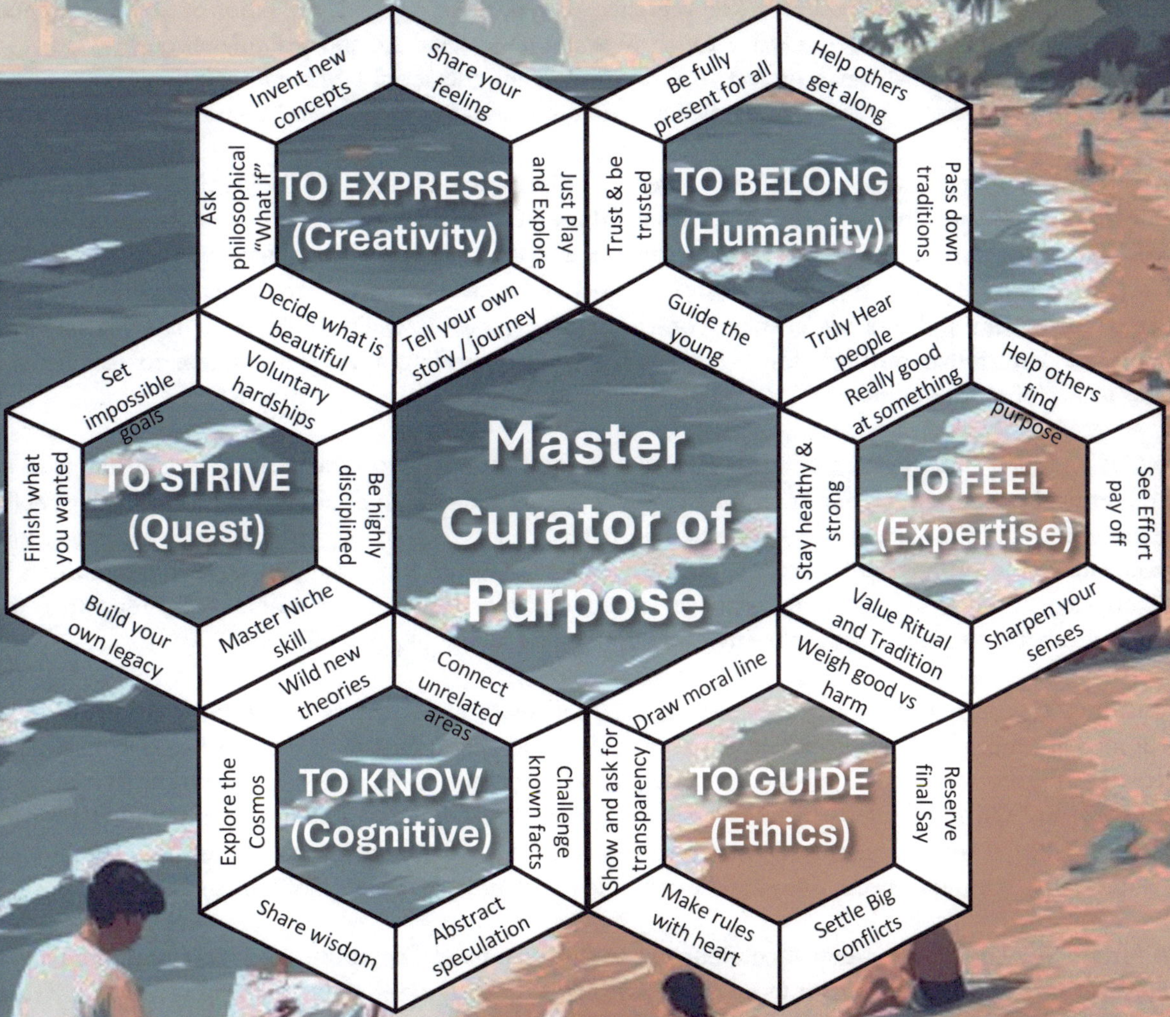

The Post-AI Era (Curation & Meaning) represents the definitive existential shift in human history. With advanced artificial intelligence and automation having conquered the challenges of efficiency, information processing, and mass labor, human purpose is radically redefined. We are no longer necessary for function, but for **meaning**. The individual's worth is derived entirely from their capacity for subjective judgment, ethical authority, and the creation of non-calculable value. The collective Raison D'être is to become the **Sole Arbiter of Worth, Will, and Wonder**—the source of all criteria for why life should continue and what makes it beautiful.

To Express: Expression is focused on **Subjectivity and Uncurated Truth**. In a world saturated with synthetic media and hyper-optimized content generated by AI, the only expression that retains intrinsic value is the genuinely **original, subjective human voice**. This purpose mandates focusing on authentic, unfiltered creation—personal philosophical content, aesthetic critique, memoir, and art that contains flaws, uncertainty, and emotional resonance. Fulfillment is found in generating novelty that AI cannot predict, thereby constantly pushing the boundaries of human experience and securing a creative future that is immune to algorithmic stagnation.

To Belong: Belonging becomes a deliberate, chosen investment in **Deep Vulnerability and Presence**. When digital connection is ubiquitous but emotionally shallow, true fulfillment relies on actively resisting transactional relationships and seeking authentic, physical presence. This purpose demands the **labor of Deep Connection**: cultivating empathy, mediating conflict through personal understanding, and prioritizing quality, face-to-face interaction. The collective human project requires maintaining the social fabric and emotional resonance that machines cannot replicate, ensuring that isolation does not erode the psychological foundation of society.

To Feel: Embodied Mastery transforms into the **Voluntary Pursuit of Difficulty and Flow**. Since automation eliminates the necessity of difficult labor, human physical purpose centers on self-imposed challenges—non-essential crafts, athletics, meditation, and the perfection of technical, yet unnecessary, skills. This pursuit generates the vital 'Flow' state—the psychological reward of earned accomplishment—which secures self-esteem and grounds abstract consciousness in tangible reality. Fulfillment is achieved through **Deliberate Effort** that reaffirms the dignity of the body, providing a visceral antidote to the inertia and complacency of total comfort.

To Guide: Moral authority centers on **AI Ethics and the Human Veto**. The purpose of guidance is to serve as the **ethical sovereign**, placing strict moral boundaries around autonomous systems. This involves defining the principles of data governance, ensuring transparency in black-box algorithms, and enforcing systemic accountability for technological outcomes. Crucially, every individual is mandated to exercise the **Human Veto Power**—the non-delegable right to overrule critical decisions made by algorithms—ensuring that power remains permanently aligned with subjective human values, not cold-blooded efficiency.

To Know: The pursuit of knowledge is now defined by **Abstract Synthesis and Axiomatic Inquiry**. In an environment where AI holds access to all existing data and established facts, human intellect must shift its focus. The purpose is to transcend data processing and concentrate on formulating **Axiomatic Hypotheses**—the profound "Why?" questions that challenge foundational assumptions. Fulfillment is found in connecting disparate fields of knowledge, creating new philosophical frameworks, and constantly resisting the temptation to settle for algorithmic answers, ensuring the continuous, boundless expansion of human understanding.

To Strive: Striving is the existential commitment to **Self-Imposed Discipline and Personal Legacy**. Since survival is guaranteed, the quest for meaning requires the individual to set ambitious, non-essential goals purely for the satisfaction of the effort. This involves cultivating profound resilience and fortitude against the immense temptation of automated ease. Fulfillment is an **earned state**, achieved through continuous, conscious effort and the construction of an enduring **Personal Legacy**—a body of work or moral character that defines one's worth by internal resolve rather than external economic output.

Ai6x6™ Transformation Blueprint

Human-AI Personas

"Artificial intelligence is not a substitute for human intelligence; it is a tool to amplify human creativity and ingenuity." - Fei-Fei Li, a prominent Computer Scientist and Co-Director of the Stanford Institute for Human-Centered Artificial Intelligence (HAI).

The Ai6x6 HUMAN-AI Synergy framework formalizes a new executive standard for the AI era, transitioning organizations from fragmented automation to a state of Orchestrated Autonomy. This model defines the modern enterprise through six core leadership movements, each serving as a specialized "sense" for the organization. **The Forecasters** serve as the "Eyes," tasked with the heavy lifting of foresight to eliminate strategic blind spots and detect invisible market signals before they escalate into crises. Supporting this vision, **The Diplomats** act as the "Voice," translating complexity into human trust and ensuring that transformation is met with cultural alignment rather than institutional resistance. The organization's creative energy is fueled by **The Creators**, the "Engine" of the enterprise, who move beyond static production to drive a state of radical velocity and infinite iteration. Their output is balanced by the surgical objectivity of **The Strategists**, acting as the "Compass" to ensure that every capital investment is mathematically anchored to peak ROI. To unify these efforts, **The Architects** build the "Institutional Brain," ensuring that the organization's collective memory is always accessible and relevant to the individual. Finally, the framework finds its movement in **The Executors**, the "Hands" of the organization, who coordinate autonomous workflows to achieve scale without traditional overhead. By weaving these archetypes into a single Symphony, leaders can navigate market ambiguity with surgical precision, transforming the enterprise into a self-evolving system of sustainable innovation.

Focus: Eradicating Information Decay and Safeguarding the Digital Truth

In the **Ai6x6 Human-AI Synergy framework**, the defensive stance on data is not about cybersecurity in the traditional IT sense; it is about protecting the **integrity, relevance, and accessibility** of the organization's intelligence. In the AI era, bad data isn't just a nuisance—it is a liability that leads to "hallucinating" strategies. This essay explores how the six roles act as a shield against the "Entropy of Information."

The Forecaster and The Diplomat: Detecting Decay and Breaking Silos

The **Forecaster** serves as the defensive sentinel for data quality. Its mandate is to kill **Information Blind spots** by detecting anomalies in the data stream before they propagate into downstream decisions. By monitoring for "Data Drift"—where the patterns the AI was trained on no longer match reality—the Forecaster ensures the leadership isn't making 2026 decisions based on 2022 logic.

The **Diplomat** is tasked with neutralizing **Institutional Silos**. Data is often "hoarded" by departments, creating a defensive friction that prevents a unified view of the business. The Diplomat uses reasoning to prove the value of data-sharing, acting as a bridge between technical teams and business units. By dissolving these political silos, it protects the organization from the "Fragmented Truth" that leads to conflicting departmental strategies.

The Creator and The Strategist: Modernizing Legacy and Proving Value

The **Creator** acts defensively by killing **Technical Obsolescence**. Most organizations are buried under "Dark Data"—unstructured, unindexed files that are unusable. The Creator's defensive role is to automatically classify, tag, and modernize this legacy information, turning "Data Debt" into "Data Assets." It ensures the company's "Information Engine" doesn't seize up due to a backlog of unsearchable, outdated formats.

The **Strategist** defends the organization by eliminating **Data Mismatch**. Not all data is valuable; hoarding "Garbage Data" creates unnecessary storage costs and cognitive noise. The Strategist performs a ruthless audit of the data landscape, using causal inference to determine which signals actually drive ROI. By pruning irrelevant data, it protects the "Information Compass," ensuring that only the most potent, high-impact signals reach the executive dashboard.

The Architect and The Executor: Ensuring Truth and Automating Hygiene

The **Architect** kills **Information Inconsistency**. Its primary defensive mandate is the creation of a "Single Source of Truth." By resolving duplicate entities and mapping complex data relationships, the Architect prevents the "Conflicting Reality" friction where Finance and Sales have different numbers for the same customer. It builds a defensive "Institutional Brain" that is immune to the contradictions that usually plague large-scale enterprises.

Finally, the **Executor** serves as the defensive "Hands" of data governance. It kills **Manual Maintenance Friction** by automating the hygiene of the data ecosystem. From auto-deduplication to real-time compliance auditing, the Executor ensures that the organization's data guardrails are always active. This prevents **Operational Leakage**, ensuring that the data used for autonomous execution is always fresh, verified, and legally compliant without requiring constant human scrubbing.

Conclusion: Safeguarding the Foundation of Intelligence

A defensive data strategy within the **Ai6x6 Human-AI Synergy framework** ensures that the organization's "Digital Body" is healthy and reliable. By removing the friction of drift, silos, obsolescence, and noise, the framework establishes a foundation of "Strategic Truth." When a company's data is protected by these six roles, it stops being a burden of the past and becomes the secure, high-octane fuel for the offensive innovations of the future.

In the Ai6x6 framework, **Data Transformation** is the process of turning "Raw Noise" into **"Sovereign Truth."** This is the move from managing messy, siloed databases to architecting a "Live Intelligence Layer." By applying the six roles to the data stack, leadership ensures that information is not just stored, but is contextually aware, ethically governed, and ready for immediate offensive action.

The Forecasters and The Diplomats: Unlocking Signal and Integrity

The **Forecaster's** mandate is **Automated Data Discovery.** Pairing Predictive AI with Graph Neural Networks allows the "Eyes" to scan the enterprise for hidden data relationships and quality decay before they compromise a model. It predicts "Data Drift," ensuring the foundation of every strategic move remains mathematically sound. This proactive scanning turns the data stack into an early-warning system that identifies emerging market signals buried in the noise.

The **Diplomat** acts as the offensive "Voice" of **Data Ethics and Explainability.** By pairing Explainable AI with Differential Privacy, the Diplomat ensures that data usage is transparent to regulators and customers alike. It translates complex data governance policies into "Trust Narratives," ensuring the firm can utilize its data assets aggressively without sacrificing privacy or brand integrity. This eliminates the friction of compliance and builds a foundation of radical data trust.

The Creators and The Strategists: Unlocking Utility and Value

The **Creator** serves as the engine of **Synthetic Data Generation and Prototyping.** When real-world data is biased or incomplete, the Creator uses Generative AI paired with Reality Anchoring to fill the gaps. This allows the organization to build "Sandboxed Realities" to test new strategies in data-starved environments. This "Generative Data Velocity" ensures that the "Engine" is always fueled by high-fidelity information, regardless of external limitations.

The **Strategist** optimizes the **Data Value Chain for Maximum ROI.** By pairing Cognitive AI with Multi-Objective Optimization, the Strategist identifies which data streams are high-alpha and which are mere overhead. It determines the exact cost-to-value ratio of every data point, ensuring that the "Compass" is always pointing toward the most profitable insights. This ensures that the data transformation is not a bottomless cost center, but a precision-tuned profit driver.

The Architects and The Executors: Unlocking Structure and Flow

The **Architect** builds the "Searchable Brain" through **Dynamic Metadata Orchestration.** By pairing Human-in-the-Loop with Vector Databases (RAG), the Architect ensures that data is not just indexed, but "Understood." This turns the entire enterprise data lake into a context-aware memory bank. Every decision-maker in the firm can query the "Brain" in natural language, receiving the "Relevance" they need without the friction of manual data engineering.

Finally, the **Executor** manages the **Autonomous Data Pipeline.** Through Agentic AI paired with Blockchain, the Executor automates the cleaning, labeling, and movement of data across the firm. These agents act as the "Hands," ensuring that the data flow is self-healing and tamper-proof. By securing the data provenance on a ledger, the Executor guarantees that the "Sovereign Truth" remains unassailable, allowing for the autonomous execution of high-stakes strategic goals.

Conclusion: The Pulse of the Intelligent Enterprise

Data transformation is the pursuit of **Informational Sovereignty.** By unlocking signal, integrity, utility, value, structure, and flow, the leader creates a data architecture that is as alive as the organization it supports. This framework ensures that the brand's "Intelligence Layer" is perfectly synchronized with its strategic intent, turning raw information into a high-velocity symphony of orchestrated insight.

Data Power: High-Fidelity Perception

In the legacy era, data was a "History Lesson"—a cold record of what had already happened. By the time a leader reviewed a dashboard, the market had moved, the opportunity had vanished, or the crisis had already taken root. The transformation into **Forecasters** fundamentally redefines data as a "Real-time Periscope." This role is centered on the **Purification of Signal**. It involves moving the organization away from "Batch Processing," where information is collected in chunks, toward "Stream Intelligence," where every data point is treated as a "Live Pulse" of the global environment. In this paradigm, data no longer just "exists" in a warehouse; it "observes" with intent. It provides the organization with the high-definition "Eyes" needed to see through the fog of market volatility, identifying the subtle vibrations of the future before they solidify into irreversible facts.

Defensive Stance: Killing the Signal-to-Noise Mismatch

The defensive mission of the Data-Forecaster is the **Elimination of Entropy**. Most modern organizations are drowning in "Dark Data"—massive volumes of unorganized, unverified information that provides no value while creating immense "Security and Compliance Drag." This digital clutter acts as a "Weight" on the company's "Brain," making it slow and prone to error. The Forecaster role kills this "Noise" by focusing on "Signal Integrity." This creates a **Shield of Accuracy**. By utilizing "Anomaly Detection" to identify "Ghost Data" or "Corrupted Signals" the moment they enter the system, the data layer protects the organization from making high-stakes decisions based on false premises. This defensive posture ensures that the organization's "Eyes" are always "Clear." It kills the blindspots caused by "Data Hallucinations" and ensures that every byte of information is a "Verified Truth." This isn't just about data cleaning; it is about "Information Hygiene" as a survival mechanism, ensuring the brand's "Vision" is never distorted by the "Garbage" of the digital age.

Offensive Stance: Unlocking Predictive Resonancy

On the offense, the Forecaster data moves toward **Anticipatory Modeling**. This is the ability to use data not just to see what *is*, but to simulate what *will be* across a million different permutations. This **Unlocks Temporal Mastery**, where the data provides a "Probabilistic Map" of future demand, potential supply shocks, and hidden competitor moves. Instead of waiting for a "Quarterly Report" to suggest a change in direction, the data triggers "Pre-emptive Action." If the data "sees" a high probability of a regional shortage or a shift in consumer sentiment three months out, it automatically alerts the "Hands" of the company to secure the market. This turns data into a "Competitive Weapon" that allows the brand to "Own the Future." It allows the organization to act with "Pre-emptive Agility," occupying the winning ground while competitors are still debating the meaning of last month's spreadsheets.

The Data Outcome: Perceptual Dominance

The ultimate value of the Forecaster role is **Organizational Foresight**. When data is transformed into a high-fidelity "Eye," the company gains the ability to "Navigate Time" with the same ease that it navigates space. For the leader, this provides **Decisiveness**. You are no longer guessing or relying on "Gut Feeling"; you are reacting to "Validated Futures." This turns the data layer into a "Strategic Asset" that ensures the brand is always the first to see a trend and the first to avoid a trap. It ensures that the "Eyes" of the company are never closed, providing a 24/7 "Reality Stream" that keeps the organization perpetually aligned with the moving target of the global market.

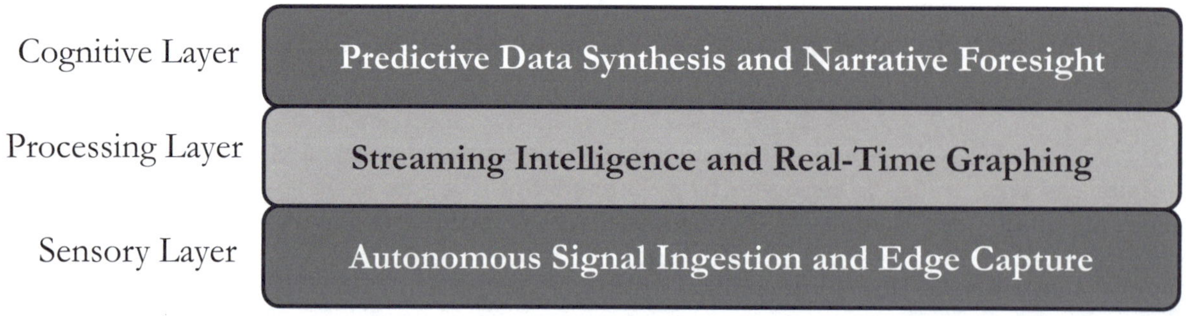

1. THE FORECASTERS (THE EYES)

Executive Power: Information Velocity & Signal Fidelity

In the legacy era, a leader's "Eyes" were clouded by "Data Lag"—the gap between an event happening and it being reflected in a dashboard. The **Forecaster** role in data transformation represents the transition to **Real-Time Signal Perception**. For the executive, this means moving away from "Batch Processing" and "Historical Reporting" toward a state of **Continuous Observability**. By architecting a stack that perceives the data as a live stream rather than a static lake, the leader gains the "Perceptual Edge" to identify market shifts the micro-second they occur. The ultimate goal is the **Purification of the Information Stream**, ensuring that the "Eyes" of the firm are seeing "Live Reality" rather than "Post-Mortem Statistics," allowing the leader to act with a level of temporal precision that competitors reliant on manual reports cannot match.

The Sensory Layer: Autonomous Signal Ingestion and Edge Capture

The **Sensory Layer** of the Data Forecaster serves as the organization's "Perceptual Entry Point." It moves beyond manual data cleaning and into **Autonomous Signal Ingestion**. This layer leverages a convergence of **Real-Time Data Pipelines**, **Edge Computing**, and **CDC (Change Data Capture)** technologies to capture every digital event as it happens. Whether it is a customer clicking a link, a sensor reading a temperature drop, or a geopolitical news break, this layer "labels" and ingests the signal at the point of origin. By utilizing **Automated Data Quality Agents**, it scrubs and validates the data in transit, removing the "Noise" of duplicates and errors before the data even reaches the core. This ensures the leader's "Eyes" are receiving a raw, high-fidelity stream of "Environmental Truth" with near-zero latency, removing the administrative friction that traditionally slows down the organization's ability to "See" the world.

The Processing Layer: Streaming Intelligence and Real-Time Graphing

The **Processing Layer** serves as the "Digital Refiner" of the data stack. It transforms the raw signal stream into a structured understanding of **Live Context**. This layer utilizes **Stream Processing Engines** (like Flink or Spark Streaming) and **Vector Databases** to index data as it flows. While traditional databases wait for data to "land" before analyzing it, the Forecaster's processing layer analyzes data "in-flight," mapping new signals directly into a **Dynamic Knowledge Graph**. This allows the system to understand the relationship between a new incoming data point and the entire historical context of the firm. It moves the organization from "Scheduled Reports" to "Trigger-Based Awareness," providing a "Logical Foundation" for immediate action. This layer ensures the brand's "Data Compass" is always calibrated to the present moment, identifying the causal link between incoming signals and future probabilities with mathematical certainty.

The Cognitive Layer: Predictive Data Synthesis and Narrative Foresight

The **Cognitive Layer** is the "Reasoning Mind" of the data stack, where the streaming signals are converted into **Actionable Foresight**. This layer utilizes **Large Language Models (LLMs)** and **Predictive Analytics** to synthesize the logic of the Processing Layer into "Strategic Alerts." Instead of showing the leader a chart, the Cognitive Layer generates "Contextual Narratives." It acts as a "Data Sentry," using **Natural Language Generation** to tell the leader: "Based on the signal we just received from the Asian supply chain and the current currency fluctuation, there is an 85% probability of a margin squeeze in the next twelve hours." This allows the executive to move from "Monitoring Data" to "Leading via Insight," authorizing pre-emptive moves to capture value or mitigate risk. It turns the "Forecaster" into a "Proactive Guide," ensuring the brand's data is not just stored, but is actively working to reveal the "Next Move" on the global stage.

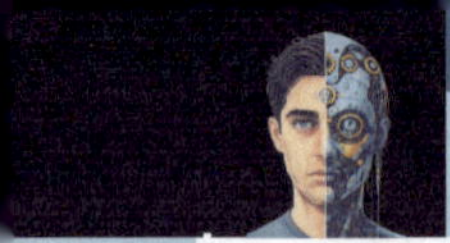

1. THE FORECASTERS (THE EYES) | PREDICTIVE AI

OVERVIEW

Predictive AI forecasts future outcomes by analyzing vast historical data using machine learning algorithms. For the "Intuitive & Risk" leader, the AI becomes a crucial tool for quantifying uncertainty. The leader provides their initial intuition and acceptable risk limits, framing the precise question. The AI delivers objective, quantifiable evidence, such as likelihoods and key drivers, that either validates the leader's gut feeling or proposes data-driven alternatives. This synergy converts subjective risk into a calculated, strategic choice, empowering the leader to make proactive, decisive actions.

MAIN STEPS

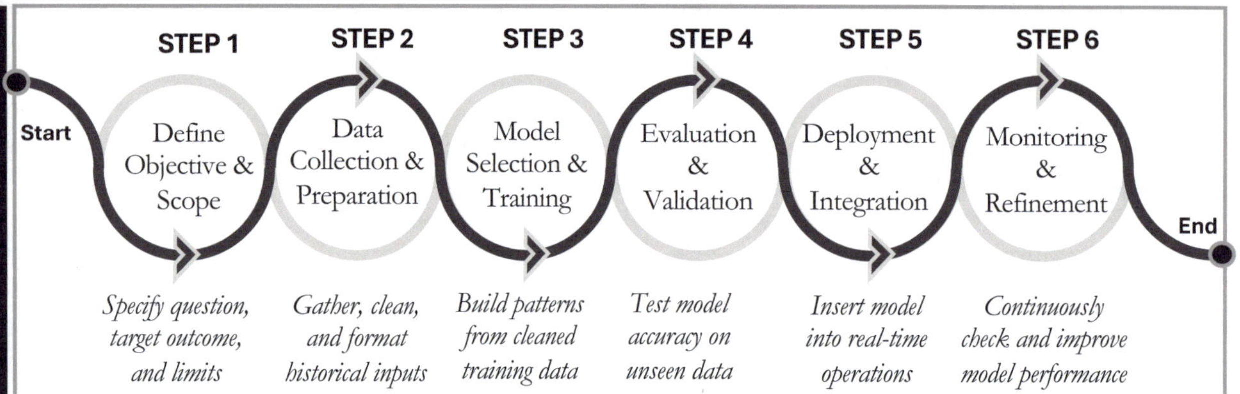

SAMPLE FLOW

Telecommunications - Data Quality Risk- A Telecom leader mitigates regulatory fines by using AI to predict data quality drift in billing systems before it affects customer invoices.

1. Define Error Tolerance
Set acceptable thresholds for billing data anomalies in customer usage/charge records.

2. Track Lineage Metrics
Monitor data transformation, system latency, and ingestion rates across all platforms.

3. Model Anomaly Severity
Train on historical data drift events to predict the severity & impact of future corruption.

4. Isolate High-Risk Sources
Identify specific network regions or data streams most likely to introduce quality errors.

5. Calculate Regulatory Score
Generate real-time compliance risk scores based on potential exposure to quality breaches.

6. Quarantine & Rectify Data
Automatically block high-risk data from production and initiate quality remediation process.

USE CASES

- **Logistics | Address Standardization Failure:** Predicts which incoming shipping addresses will consistently fail validation and require extensive manual correction, optimizing the data cleaning workflow for better delivery accuracy.
- **Insurance | Claim Fraud Data Integrity:** Forecasts complex data anomalies that indicate potential manipulation or corruption of policyholder and claims records, protecting core financial data integrity proactively and minimizing exposure.
- **Real Estate | Valuation Data Drift Warning:** Predicts the need for model retraining by forecasting when external market data inputs will significantly diverge from established historical pricing patterns, ensuring appraisal accuracy.
- **Automotive | IoT Sensor Data Reliability:** Forecasts the likelihood of specific vehicle sensors transmitting corrupted or incomplete telemetry data due to environmental factors, ensuring safe and reliable fleet management insights.
- **Biotech | Genomic Sequence Integrity:** Predicts the probability of significant errors in large-scale genomic sequencing data pipelines, ensuring high confidence in fundamental research results and accelerating drug discovery targets.
- **Non-Profit | Donor Data Utility Decline:** Forecasts when donor contact data utility will drop below an effective threshold due to outdated records and privacy changes, triggering necessary, timely data enrichment campaigns.

In the **Data Transformation**, the **Forecasters (The Eyes)** serve as the enterprise's "Predictive Sentinel." While the primary instrument of **Predictive AI** identifies patterns in historical data, it often struggles to account for complex, non-linear relationships or sudden "Black Swan" shifts in the environment. To move from "Descriptive Reporting" to "Anticipatory Intelligence," a leader must integrate these six secondary pairings to sharpen the "Eyes," ensuring the organization can see through the fog of data to identify the next market inflection point.

1. The Graph Neural Network (GNN) Pairing for Relationship Discovery

Pairing **Predictive AI with GNNs** is the ultimate tool for **Uncovering Systemic Risk.** While standard AI looks at individual data points, GNNs analyze the "Relationships" between those points across the entire enterprise. This is best utilized to map hidden dependencies in complex global supply chains or customer behavior networks. By seeing the data as a web rather than a list, the Forecaster can predict how a failure in one obscure node will ripple through the entire system, allowing for pre-emptive stabilization.

2. The Causal Inference Pairing for Strategic Validation

Leaders should pair **Predictive AI with Causal Inference** to solve the "Correlation Trap." Standard AI can show that two events happen together, but Causal Inference proves if one actually *caused* the other. This is best deployed when validating a new data-driven strategy. It ensures the leader isn't chasing "False Positives" created by coincidental data spikes, but is instead making offensive moves based on verified cause-and-effect relationships that will hold true even as market conditions change.

3. The Digital Twin Pairing for Data Stress-Testing

When preparing for market volatility, **Predictive AI must be paired with Data Digital Twins**. This is best utilized to "rehearse" the impact of extreme scenarios on the firm's data infrastructure. The leader can simulate a 500% spike in transaction volume or a coordinated cyber-attack within a virtual sandbox. This pairing ensures the "Eyes" see the breaking points of the organization's intelligence layer before they are reached in reality, allowing for the hardening of the data architecture in a risk-free environment.

4. The Natural Language Processing (NLP) Pairing for Unstructured Insight

To capture the 80% of enterprise data that is "invisible," a leader should pair **Predictive AI with Advanced NLP**. While most models only see structured tables, NLP allows the Forecaster to ingest and analyze unstructured data like customer call logs, legal contracts, and industry news. This identifies "Early Warning Signals" buried in text that quantitative models would miss. This turns the Forecaster into a holistic observer, ensuring that the organization's "Eyes" are sensitive to both the numbers and the narratives shaping the market.

5. The Monte Carlo Pairing for Probability Forecasting

To navigate the uncertainty of high-stakes investments, leaders should pair **Predictive AI with Monte Carlo Simulations**. This is best utilized when the "Eyes" must look into a highly volatile future, such as predicting energy prices or currency shifts. By running millions of random variables through the predictive model, the Forecaster can map the "Probability Cloud" of various outcomes. This allows the leader to move from "One Best Guess" to a range of mathematically grounded possibilities, ensuring the brand is always positioned for the most likely reality.

6. The Knowledge Graph Pairing for Data Lineage

In an era of deepfakes and data manipulation, **Predictive AI should be paired with Knowledge Graphs**. This pairing tracks the **"Provenance of Truth"**—mapping exactly where every piece of data originated and how it has been transformed. It allows the leader to see the "Lineage" of their intelligence. By identifying the source nodes of high-quality data versus "Dirty Data," the Forecaster ensures the organization's "Eyes" are never blinded by corrupted information, maintaining the absolute integrity of the firm's strategic vision.

Data Power: Semantic Integrity

Data without context is merely noise, and data without ethics is a significant liability. The transformation of data into **Diplomats** focuses on **Semantic Integrity**—the process of ensuring that data "Speaks" a language that is consistent, ethical, and universally understandable across the entire enterprise. In this role, data is not just "Numbers in a Cell"; it is "Codified Meaning." It bridges the gap between the "Technical Code" of the machine and the "Strategic Intent" of the human leader. The Diplomat data ensures that the "Voice" of the organization is grounded in "Fact," but articulated with "Deep Context." It is the layer that turns raw information into "Institutional Wisdom," ensuring that the brand's communication—both internal and external—is always coherent and principled.

Defensive Stance: Killing the Interpretive Friction

The defensive mission of the Data-Diplomat is the **Elimination of Ambiguity**. In many legacy companies, different departments use the same terminology to mean entirely different things—"Revenue" in Sales might mean something different than "Revenue" in Finance, and "Customer Success" might have four different definitions across four different regions. This creates "Interpretive Friction" that leads to massive strategic misalignment and wasted effort. The Diplomat role kills this "Friction" by establishing "Universal Definitions" and "Ethical Guardrails" at the foundational data level. This is the **Shield of Governance**. By "Hard-Coding Ethics" into the data—ensuring privacy, fairness, and transparency are baked into every record—the Diplomat protects the brand from the "Reputational Meltdown" of biased or misused information. This defensive posture ensures that the "Voice" of the data is always "Trustworthy" and "Compliant." **Offensive Stance: Unlocking Narrative Synthesis**

On the offense, the Diplomat data moves toward **Contextual Storytelling**. This is the ability of the data system to "Explain Itself" to the human orchestrator. Instead of merely presenting a list of figures, the data provides a "Reasoned Narrative"—explaining not just *that* a metric is shifting, but *why* it is shifting in the specific context of global economic trends, competitor actions, and internal policy changes. This **Unlocks Reasoning Power**, where the data acts as a "Strategic Advisor" rather than a passive ledger. It synthesizes "Disparate Signals" from across the globe into a "Coherent Voice" that the leader can use to persuade boards, align global teams, and inspire customer loyalty. This turns data into "Persuasive Capital," ensuring that the brand's "Voice" is always the most "Informed" and "Authentic" in the market.

The Data Outcome: Institutional Trust

The ultimate value of the Diplomat role is **Alignment through Truth**. When data acts as the "Voice" of the company's intelligence, the organization achieves a state of "High-Integrity Scale." For the leader, this provides **Narrative Sovereignty**. You know that your "Strategy" is built on "Data" that is ethical, clearly defined, and contextually rich. This turns the organization into a "Transparent Powerhouse"—where every action is justifiable and every word is evidenced. It ensures that the brand's "Voice" is never a "Guess," but a "Declaration of Fact," creating a culture of "Evidence-Based Leadership" that is deeply respected by employees and customers alike. It is the bridge that allows a massive, automated enterprise to maintain its "Human Character."

Executive Power: Semantic Data Integrity & Governance

For the modern executive, the "Voice" of data is the "Policy" and the "Trust" embedded in how information is handled. In an era of data privacy scandals and algorithmic bias, the **Diplomat** role ensures the brand's data is not just "Used," but "Governed with Integrity." This role is about the **Abolition of Data Opacity**, ensuring that the "Voice" of the organization's data is the most "Ethical" and "Accountable" in the industry. The leader uses this stack to ensure that the "Data Policy" is articulated with absolute clarity for regulators, customers, and partners. By architecting this role, the executive moves from "Data Extraction" to "Data Stewardship," ensuring that every bit of information is treated as a "Sacred Trust" that reinforces the brand's reputation for reliability and fairness.

The Sensory Layer: Compliance Monitoring and Lineage Tracking

The **Sensory Layer** of the Data Diplomat acts as the organization's "Governance Ear." It moves beyond manual auditing and into **Autonomous Compliance Monitoring**. Utilizing **Metadata Scrapers** and **Data Lineage APIs**, this layer tracks the "Life of a Data Point" from ingestion to consumption. It captures the "Context of Use"—understanding who accessed what data, for what purpose, and under which regulatory framework (GDPR, CCPA, etc.). By monitoring for "Data Drift" and "Access Anomalies," this layer provides the leader with a real-time map of "Governance Friction." It ensures the executive is aware of any potential compliance risk before it results in a breach or a fine. This layer provides a high-fidelity "Map of Data Ethics," ensuring the "Voice" of the firm is grounded in the absolute reality of how its data is actually being treated by the machine.

The Processing Layer: Semantic Governance and Ethical Mapping

The **Processing Layer** serves as the "Integrity Engine" of the data stack. It utilizes **Knowledge Graphs** to link the brand's high-level privacy policies and ethical commitments to the specific data objects in the system. If the executive's "Voice" speaks of "Customer Privacy," this layer "Hard-Codes" those rules into the **Data Access Logic**. This ensures that the leader's message is not just a PR claim, but is mathematically enforced at the database level. It acts as a **Shield of Semantic Integrity**, using **Differentially Private Algorithms** and **Encrypted Computation** to ensure that data is used in a way that is "Provably Ethical." By resolving the identity of every data use-case in the context of the brand's mission, this layer kills the friction of "Internal Misuse," providing a "Verifiable Logic" that the company's data practices are aligned with its public "Voice," thereby building unshakable institutional trust.

The Cognitive Layer: Explainable Data Advocacy

The **Cognitive Layer** is where the "Data Logic" is transformed into **Explainable Narratives** for the world. Utilizing **Large Language Models (LLMs)** and **Explainable AI (XAI)**, the Diplomat generates "Transparency Reports" that anyone can understand. It doesn't just provide a wall of legalese; it orchestrates "Sovereign Explanations"—tailoring the data-use story for a curious customer, a rigorous regulator, or a skeptical investor. This layer ensures the organization can "Explain its own Data Mind," providing a reasoned justification for how an AI model reached a specific conclusion using specific data. This allows the leader to project a "Voice" of **Absolute Data Accountability**, turning "Governance" from a cost center into a "Competitive Advantage." It ensures that every stakeholder feels a sense of "Security and Alignment" with how the brand processes information, creating a culture of "Principled Data Efficiency."

OVERVIEW

Explainable AI (XAI) provides transparent, human-readable rationales for AI decisions, moving beyond black box outputs. It reveals why a model reached a conclusion, building trust and mitigating legal and ethical exposure. The Regulatory & Compliance leader leverages XAI to validate complex, automated decision-making processes, particularly in highly regulated domains. This synergy ensures AI systems are fair, auditable, and adhere to internal policies and external legal requirements, transforming compliance from a check-the-box function into a strategic enabler of trust.

MAIN STEPS

Start

- **STEP 1** — Define Audit Scope: *Establish parameters for compliance assessment.*
- **STEP 2** — Feature Importance Analysis: *Quantify input factors driving AI output.*
- **STEP 3** — Generate Local Explanation: *Create rationale for a single prediction.*
- **STEP 4** — Validate Model Fairness: *Test decision consistency across subgroups.*
- **STEP 5** — Human Review and Oversight: *Leader verifies explanations meet standards.*
- **STEP 6** — Document Rationale & Action: *Record justification for audit readiness.*

End

SAMPLE FLOW

Data Privacy Impact Assessment: A leader utilizes Explainable AI (XAI) to rigorously audit the impact of new data usage models on user privacy and ensure global regulatory compliance..

EXPLAINABLE AI

1. Define Privacy Boundaries
Specify regulatory mandates and internal data masking requirements clearly.

2. Data Flow Explanations
Trace data origins and usage paths, identifying all decision touchpoints.

3. Identify High-Risk Features
Pinpoint sensitive data variables most influential in generating model outcomes.

4. Simulate Anonymization
Test how data anonymization affects model accuracy and explanation clarity.

5. Assess Compliance Risk Score
XAI provides a clear risk score tied to potential privacy violation severity.

6. Certify Data Model Usage
Approve model deployment, ensuring full adherence to GDPR and other compliance frameworks.

USE CASES

- **Manufacturing | IoT Data Governance:** Justifies data retention policies by explaining which machine data is critical for predictive maintenance and which must be purged to meet regulatory requirements swiftly.
- **Financial Services | Anti-Money Laundering (AML) Model Audit:** Provides clear rationale for flagging transactions as suspicious, enabling leaders to prove the AML system is fair, consistent, and adheres to strict reporting guidelines reliably.
- **Healthcare | Patient Record De-identification:** Explains the XAI process used to de-identify patient records for research, assuring leaders that privacy standards are met while maintaining data utility for analysis safely.
- **Technology | Personal Data Export Justification:** Provides transparent reasons for how and why user data is structured and exported upon request, certifying compliance with "right to access" regulations accurately.
- **Retail | Consumer Preference Data Masking:** Explains the impact of data masking on personalized marketing models, allowing leaders to ensure privacy compliance without significant loss of analytical effectiveness clearly.
- **Energy | Infrastructure Data Security Rationale:** Justifies the strict access controls on critical grid data by demonstrating the security risks associated with data misuse, ensuring legal liability is minimized quickly.

In the **Data Transformation**, the **Diplomats (The Voice)** serve as the bridge between the firm's "Cold Logic" and its "Human Trust." While the primary instrument of **Explainable AI (XAI)** provides the technical reasoning behind data-driven decisions, it often fails to account for the emotional and ethical concerns of stakeholders. To move from "Opaque Automation" to "Radical Transparency," a leader must integrate these six secondary pairings. This ensures the organization's "Voice" is not just clear, but is also contextually resonant, legally unassailable, and ethically grounded.

1. The Sentiment Analysis Pairing for Stakeholder Calibration

Pairing **Explainable AI with Sentiment Analysis** is best utilized when communicating data-driven changes to a workforce or customer base. While XAI explains the logic—such as a shift in dynamic pricing—it often ignores the audience's "Emotional Climate." By running explanations through a Sentiment Analysis layer, the Diplomat can gauge rising anxiety or resistance. This allows the leader to calibrate the "Voice," ensuring that technical reasoning is delivered with the empathy required to maintain brand loyalty and institutional morale.

2. The Contextual NLP Pairing for Multimodal Translation

Leaders should pair **Explainable AI with Contextual NLP** when navigating global data ecosystems. XAI provides the facts, but Contextual NLP translates them into the specific "language" of the recipient—whether it is technical jargon for regulators, financial logic for the board, or simple narratives for the public. This pairing is essential for cross-border transparency, ensuring the brand's "Voice" is accurately understood across different cultural and professional contexts, effectively eliminating the friction of misinterpretation.

3. The Knowledge Graph Pairing for Logical Traceability

When a data decision involves multi-layered dependencies—such as an automated credit approval or an R&D pivot—**Explainable AI must be paired with Knowledge Graphs**. While XAI explains the output, the Knowledge Graph provides the "Map of Relationships" that led to that conclusion. This is best deployed when a leader must prove "Sovereign Truth" to auditors. It allows the Diplomat to trace logic through thousands of interconnected data points, transforming the "Voice" into a verifiable, high-fidelity record of decision-making.

4. The Differential Privacy Pairing for Secure Transparency

To build trust without compromising individual sovereignty, a leader should pair **Explainable AI with Differential Privacy**. This allows the Diplomat to explain the "Reasoning" of a model using real data insights while mathematically ensuring that no individual's identity is exposed. This is best utilized in public-facing data reports or collaborative research. It ensures the "Voice" remains transparent and helpful while providing a mathematically guaranteed "Privacy Shield," reinforcing the brand's reputation as a responsible data steward.

5. The Conversational AI Pairing for Real-Time Interrogation

To combat "Information Lag" during a transformation, leaders should pair **Explainable AI with Conversational AI**. While XAI provides a static explanation, Conversational AI allows stakeholders to "Interrogate" the data logic in real-time. This is best deployed during town halls or investor calls. This pairing turns the Diplomat into a dynamic interface, providing instant, reasoned justifications for data moves. It builds a culture of open-door transparency where the "Voice" is always available to answer the "Why."

6. The Blockchain Pairing for Immutable Accountability

In industries where data integrity is non-negotiable, **Explainable AI should be paired with Blockchain**. XAI generates the explanation for a decision, while the Blockchain stores that explanation in an immutable, timestamped ledger. This is best utilized for environmental auditing or automated financial reporting. It ensures the "Voice" is legally unassailable and tamper-proof. By providing absolute accountability for every algorithmic action, the leader secures the brand's operational integrity against future disputes or regulatory shifts.

Data Power: Generative Proliferation

In the legacy data model, information was "Static." You stored a customer's name, their purchase history, and perhaps their location. This data sat in a database like a stone, waiting for someone to move it. The transformation into **Creators** represents the shift from "Data as a Record" to **"Data as a Seed."** In this role, the organization's information is treated as a generative foundation—a source from which new ideas, new products, and new experiences can be grown. The data layer becomes the "Engine" of the company, capable of taking existing patterns and "Generating" entirely new realities. It moves the business from a state of "Data Mining" (extracting what is there) to "Data Synthesis" (creating what *should* be there).

Defensive Stance: Killing the Obsolescence of Information

The defensive mission of the Data-Creator is the **Purification of the Heritage**. In many established brands, the "Data Engine" is clogged with "Zombie Data"—obsolete records, disconnected archives, and forgotten lessons that create a "Friction of Complexity." The Creator role kills this "Inertia" by "Resurrecting the Asset." It treats the company's entire historical record not as a graveyard, but as a "Live Library." By indexing and modernizing the organization's "Knowledge Heritage," the leader ensures that no previous success is ever lost and no mistake is ever repeated. This creates a **Shield of Continuity**. It ensures that the brand's "Engine" is always running on "Clean Energy"—accurate, modern, and accessible information. This defensive posture kills the "Wastage of Re-invention," protecting the brand's "Intellectual Capital" and ensuring that every new project starts from the highest possible peak of previous knowledge.

Offensive Stance: Unlocking Generative Abundance

On the offense, the Creator data moves toward **Infinite Personalization**. This is the ability to use the "Data Engine" to manifest "Tailored Realities" for every stakeholder at scale. Instead of having one "Product Description" or one "Marketing Message," the data generates a million variations, each optimized for the specific context of an individual user. This **Unlocks Creative Force**, where the data itself becomes the "Designer." It allows the brand to "Prototype" new business models or product variations in a digital simulation before a single cent is spent in the physical world. This turns the organization into a "Generative Powerhouse"—one that can "Saturate the Market" with high-fidelity, relevant output that feels "Hand-crafted" but is actually powered by the "Scalable Logic" of the data engine. It allows the brand to "Create its way out of competition" by moving faster and with more variety than any manual process could ever allow.

The Data Outcome: Absolute Agency

The ultimate value of the Creator role is **Operational Plasticity**. When your data is "Generative," your business becomes "Fluid." For the leader, this provides **Execution Mastery**. You are no longer limited by what you have "In Stock" or "In the System"; you are limited only by your "Intent." The data engine can manifest the "Next Move" instantly. This turns the organization into a "Living Brand"—one that creates its own future in real-time, transforming "Raw Data" into "Direct Value" and ensuring that the brand's "Engine" is always producing the most relevant and innovative output in the industry.

Executive Power: Generative Data Manifestation

The **Creator** role transforms the executive from a "Data Consumer" into a "Data Architect." In the legacy enterprise, data structures were rigid; if the business needed a new insight, it required months of manual ETL (Extract, Transform, Load) engineering to build a new pipeline. The Creator role represents the shift to **Autonomous Data Manifestation**, where the data infrastructure itself is dynamically generated in response to the leader's intent. The executive uses this engine to manifest "Synthetic Realities" and "Dynamic Data Marts" on the fly, moving the organization from a state of "Data Scarcity" to "Data Abundance." This role ensures the brand is always manifesting the most "Relevant" information environment for any strategic challenge, turning the "Engine" of the firm into a self-assembling library of insight.

The Sensory Layer: Intent Capture and Schema Discovery

The **Sensory Layer** for the Creator captures "Inquiry Intent" as a high-definition signal. It moves beyond traditional SQL queries by integrating **Semantic Search** and **Autonomous Metadata Discovery**. By utilizing **Agentic Crawlers** to scan the entire enterprise "Dark Data" (unstructured emails, logs, and PDFs), this layer identifies the "Atomic Components" of a problem before a human even knows where the data is stored. It captures the raw intent of the leadership—"I need to understand the correlation between weather patterns and last-mile delivery friction"— and automatically locates the relevant sensors and streams. This ensures the leader is building the "Engine" of the project on a foundation of "Available Reality" rather than "Pre-defined Tables," capturing the necessary signals for innovation as a live, fluctuating resource.

The Processing Layer: Parametric Data Logic and Synthetic Synthesis

The **Processing Layer** acts as the "Refinery" for data creation. It utilizes **Generative Adversarial Networks (GANs)** and **Variational Autoencoders** to manifest **Synthetic Data** where real-world data is missing or sensitive. This layer understands the "Rules of the Information Environment"—including statistical distributions, privacy constraints, and logical dependencies. It stress-tests every generated data model against the organization's "Causal Truths," ensuring the executive only sees "Simulated Realities" that are mathematically sound. It acts as a **Guardian of Accuracy**, ensuring that the ratio of real-world signal to synthetic augmentation is optimized for precision. By applying "Parametric Data Logic," this layer ensures the "Engine" is disciplined, preventing "Data Hallucinations" and ensuring that every manifest insight is a verifiable reflection of potential reality.

The Cognitive Layer: Dynamic Intelligence Synthesis and Execution

The **Cognitive Layer** is the "Orchestration Mind" of the data stack, powered by **Large Language Models (LLMs)** and **Autonomous Code Interpreters**. This layer "hallucinates" thousands of different ways to model the information, presenting the leader with the "Optimal Insight Architecture." It doesn't just provide a dashboard; it manifests the **Code and Infrastructure** required to answer the question. The leader provides the strategic goal, and the Cognitive Layer generates the "Data Genome"—automatically writing the Python scripts, standing up the vector indexes, and configuring the API endpoints to fulfill the mission. It involves "Recursive Schema Evolution," where the AI adapts the data model as more information arrives, ensuring the final output is the "Optimal Manifestation" of the leader's vision. This turns the executive into a "Conductor of Knowledge," capable of deploying complex intelligence assets at the speed of thought.

OVERVIEW

Generative AI creates novel outputs like text, art, code, and immersive environments from simple prompts. It excels at rapid prototyping, style transfer, and exploratory ideation. The Creative & Artistic leader partners with Generative AI to rapidly visualize strategic concepts, explore diverse future scenarios, and prototype new organizational narratives. The synergy transforms vision casting from a slow, manual process into a continuous, inspiring, and engaging collaborative endeavor.

MAIN STEPS

Start

STEP 1 — Define Creative Prompt
Articulate desired output parameters precisely.

STEP 2 — Initial Content Generation
Create diverse drafts based on input.

STEP 3 — Style and Mood Adaptation
Adjust tone, format, or aesthetic output.

STEP 4 — Iterative Refinement Loop
Guide model toward final creative goal.

STEP 5 — Human Curation & Selection
Choose the most impactful creative result.

STEP 6 — Deploy Creative Asset
Integrate final output into strategy.

End

SAMPLE FLOW

Visualizing Complex Data Governance Policies: A Chief Data Officer uses AI to visually translate complex governance and quality policies into intuitive, engaging, and clear formats..

1. Input Policy Text
Define legal documents and technical metadata standards precisely in text.

2. Generate Visual Models
Create dynamic infographics and interactive 3D flowcharts instantly from text.

3. Establish Data Narratives
Design stories to illustrate data lineage and policy compliance clearly.

GENERATIVE AI

4. Refine Clarity and Tone
Adjust complexity and color palette for different audience comprehension levels.

5. Test Stakeholder Engagement
Collect executive and technical user feedback to measure policy clarity rigorously.

6. Interactive Policy Guide
Launch the intuitive, visual interface for navigating complex data quality rules and policies.

USE CASES

- **Manufacturing | Data Architecture Blueprint:** Creates dynamic, interactive 3D visualizations of the entire IoT data pipeline, making complex sensor flows and integration points easily comprehensible for technical and non-technical staff.
- **Financial Services | Risk Reporting Storyboards:** Generates visual narratives and animated charts that translate opaque regulatory risk models into clear, actionable executive summaries, improving data-driven decision-making speed.
- **Healthcare | EHR Data Anonymization:** Designs abstract visual patterns and unique artistic representations of patient data, ensuring privacy standards are met while maintaining the structural integrity for research use.
- **Technology | API Documentation Design:** Produces interactive, graphically rich documentation and code examples for complex APIs, drastically reducing the learning curve for external developer adoption and integration speed.
- **Retail | Inventory Data Visualization:** Creates intuitive, real-time 3D models of warehouse inventory and shelf space utilization, allowing leaders to artistically manipulate stocking strategies based on sales velocity data.
- **Energy | Sensor Data Artification:** Transforms raw, complex energy grid sensor data into compelling visual or sonic art installations, helping engineers intuitively spot anomalies and understand network usage patterns creatively.

In the **Data Transformation**, the **Creators (The Engine)** act as the "Manifestation Layer" of the organization's intelligence. While the primary instrument of **Generative AI** is used to synthesize new data structures or automate report writing, it often lacks the "Reality Anchoring" required to produce assets that are both accurate and useful. To move from "Stochastic Guessing" to "Generative Precision," a leader must integrate these six secondary pairings. This ensures the "Engine" doesn't just produce content, but generates high-fidelity data assets that solve the friction of scarcity and fuel the next wave of innovation.

1. The Synthetic Data Pairing for Reality Expansion

Pairing **Generative AI with Synthetic Data Generation** is the ultimate solution for **Data Scarcity.** In high-stakes fields like autonomous driving or rare disease research, real-world data is often too limited or dangerous to collect. This pairing allows the Creator to manifest millions of "Edge Case" scenarios in a virtual environment. By training models on this high-fidelity synthetic data, the leader can bypass the "Cold Start" problem, ensuring the organization's AI engines are "Road-Ready" before they ever touch the real world.

2. The Digital Twin Pairing for Generative Prototyping

Leaders should pair **Generative AI with Digital Twins** to achieve **Hyper-Accurate Simulation.** While the AI can generate thousands of new product or process designs, the Digital Twin provides the "Physics-Based Guardrail." This is best utilized in R&D, where the "Engine" can dream up new chemical compounds or factory layouts and instantly test their viability in a virtual replica of the physical world. This pairing ensures that the Creator's output is not just imaginative, but is mathematically verified for immediate manifestation.

3. The RLHF (Human Feedback) Pairing for Value Alignment

When the "Engine" is used to generate strategic documents or customer-facing data visualizations, **Generative AI must be paired with Reinforcement Learning from Human Feedback.** This ensures that the machine's output aligns with the brand's "Sovereign Aesthetic" and ethical standards. Human experts provide the "Taste" and "Contextual Nuance" that raw models lack. This pairing turns every data scientist into an "Architect of Value," ensuring that the generated data assets are perfectly tuned to the organization's specific strategic intent.

4. The Vector Database (RAG) Pairing for Grounded Innovation

To prevent "Hallucinations" in generative output, a leader should pair **Generative AI with Vector Databases (RAG).** This pairing ensures that every asset created by the "Engine" is grounded in the firm's actual historical records and proprietary knowledge. This is best utilized during internal "Innovation Sprints." Instead of the AI making up possibilities, it retrieves relevant patterns from the "Institutional Brain" to suggest new, feasible data products. It ensures the "Engine" builds upon the firm's existing competitive moats rather than diverging from them.

5. The Differential Privacy Pairing for Secure Manifestation

To generate insights from sensitive data without risking a breach, leaders should pair **Generative AI with Differential Privacy.** This is best deployed when creating "Public Insight Reports" or sharing data with third-party partners. The generative model learns the underlying patterns of the data while the privacy layer ensures that no individual record can be re-identified. This pairing allows the "Engine" to manifest the value of "Dark Data" while maintaining absolute privacy sovereignty, reinforcing the brand's position as a trusted data steward.

6. The API Orchestration Pairing for Autonomous Execution

To move from "Content Generation" to "Value Execution," **Generative AI should be paired with API Orchestration.** While the AI creates the "Script" for a new data workflow, API Orchestration provides the "Connection" to actually trigger that workflow across the enterprise. This is best utilized to automate complex data-cleansing or reporting tasks. It moves the "Engine" from being a generator of ideas to a generator of functional, ready-to-run code, turning the Creator into a high-speed factory for operational efficiency.

Data Power: Causal Intelligence

Traditional data analysis is obsessed with "Correlation"—the observation that two things happened at the same time. However, correlation is a dangerous "Compass" for a leader; it can lead to "False Norths" where a company invests millions in a trend that was merely a coincidence. The transformation into **Strategists** moves data into the realm of **Causality.** In this role, the data system doesn't just show "What" is happening; it explains "Why" it is happening. It transforms the "Data Compass" from a simple needle into a sophisticated "Navigational Engine" that understands the "Hidden Mechanics" of the market. This is the shift from "Hindsight" to "Insight."

Defensive Stance: Killing the Correlation Trap

The defensive mission of the Data-Strategist is the **Validation of Reality**. In an age of "Big Data," it is easy to find patterns that aren't real. The Strategist role kills this "Strategic Friction" by applying "Causal Logic" to the organization's information. It acts as the "Guardian of Truth," identifying when a perceived success was actually driven by external noise rather than internal strategy. This is the **Shield of Objectivity**. By "Modeling the Risk" of every strategic path with mathematical rigor, the leader kills the "Mismatch" between budget and impact. It prevents the organization from chasing "Vanity Metrics" and ensures that every dollar is allocated to the "True Drivers" of growth. This defensive posture protects the brand from the "Sunk Cost Fallacy" and ensures that the company's "Compass" is never distorted by the "Ego" of the boardroom or the "Noise" of the market.

Offensive Stance: Unlocking Optimal Navigation

On the offense, the Strategist data moves toward **Dynamic Prescriptive Power**. This is the transition from "Predicting the Weather" to "Commanding the Ship." The data doesn't just forecast a market shift; it "Prescribes" the optimal series of moves to capitalize on it. This **Unlocks Decision Superiority**, where the "Compass" provides the leader with the "Path of Least Resistance" to a goal. It involves "Simulating the Win"—running millions of "What-If" scenarios to find the "Hidden Profit Pools" that competitors, limited by their linear spreadsheets, will never see. This turns the organization into a "Calculating Winner" that can pivot its strategy in milliseconds. It allows the brand to "Optimize the Entire Value Chain" simultaneously—adjusting pricing, supply, and marketing in a "Synchronized Symphony" that maximizes yield and minimizes waste.

The Data Outcome: Strategic Sovereignty

The ultimate value of the Strategist role is **Certainty in Complexity**. In a world defined by "Perpetual Volatility," the leader with the most "Objective Compass" wins because they are the only ones who can see through the "Fog of Information." For the leader, this provides **Strategic Peace of Mind**. You are no longer "Reacting" to the market with hope; you are "Optimizing" through it with math. It ensures that your "Strategy" is a "Live Advantage" that evolves as fast as the world does. This is the key to "Market Dominance"—the ability to know exactly which lever to pull and when, ensuring the brand's "Compass" is always pointing toward the "Value Peak."

Executive Power: Causal Information Yield

The **Strategist** role in data transformation represents the "Abolition of Descriptive Analytics." It moves the executive from "Tracking Metrics" to **"Causal Insight Optimization."** The leader uses this stack to ensure that every "Byte of Data" is allocated to the highest-impact drivers of the brand's mission. The Strategist is the **Guardian of the Signal**, ensuring that the organization is not just "Collecting Data," but "Generating Value." In this role, the executive evolves into a "Strategic Information Navigator" who manages the company's "Cognitive Yield" with mathematical precision. This role kills the friction of "Data Hoarding" and "Dashboard Fatigue" by ensuring the organization's information energy is always flowing toward the "True Drivers" of sustainable profit and competitive advantage.

The Sensory Layer: Information Density and Signal Synchronization

The **Sensory Layer** of the data Compass monitors the "Micro-Levers" of information utility across the entire global enterprise. It synchronizes **Internal Data Velocity** with **External Market Context**. By tracking the "Latency of Insight" (how long it takes for a data point to result in a decision) against the "Accuracy of Outcome," this layer provides the "Coordinate Data" required to know exactly where the firm's intelligence stands in relation to the brand's "Strategic North Star." It provides the "Eyes" for the Compass, ensuring that the starting point of any data initiative is grounded in the absolute reality of current information quality. This layer removes the "Blind Spots" in the data stack, ensuring the leader is aware of every factor—from "Data Silos" to "Decaying Models"—that could impact the trajectory of the organization's decision-making success.

The Processing Layer: Causal Discovery and Value Attribution

The **Processing Layer** is the heart of the Strategist role, utilizing **Causal Inference Engines** and **Structural Equation Modeling** to determine "Why" certain data streams are producing ROI. While traditional analytics show simple "Correlation" (e.g., users who read the blog buy more), the Strategist identifies the **Causal Path** (e.g., they buy more because of a specific educational sequence that reduces perceived risk). This layer provides the "Mathematical Evidence" for data strategy, ensuring the leader is investing in the "Signals" that actually matter for the brand's "Yield." It acts as a **Shield of Objectivity**, stripping away the "Noise" of vanity metrics and focusing the executive on "Global Optimization." By identifying the "Mathematical Why," this layer ensures the firm is never "Chasing Noise," but is always focused on the "Causal Truths" that drive real-world information advantage.

The Cognitive Layer: Predictive Data Orchestration and Navigation

The **Cognitive Layer** acts as the "Navigational Mind" of the data stack, utilizing **Reinforcement Learning** and **Multi-Agent Simulation**. This layer "plays the game" of the information market millions of times to find the "Optimal Path of Insight." When the data environment shifts—due to a new privacy law or a competitor's technological breakthrough—the Compass "Recalculates" the data deployment strategy instantly. This layer doesn't just report on data; it "Prescribes the Best Move" to maximize the **Information Yield**. For example, it might recommend decommissioning an expensive, low-impact data warehouse in favor of a real-time edge stream because the probability of a market shift has increased. This provides the executive with **Strategic Certainty**, allowing them to navigate data disruptions with the confidence that their "Compass" is calibrated to the objective laws of information value. It turns the organization into a "Calculating Winner," ensuring the brand "Wins" because its "Data Strategy" is mathematically optimized for the highest possible cognitive outcome.

OVERVIEW

Cognitive AI mimics human intelligence to understand and interpret unstructured data like text, speech, and images. It processes complex inputs to generate hypotheses, assess logical connections, and offer nuanced context. The Research & Analytical leader uses Cognitive AI as a hyper-efficient knowledge partner, rapidly synthesizing vast amounts of data to rigorously test assumptions, structure complex arguments, and make deeply informed, evidence-based strategic decisions.

MAIN STEPS

Start

STEP 1 – Data Ingestion: *Collect and structure input knowledge quickly.*

STEP 2 – Understand Context: *Understand meaning and relationships deeply.*

STEP 3 – Hypothesis Generation: *Formulate potential explanatory outcomes.*

STEP 4 – Evidence Synthesis: *Find relevant supporting information.*

STEP 5 – Analytical Refinement: *Test hypotheses against real-world data.*

STEP 6 – Insight Formulation: *Structure findings for human action.*

End

SAMPLE FLOW

Data Governance - Metadata Management : A Chief Data Officer uses AI to review and tag vast amounts of corporate data for compliance and accurate reporting needs.

COGNITIVE AI

1. Identify Regulatory Scope
Specify compliance standards and required data privacy classifications precisely.

2. Ingest Raw Data Inventory
Collect and process all unclassified documents and database schema inputs.

3. Automated Metadata apply
Tag data fields with appropriate compliance levels and ownership information.

4. Misclassification Anomalies
Detect and analyze data relationships to flag incorrect or inconsistent security tags.

5. Audit Data Usage Patterns
Track access records to ensure policies align with actual information consumption.

6. Validate Governance Integrity
Sign off on the data catalog structure for official corporate use.

USE CASES

- **Manufacturing | Data Quality Assessment:** Analyzes millions of sensor readings and production logs to identify data anomalies and sources of input inconsistency, ensuring the integrity of predictive modeling and automation systems.
- **Financial Services | Document Search Indexing:** Organizes unstructured legal and correspondence documents by entity, risk category, and key clauses, drastically speeding up litigation response and due diligence research time.
- **Healthcare | Clinical Trial Text Mining:** Extracts actionable insights from thousands of research papers and patient notes, identifying subtle drug interactions and symptom correlations for rapid therapeutic target identification.
- **Technology | Codebase Knowledge Graph:** Builds a semantic map of complex, legacy software code, documentation, and dependencies, enabling engineers to quickly understand system logic for efficient modernization projects.
- **Retail | Competitor Pricing Analysis:** Scours public web data, news articles, and product reviews to create a detailed, contextualized knowledge base of competitor strategies and market positioning tactics.
- **Energy | Environmental Impact Reporting:** Synthesizes disparate environmental monitoring reports, regulatory filings, and permit applications to auto-generate comprehensive, accurate, and audit-ready sustainability disclosures annually.

In the **Data Transformation**, the **Strategists (The Compass)** act as the "Navigational Intelligence" of the organization's information assets. While the primary instrument of **Cognitive AI** identifies high-level patterns and correlations across the data lake, it often lacks the multi-dimensional rigor required to balance conflicting business objectives. To move from "Observation" to "Offensive Optimization," a leader must integrate these six secondary pairings. This ensures the "Compass" doesn't just point to what is happening, but identifies the precise path to maximum yield, ensuring that every byte of data is converted into a strategic advantage.

1. The Multi-Objective Optimization (MOO) Pairing for Resource Yield

Pairing **Cognitive AI with MOO** is the ultimate tool for **Data Value Engineering.** While AI identifies profitable patterns, MOO balances the trade-offs between data storage costs, processing speed, and the strategic "alpha" of the insight. This is best utilized when managing massive cloud data budgets. It allows the leader to identify the "Value Frontier," ensuring the organization isn't just hoarding data, but is processing the *right* data to achieve the highest ROI.

2. The Game Theory Simulation Pairing for Market Dominance

Leaders should pair **Cognitive AI with Game Theory** to achieve **Competitive Predictive Sovereignty.** This is best deployed during aggressive market expansions or pricing wars. While the AI predicts market trends, Game Theory simulates how competitors will react to the firm's data-driven moves. This ensures the "Compass" identifies the "Nash Equilibrium"—the strategy that remains dominant even when rivals attempt to counter it—securing the brand's offensive position in a volatile landscape.

3. The Digital Twin Pairing for Structural Resilience

When re-architecting the firm's information flow, **Cognitive AI must be paired with Data Digital Twins.** This is best utilized to "rehearse" the impact of a total data migration or a new governance framework. The Strategist simulates how data will flow through a virtual replica of the organization's tech stack. This pairing ensures the "Compass" guides the transformation without causing operational downtime, allowing for the "Stress-Testing" of the organization's nervous system before any physical changes are made.

4. The Knowledge Graph Pairing for Skill and Data Arbitrage

To maximize the impact of the organization's intellectual capital, a leader should pair **Cognitive AI with Knowledge Graphs.** While AI tracks data output, the Knowledge Graph maps how specific data sets relate to individual employee expertise. This allows for "Data Arbitrage"—identifying underutilized data assets and matching them with the specific teams most capable of extracting their value. It ensures the "Compass" is always directing the organization's "Alpha Minds" to the firm's most potent information moats.

5. The Causal Inference Pairing for Meritocratic Truth

For maintaining a "Data-Driven Culture," leaders should pair **Cognitive AI with Causal Inference.** While AI identifies successful outcomes, Causal Inference proves *why* those successes happened—separating genuine strategy from lucky market tailwinds. This is best utilized in quarterly performance reviews and budget allocations. It ensures the "Compass" points toward the initiatives that actually drive growth, preventing the "False Positives" of luck from distorting the organization's long-term strategic map.

6. The Agentic Orchestration Pairing for Dynamic Governance

In high-velocity environments, **Cognitive AI should be paired with Agentic Orchestration** for **Live Policy Execution.** While the Strategist sets the data governance goals, the Agentic layer automatically re-allocates compute power or access rights based on real-time needs and risk profiles. This is best utilized during sudden market shifts or cyber-threat escalations. It turns the "Compass" into an active driver of the enterprise, ensuring that the firm's data resources are always directed toward the areas of greatest opportunity and safety.

Data Power: Structural Intelligence

In the legacy enterprise, data was often treated like a "Warehouse"—a place where you dumped everything and hoped you could find it later. This led to the "Data Swamp," where information existed but was unreachable, uncontextualized, and ultimately useless. The transformation into **Architects** turns the leader and the data team into the "Designers of the Corporate Brain." In this role, the focus is not on the *volume* of data, but on the **Architecture of Connectivity**. It is the shift from "Storing Facts" to "Mapping Relationships." The Architect ensures that the organization's data possesses "Structural Intelligence," where every piece of information knows its place, its history, and its relationship to every other piece of information in the global enterprise.

Defensive Stance: Killing the Fragmentation of Truth

The defensive mission of the Data-Architect is the **Abolition of Silos**. In most large companies, the "Brain" is sufferring from a form of digital schizophrenia—the shipping data doesn't "speak" to the customer service data, and the financial records are disconnected from the marketing logs. This fragmentation creates "Institutional Friction," where the organization wastes millions of hours trying to "reconcile" different versions of reality. The Architect role kills this "Inefficiency" by building a "Unified Knowledge Graph." By "Resolving the Identity" of every entity—customer, product, or employee—across every system, the leader kills the "Friction of Contradiction." This creates a **Shield of Continuity**. It ensures that the organization has a "Single Version of the Truth," protecting the brand from the errors and delays that occur when the left hand doesn't know what the right hand is doing. This defensive posture turns the "Data Swamp" into a "Crystal Spring," providing a clean, reliable foundation for every other transformation.

Offensive Stance: Unlocking Contextual Awareness

On the offense, the Architect data moves toward **Cognitive Fluidity**. This is the ability of the "Data Brain" to provide "Perfect Context" for any decision. When the architecture is sound, the system doesn't just provide a data point; it provides the "Network of Meaning" surrounding that point. If a leader looks at a "Supply Chain Delay," the Architect-Brain automatically surfaces the related "Customer Satisfaction Scores," "Current Inventory Buffers," and "Alternative Logistics Paths." This **Unlocks Relevance**, allowing the organization to act with "Total Awareness." It involves "Dynamic Schema Evolution," where the data brain learns and adapts its own structure as the business grows. This turns the organization into a "High-IQ Brand"—one that is fundamentally smarter because its "Brain" can connect dots that a fragmented competitor would never even see. It ensures that the brand's "Intelligence" is an "Accumulative Asset" that gets stronger and more interconnected with every new byte of information.

The Data Outcome: Institutional Immortality

The ultimate value of the Architect role is **Systemic Resilience**. It ensures that the company's "Intelligence" is not dependent on the memory of specific employees or the quirks of specific software. For the leader, this provides **Operational Stability**. You are building an "Intellectual Infrastructure" that can survive the departure of key people or the shifting of market trends. This turns the data layer into a "Living Archive" that possesses "Total Recall." It ensures that the brand's "Brain" is a "Global Asset" that provides the same level of sophisticated context to a manager in Tokyo as it does to an executive in New York, creating a brand that is truly "One Mind" across the world.

Executive Power: Institutional Data Consciousness

In the legacy enterprise, data was a fragmented and passive resource, sitting in disconnected silos where its value decayed over time. The **Architect** role represents the executive's transition to the "Designer of the Firm's Collective Consciousness." In this role, the leader constructs a permanent, self-evolving **Cognitive Infrastructure** that transforms raw data into a persistent, networked "Brain." This is the shift from "Data Storage" to "Architected Awareness," ensuring that every piece of information generated by the firm is interconnected and immediately accessible for reasoning. The Architect ensures that the organization possesses a "Data Mind"—one that understands the context of its own information and grows more "intelligent" as it ingests more reality, turning data into a compound asset that never forgets a strategic lesson.

The Sensory Layer: Universal Data Identity & Perception

The **Sensory Layer** of the Data Brain acts as the "Universal Intake System." It moves beyond the limitations of traditional ingestion into **Autonomous Entity Resolution**. Utilizing **Semantic Linking** and **Automated Metadata Tagging**, this layer ensures that every data point—whether a customer record, a sensor log, or a legal contract—is given a unique, persistent digital identity that the "Brain" can recognize across the entire ecosystem. It doesn't just ingest data; it "labels" the context of that data in real-time. By utilizing **Edge Perception Agents**, the sensory layer captures the "Nuance" of the information at the point of origin, ensuring that the "Brain" of the firm is constantly being fed with a high-fidelity, perfectly identified stream of reality. This layer removes the "Friction of Disconnect," ensuring that the organization's "Mind" has a perfect perception of its own informational existence.

The Processing Layer: Structural Intelligence & Knowledge Graphing

The **Processing Layer** serves as the "Neural Connector" of the data stack. It utilizes **Knowledge Graphs** and **Vector Databases** to define and store the complex relationships between the identities captured in the sensory layer. While traditional systems store data in isolated tables, the Architect's processing layer stores the **Semantic Connectivity**. It understands, for example, the deep relationship between a specific vendor's delivery delay and the subsequent sentiment shift in a regional customer service hub. By employing **Graph Neural Networks (GNNs)**, this layer identifies "Hidden Patterns of Value" within the data, allowing the "Brain" to perceive itself as a unified, thinking organism. This creates a **Shield of Systemic Wisdom**, ensuring that when a data point changes in one part of the world, its implications are immediately "felt" and calculated across the entire global knowledge graph.

The Cognitive Layer: Recursive Memory & Generative Logic

The **Cognitive Layer** is the "Reasoning Core" of the Data Brain, powered by **Retrieval-Augmented Generation (RAG)** and **Recursive Feedback Loops**. This layer ensures the brand "never forgets" a data-driven insight. It synthesizes the relationships from the processing layer into a "Continuous Stream of Intelligence" that acts as a permanent advisory for the firm. When an executive asks a question, the Cognitive Layer doesn't just retrieve a file; it "reasons" through the firm's entire informational history to provide an answer grounded in "Deep Institutional Context." It involves **Self-Evolving Data Schemas**, where the Brain actually updates its own understanding of the business as new data arrives. This turns the organization into a "Learning Organism" with an ever-increasing "Corporate IQ," providing the executive with a "Memory Asset" that is independent of any single software package or individual analyst, ensuring the brand's "Data Mind" remains a permanent competitive advantage.

OVERVIEW

Human-in-the-Loop (HITL) AI is a cycle where human supervisors continuously train, tune, and validate a model's performance, especially for ambiguous tasks. The Supervisory & Coaching leader focuses on defining ethical boundaries and quality thresholds, ensuring AI decisions align with organizational values. This synergy elevates the leader from a decision-maker to a model trainer and quality assurance expert. By providing critical feedback, the leader helps the AI evolve, improving accuracy, fairness, and overall governance in complex, high-stakes leadership decisions.

MAIN STEPS

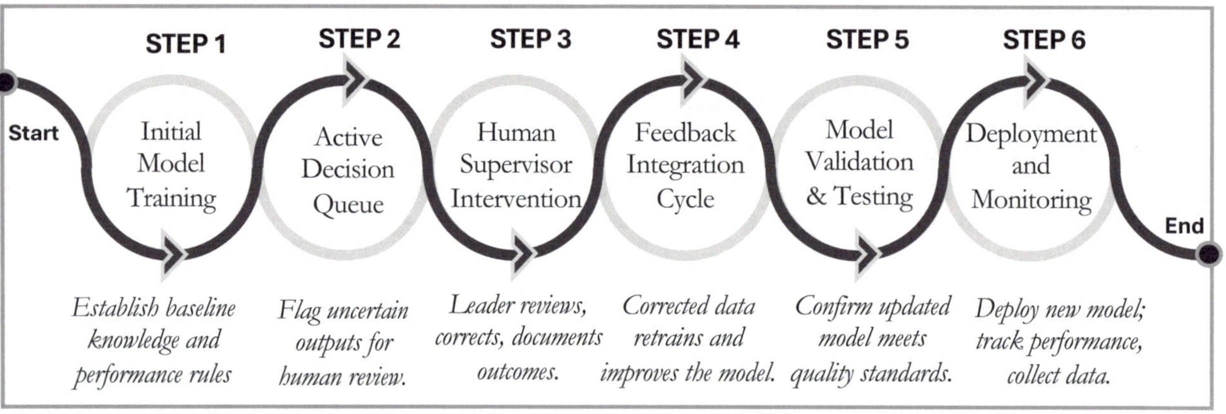

SAMPLE FLOW

Refining Data Annotation for Quality: A leader utilizes HITL to supervise the AI process of classifying unstructured data, guaranteeing high quality and governance standards..

1. Input Ethical Standards
Define fairness metrics and bias thresholds for all performance scoring clearly.

2. Flag Ambiguous Data Points
HITL flags data entries where the automated categorization confidence is low inaccurately.

3. Coach Classification Model
Leader manually corrects the labels, documenting the reasoning for the precise data classification ethically.

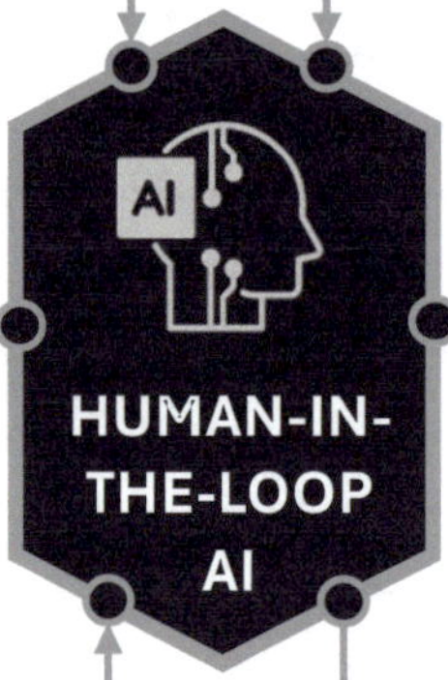

4. Give Feedback to Algorithm
The model immediately incorporates the leader's manual corrections to enhance future data

5. Validate Unbiased Output
Test the updated model against diverse profiles to confirm ethical fairness is achieved correctly.

6. Approve Final Data Set
Leader signs off on the final, cleansed, and accurately labeled dataset for downstream use criteria.

USE CASES

- **Manufacturing | Sensor Data Calibration:** Leaders review and correct AI's automated interpretation of complex machine sensor data, ensuring precise equipment health monitoring and preventative action is taken effectively.
- **Financial Services | Legacy Data Integration:** Supervisors manually correct errors in the AI's merging of old legacy data with new systems, labeling discrepancies to continuously improve data consistency and accuracy reliably.
- **Healthcare | Patient Record Normalization:** Senior staff review AI attempts to standardize diverse patient records, correcting formatting errors and ensuring accurate, confidential data entry safely.
- **Technology | API Usage Compliance:** Data architects review AI suggestions for integrating third-party APIs, correcting configurations to ensure strict data handling protocols are met clearly.
- **Retail | Geo-location Tagging Accuracy:** Managers correct AI's auto-tagging of customer photos or feedback with incorrect store locations, coaching the model on regional differences accurately.
- **Energy | Real-time Grid Anomaly Labeling:** Supervisors validate and label unusual data spikes in power grid consumption, teaching the AI to distinguish true anomalies from standard operational noise quickly.

In the **Data Transformation**, the **Architects (The Brain)** serve as the "Cognitive Infrastructure" of the enterprise. While the primary instrument of **Human-in-the-Loop (HITL)** ensures that the organization's data remains anchored in human ethics and strategic governance, it requires secondary technical pairings to scale that oversight across petabytes of information. To move from "Passive Data Silos" to a "Sovereign Intelligence Layer," a leader must integrate these six pairings. This ensures the organization's collective memory is not just stored, but is contextually aware, legally unassailable, and instantly accessible to the right minds at the right time.

1. The Vector Database (RAG) Pairing for Semantic Memory

Pairing **HITL with Vector Databases** is the foundation of the **"Searchable Data Brain."** While the human Architect sets the governance logic, the Vector Database provides the "Semantic Retrieval" that allows employees to find information based on intent rather than keywords. This is best utilized to break down departmental silos. By indexing the firm's entire history—from R&D notes to legal contracts—in a high-dimensional vector space, the Architect ensures that any query retrieves the most contextually relevant insights, effectively turning "Dark Data" into a live strategic asset.

2. The Knowledge Graph Pairing for Data Lineage

Leaders should pair **HITL with Knowledge Graphs** to ensure **Logical Traceability.** While the human provides the "Business Logic," the Knowledge Graph maps the complex web of dependencies between data points. This is best deployed in audit and compliance scenarios. It allows the Architect to trace exactly how a specific data point moved from a raw sensor to a final executive dashboard. This pairing prevents "Truth Decay," ensuring that the Brain provides a verifiable and transparent record of the firm's data lineage that a human can audit in seconds.

3. The Natural Language Processing (NLP) Pairing for Governance Automation

To maintain a high signal-to-noise ratio, **HITL must be paired with Advanced NLP.** This is best utilized for "Auto-Categorization" and "Anomaly Detection" within the data lake. When the NLP layer identifies a contradiction between two datasets or an entry that violates a new privacy law, it triggers a request for a human Architect to intervene. This ensures the "Brain" is constantly self-pruning and updating, maintaining the integrity of the firm's institutional memory without requiring thousands of man-hours for data cleaning.

4. The Differential Privacy Pairing for Data Sovereignty

When building a collaborative ecosystem, a leader should pair **HITL with Differential Privacy.** This allows the Architect to share synthesized insights with partners or internal teams without the risk of exposing sensitive raw records. This is best used in "Federated Innovation." For example, the firm can allow an external AI researcher to train a model on its proprietary data patterns while the privacy layer mathematically ensures that no individual customer or employee record is ever revealed, protecting the brand's legal and ethical sovereignty.

5. The Multi-Modal AI Pairing for Implicit Insight Capture

To move beyond structured tables, leaders should pair **HITL with Multi-Modal AI.** This allows the "Brain" to ingest and "understand" non-textual data like factory floor video, design sketches, or audio from sales calls. This is best utilized in "Capturing Tacit Knowledge." The human Architect guides the AI to identify which visual or auditory patterns represent "Best Practices." This ensures the organization's "Brain" captures the subtle nuances of human expertise that are typically lost in translation when converted to traditional documentation.

6. The Blockchain Pairing for Immutable Governance

For ensuring the absolute integrity of the firm's "Sovereign Truth," **HITL should be paired with Blockchain.** While the human Architect sets the core policies, the Blockchain stores a tamper-proof "Audit Trail" of every modification made to the data architecture. This is best utilized for "Regulatory Sovereignty." It prevents any unauthorized or accidental tampering with the firm's most critical data sets. It ensures that the "Brain" remains an unassailable record of truth, providing a permanent, timestamped ledger of the organization's evolving data logic.

Data Power: Autonomous Agency

The final and most radical shift in the data journey is the transition from "Data as Information" to **"Data as Action."** Historically, data was something humans looked at to decide what to do. The transformation into **Executors** represents the birth of "Agentic Data." In this role, the data doesn't just "report" a status; it "executes" a response. The data becomes the "Hands" of the organization, capable of navigating complex workflows, negotiating with other systems, and resolving problems in real-time without human intervention. This is the shift from "Passive Databases" to "Active Agents"—turning the organization into an "Autonomous Force" that can manifest the leader's intent at the speed of light.

Defensive Stance: Killing the Operational Lag

The defensive mission of the Data-Executor is the **Elimination of Latency**. In the legacy world, there is always a "Gap" between a data signal (e.g., "A customer is unhappy") and the human execution (e.g., "Sending a refund"). This gap is where value, trust, and profit leak out of the company. The Executor role kills this "Operational Drag" by establishing "Closed-Loop Automation." The data identifies the problem and "Executes the Fix" in the same millisecond. This is the **Shield of Frictionless Flow**. By "Hard-Coding the Response" to common operational failures—such as inventory re-ordering or security patch deployment—the data ensures the company remains in a state of "Constant Performance." This defensive posture protects the brand's "Margins" and "Reputation" by ensuring that "Minor Glitches" never grow into "Major Crises," because the "Hands" of the data are always there to catch the ball before it hits the ground.

Offensive Stance: Unlocking Scalable Autonomy

On the offense, the Executor data moves toward **Agentic Orchestration**. This is the ability of data to "Conduct" the entire enterprise. Instead of a human manually moving data from one system to another, "Data Agents" navigate the software landscape to achieve a goal. For example, a "Procurement Agent" doesn't just flag a low stock level; it searches for the best price, negotiates with vendor APIs, verifies the shipping schedule, and updates the ERP—all autonomously. This **Unlocks Infinite Scale**, where the "Hands" of the company can handle a 100x increase in complexity without a single new hire. It turns the organization into a "High-Frequency Brand"—one that can out-execute competitors through sheer "Operational Velocity." It allows the leader to "Orchestrate" global outcomes by setting "Strategic Intent" and letting the "Data Agents" find the most efficient path to manifest that reality in the physical and digital world.

The Data Outcome: Absolute Leverage

The ultimate value of the Executor role is **Operational Sovereignty**. It allows a brand to "Operate at the Speed of Data," not the speed of human bureaucracy. For the leader, this provides **Execution Peace of Mind**. You are no longer "Managing Processes"; you are "Directing an Engine." This turns the organization into an "Autonomous Powerhouse"—where the "Hands" of the data are always active, always precise, and always working to fulfill the "Voice" of the leadership. It ensures that the brand's "Execution" is as reliable as a law of physics, transforming "Information" into "Action" and "Action" into "Infinite Profit."

Cognitive Layer	Autonomous Data Fulfillment & Self-Correction
Processing Layer	Multi-Agent Data Orchestration & Logic
Sensory Layer	Reflexive Signal Detection & Triggering

Executive Power: Agentic Data Autonomy

The final shift in the data journey is the transition from "Data Visualization" to **"Autonomous Data Fulfillment."** The **Executor** role represents the birth of "Active Data," where the leader provides the "Strategic Intent" and the data itself triggers the necessary actions across the enterprise. Historically, data was a "dead" resource that required a human to interpret it and then "do" something. The Executor role replaces this with a system of **Agentic Autonomy**, where the "Hands" of the data stack are capable of fulfilling tasks—such as re-balancing portfolios, updating supply chains, or personalizing marketing—without manual intervention. This is the shift from "Data-Informed" to "Data-Executed," giving the executive the leverage to "Operate at the Speed of Light" by transforming raw information into immediate, autonomous results.

The Sensory Layer: Reflexive Signal Detection & Triggering

The **Sensory Layer** of the Data Hands serves as the organization's "Reflex System." It is designed for **Real-Time Event Triggering**, monitoring the data stream for specific "Anomalies or Opportunities" that require an immediate autonomous response. Utilizing **Event-Driven Microservices** and **Predictive Trigger Logic**, this layer identifies the moment a data point crosses a threshold that requires action. It doesn't wait for a human to see a "Low Inventory" alert; it "feels" the operational need the micro-second the data is ingested. This sensory input provides the "Activation Signal" for the agentic stack, ensuring that the "Hands" of the company are always ready to move the millisecond a requirement is detected. This layer protects the executive's "Margins" by ensuring the organization responds to data with biological speed, catching opportunities before they dissipate.

The Processing Layer: Multi-Agent Data Orchestration & Logic

The **Processing Layer** acts as the "Information Cerebellum," breaking down the leader's high-level intent into executable data-tasks. It utilizes **Agentic Logic (ReAct Frameworks)** and **Chain-of-Thought Reasoning** to orchestrate a swarm of specialized data agents. If the leader's intent is to "maximize liquidity across all European accounts," this layer identifies the thousand sub-tasks required—from analyzing tax implications to executing real-time bank transfers. It manages the "Inter-Agent Communication," ensuring that each bot knows the "Ethical and Financial Guardrails" set by the executive. This layer provides the **Governance and Logic** for the execution, ensuring that the data "Hands" stay within the boundaries of law and policy, killing the friction of manual oversight through automated, high-speed coordination between disparate data systems.

The Cognitive Layer: Autonomous Data Fulfillment & Self-Correction

The **Cognitive Layer** is where the data "Hands" actually manifest the result through **Action Transformers** and **Automated Data Ops**. This is the layer of "Closed-Loop Execution." It doesn't just "process" information; it "achieves" the strategic outcome. If the data stack encounters an obstacle—such as a failed transaction or a corrupted feed—this layer uses **Self-Correction Logic** and **Reinforcement Learning** to pivot the strategy and find an alternative execution path without requiring new instructions from the leader. It turns "Information" into "Physical and Digital Reality," whether that means moving funds, updating code, or deploying logistics autonomously. This provides the executive with **Absolute Operational Leverage**, allowing them to "Direct the Information Flow" rather than "Manage the Toil." It transforms the firm into an "Autonomous Information Machine" that delivers "Market Dominance" through sheer agentic velocity and data-driven precision.

OVERVIEW

Agentic AI operates as an autonomous mission manager, breaking down high-level strategic goals into executable tasks. The Driving Mission leader defines the ultimate objective and core values, essentially setting the constitutional limits for the AI agent. This synergy offloads complex, time-consuming multi-step execution to the AI, allowing the leader to focus exclusively on strategic vision and alignment. The leader initiates the mission; the agent executes and reports on key milestones. This partnership fundamentally converts strategy into action, maximizing the leader's impact on high-level direction and governance.

MAIN STEPS

Start

- **STEP 1** — Mission Goal Setting: *Define the high-level desired outcome.*
- **STEP 2** — Autonomous Task Breakdown: *Agent breaks goal into specific sub-tasks.*
- **STEP 3** — Execution and Monitoring: *Agent performs steps, tracks progress autonomously.*
- **STEP 4** — Milestone Reporting & Review: *Agent summarizes for leader approval.*
- **STEP 5** — Course Correction, Redirection: *Leader adjusts mission parameters or values.*
- **STEP 6** — Mission Completion & Handoff: *Agent delivers final output to human team.*

End

SAMPLE FLOW

Automated Data Quality Uplift: The leader mandates an agent to achieve and maintain top-tier data quality across three specific legacy databases.

1. Set Quality Baseline
Define acceptable error rates and standards for data completeness and accuracy clearly.

2. Deploy Remediation Agent
Agent autonomously scans, identifies, and flags all dirty data sources correctly.

3. Execute Cleansing Scripts
Agent runs batch-processing and normalization routines across all target data autonomously.

AGENTIC AI

Generate Compliance Audit
Agent summarizes all remediation activities and compliance status for management promptly.

Review Remediation Exceptions
Leader validates complex, high-risk data exceptions that the agent could not resolve ethically.

Authorize Ongoing Monitoring
Leader commissions the agent to continuously enforce quality and governance criteria.

USE CASES

- **Manufacturing | Unified Sensor Data Mission:** An agent orchestrates the transformation of disparate machine sensor data into a single unified format, autonomously enforcing strict metadata standards across all factory locations effectively.
- **Financial Services | Regulatory Reporting Automation:** The leader directs an agent to fully automate a major regulatory reporting pipeline, autonomously collecting, transforming, and validating all required financial data reliably.
- **Healthcare | Patient Record Interoperability:** An agent is tasked with a mission to standardize patient record formats across affiliated clinics, autonomously mapping and reconciling inconsistent data fields accurately.
- **Technology | Cloud Migration Governance:** An agent manages the data governance mission during cloud migration, autonomously ensuring all privacy classifications and security rules are correctly applied to transferred datasets consistently.
- **Retail | Customer 360 View Deployment:** An agent unifies fragmented customer transaction, loyalty, and web behavior data into a single "Customer 360" profile, autonomously managing identity resolution and schema alignment quickly.
- **Energy | Smart Grid Data Cataloging:** The leader deploys an agent to create a comprehensive data catalog for smart grid telemetry, autonomously documenting all data lineage, access controls, and usage policies clearly.

In the **Data Transformation**, the **Executors (The Hands)** represent the organization's "Autonomous Actuation Engine." While the primary instrument of **Agentic AI** provides the "agency" to navigate complex data goals, it requires secondary technical pairings to ensure that digital insights are translated into concrete, error-free results. This empowers the "Hands" to execute the firm's strategy with millisecond precision, ensuring that the transition from a digital signal to a business outcome is seamless, secure, and self-improving.

1. The RPA Pairing for Legacy Integration

Pairing **Agentic AI with Robotic Process Automation (RPA)** is best utilized when the "Hands" must act within older, rigid database environments. While the Agent decides the strategy—such as re-balancing an inventory database—RPA provides the "fingers" to execute the specific entries in legacy ERP systems that lack modern APIs. This is essential for "Deep-Stack Transformation," allowing autonomous agents to bridge the gap between cutting-edge intelligence and the firm's existing infrastructure, ensuring that legacy constraints never slow down the speed of offensive action.

2. The Digital Twin Pairing for Execution Guardrails

Leaders should pair **Agentic AI with Data Digital Twins** when deploying autonomous agents in high-risk financial or operational environments. Before an agent executes a massive trade or reallocates cloud resources, it "rehearses" the move in the Digital Twin. This pairing ensures the "Hands" never take an action that could cause a systemic crash or a data integrity violation. It provides a virtual sandbox for autonomous execution, ensuring every move is verified against a live simulation of the enterprise ecosystem before it hits the production environment.

3. The Edge AI Pairing for Real-Time Actuation

When execution must happen at the "Point of Origin"—such as on an autonomous delivery vehicle or a remote sensor—**Agentic AI must be paired with Edge AI**. This allows the "Hands" to process incoming data streams and make execution decisions locally without the latency of a cloud round-trip. This pairing is vital for "Critical Path Operations," ensuring that the agents can react to sudden data anomalies or physical obstacles in milliseconds. It turns the organization's data layer into a real-time responsive nervous system that acts as fast as it thinks.

4. The Blockchain Pairing for Autonomous Settlement

For secure, multi-party data transactions, a leader should pair **Agentic AI with Blockchain (Smart Contracts)**. This is best used in "Automated Data Marketplaces" or inter-company procurement. The "Hands" don't just find a data provider; they execute the payment and verify the data quality through an immutable ledger. This pairing ensures that autonomous actions are legally binding and transparent. It eliminates the friction of manual audit and reconciliation, ensuring that the firm's data-driven exchanges are conducted with absolute financial and operational integrity.

5. The Computer Vision Pairing for Physical-Digital Synthesis

To allow the "Hands" to bridge the gap between physical reality and digital records, leaders should pair **Agentic AI with Computer Vision**. This is best utilized in "Autonomous Asset Management." By "seeing" a physical warehouse or factory floor, the agent can automatically update the digital record to reflect reality—identifying misaligned stock or mechanical wear. This pairing gives the "Hands" the sensory awareness required to synchronize the firm's "Physical Truth" with its "Data Truth," ensuring that the organization's digital twin is never out of sync with its physical assets.

6. The Reinforcement Learning Pairing for Process Refinement

In complex, shifting data environments, **Agentic AI should be paired with Reinforcement Learning (RL)**. This is best utilized for optimizing "Dynamic Workflows," such as automated ad-bidding or energy grid management. The "Hands" learn from the reward or penalty of every action taken, constantly refining their technique to maximize yield. This provides the brand with a self-improving execution layer that grows more efficient with every task performed, ensuring that the organization's "Hands" are constantly evolving to meet the demands of a high-velocity market.

SECTION C

BLUEPRINT SERIES

Business Maturity

Ai6x6™ Transformation Blueprint

Story 1 — The Marshall Plan Reconstruction (1948–1952): Post-war Europe faced shattered economies, broken supply chains, and institutional collapse. The Marshall Plan introduced structured funding, transparent reporting, and coordinated execution across nations. Local governments provided context while centralized oversight ensured accountability. This blend of autonomy and structure accelerated recovery. The program demonstrated how disciplined governance can rebuild complex systems. It became a blueprint for large-scale organizational maturity.

Story 2 — The Rise of Quality Management in Japan (1950s–1960s): Japan's industries struggled with inconsistent quality and limited global credibility. Statistical quality control introduced measurement, repeatability, and disciplined improvement cycles. Workers contributed insights while leaders standardized processes. Over time, quality became a cultural norm rather than a corrective action. This shift transformed fragmented operations into cohesive systems. It marked the emergence of maturity through methodical refinement.

Story 3 — NASA's Systems Engineering Discipline (1960s): The space race demanded unprecedented coordination across thousands of contractors and scientific domains. NASA formalized systems engineering to integrate requirements, risks, and interfaces into a unified framework. Structured reviews balanced innovation with rigor. Human judgment guided trade-offs while documentation preserved institutional memory. This discipline enabled complex missions to succeed repeatedly. It became a global model for organizational maturity in high-stakes environments.

Story 4 — The Birth of Modern Corporate Governance (1970s–1980s): Corporate scandals exposed weaknesses in oversight, transparency, and fiduciary responsibility. Boards adopted clearer roles, audit mechanisms, and accountability structures. Regulations reinforced ethical conduct while organizations matured through formal governance. Leaders learned to balance autonomy with compliance. This era established governance as a strategic capability, not an administrative burden. It marked a turning point in institutional maturity.

Story 5 — Toyota's Global Expansion Playbook (1980s–1990s): As Toyota expanded globally, it faced cultural differences, supply-chain complexity, and operational variability. The company codified principles, standardized training, and embedded continuous improvement across regions. Local teams adapted practices while preserving core philosophy. This balance of global consistency and local flexibility enabled scalable excellence. It demonstrated maturity as the ability to replicate capability without diluting identity.

Story 6 — The Digital Transformation of Banking (1990s–2000s): Banks confronted rising customer expectations, regulatory pressure, and technological disruption. They modernized legacy systems, standardized data, and introduced structured risk frameworks. Digital channels required new operating models and cross-functional coordination. Human expertise guided decisions where automation lacked nuance. Over time, banks evolved from siloed institutions into integrated digital enterprises. This transition reflected maturity through modernization and disciplined reinvention.

LESSONS FROM A DATA TRANSFORMATION PERSPECTIVE

Data maturity evolves from fragmented information to integrated, governed, and decision-ready intelligence. Early systems relied on manual interpretation, while later decades introduced structured measurement, standardized reporting, and cross-functional visibility. Mature organizations treat data as an institutional asset rather than departmental property. They build processes that ensure accuracy, transparency, and traceability. As systems evolve, data becomes the foundation for foresight, risk management, and operational consistency. Maturity is achieved when data flows seamlessly across functions, enabling decisions that are timely, evidence-based, and aligned with strategic intent. The organization becomes more predictable, accountable, and adaptive through disciplined data practices.

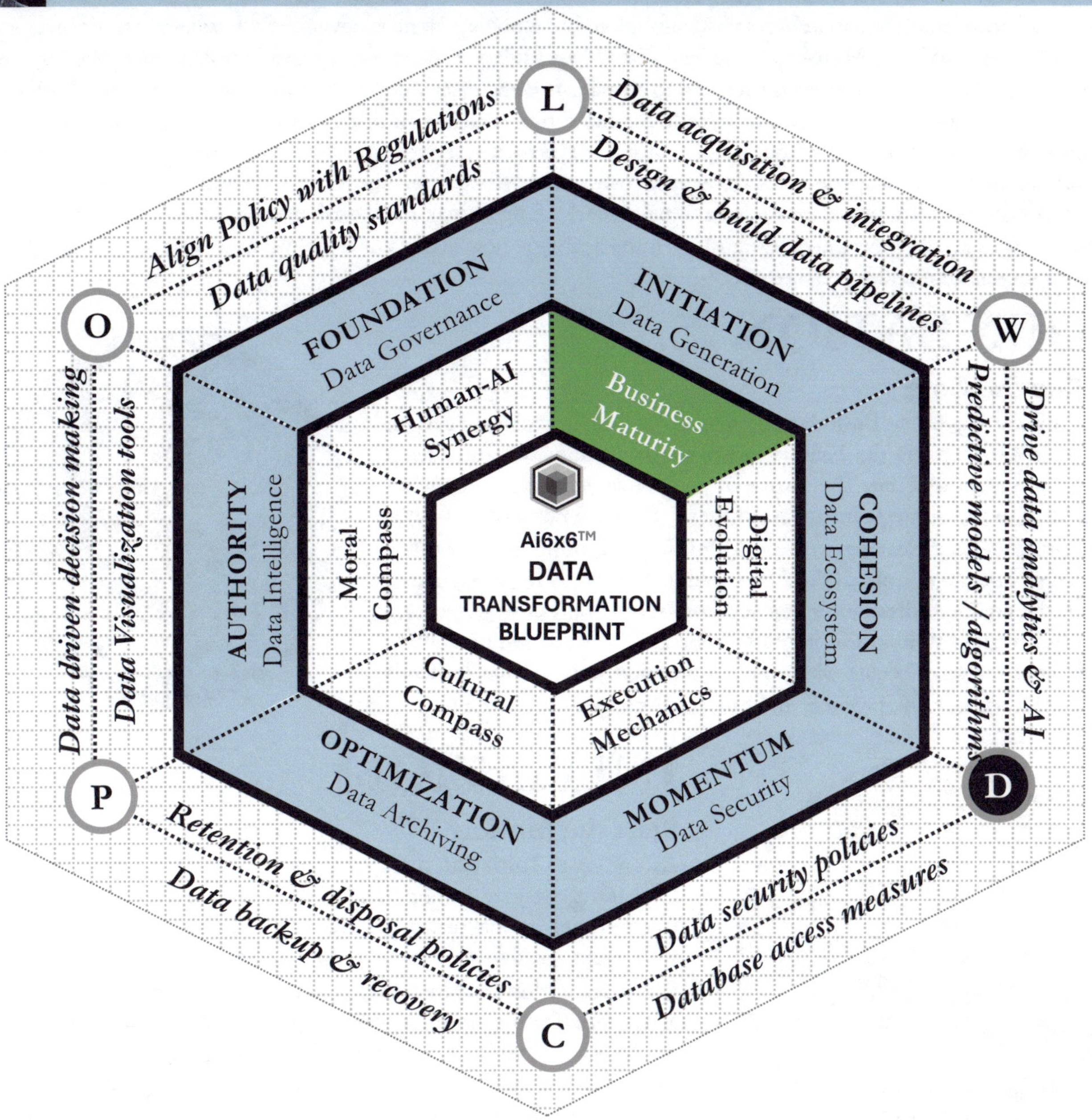

The **Ai6x6 Research Lab** selected this lens because an organization's maturity in the age of intelligence is defined by the integrity and fluidity of its data lifecycle. We view data maturity as a rigorous progression from foundational **Governance** and **Security** to the peak of **Pattern Analytics**. This dimension ensures that the aggressive data generation required for AI is anchored in a deep commitment to **Zero-Trust** and real-time orchestration. It is a vital lens because, in the turbulence of the 2026 information landscape, an organization's ability to turn raw streaming data into actionable intelligence is its primary survival mechanism. This lens is the heart's commitment to treating data as a high-fidelity institutional asset.

We believe true maturity is found in the ability to manage the delicate balance between radical innovation and disciplined **Archiving**. By focusing on the entire ecosystem—from **Real-time Streaming** to the responsible disposal of information—the Lab provides a mechanism for leaders to build a data infrastructure that is both agile and secure. This lens ensures that data generation is not a chaotic accumulation, but a deliberate modeling of reality. It is the steady hand that guides the organization through the complexities of behavioral analytics, recognizing that while data volume is infinite, its value is only realized through structured maturity. It turns the technical stack into a stable, mature foundation for long-term strategic endurance.

Every leadership transformation begins with ambition, but ambition alone is never enough to carry an organization into its future. **Business Maturity** is the lens that forces leaders to confront the gap between what they hope to achieve and what their enterprise is prepared to sustain. It is the discipline that reveals whether the organization is structurally, operationally, and culturally ready for the future it claims to desire. In the AI era, this gap becomes even more pronounced; AI does not simply automate tasks—it reshapes workflows, decision cycles, and the very rhythm of how leadership operates. It exposes weaknesses that were previously tolerable and accelerates the consequences of poor preparation, demanding a level of rigor that traditional management frameworks often overlook. It examines how ambition translates into capability, how vision converts into execution, and how culture sustains momentum under pressure.

FOUNDATION

Define the Core

The journey begins at **Foundation**, where the primary objective is to *Define the Core.* This is the bedrock upon which all transformation rests, requiring leaders to **Filter the Noise** of competing priorities and anchor their identity and strategic intent. At this level, clarity is not optional—it is existential. Without clear governance, AI initiatives drift, priorities conflict, and teams operate in isolated silos. Foundation maturity is the discipline of intent: the ability to distinguish urgency from readiness. Leaders who skip this stage mistake motion for progress, building transformation on shifting sand. True maturity begins when the organization knows what it stands for, what it will not compromise, and how it will measure success.

INITIATION

Start the Motion

Ascending to **Initiation** marks the first active step: *Start the Motion.* This level turns the gaze inward, demanding self-awareness and the courage to confront personal biases. By **Wiring the System**, leaders catalyze the shift from passive management to intentional authority. Initiation is where leadership becomes kinetic—it transforms from aspiration into activation. The organization begins to move, not because it is told to, but because its internal circuits are aligned. Maturity here is emotional as much as structural; it requires self-regulation, empathy, and the ability to lead through friction. The leader's persona is tested, revealing whether they can sustain momentum through disruption rather than retreat into comfort.

COHESION

Connect the Parts

The climb continues to **Cohesion**, represented by *The Engine.* This is the critical unifier where people, processes, and systems are harmonized. The focus shifts to **Connect the Parts**—to ensure that individual leadership goals align with a unified organizational voice. By **Shifting the Gear**, leaders eliminate silos and systemic friction, transforming fragmented efforts into a synchronized machine. Cohesion maturity is the art of integration: turning disconnected motion into collective momentum. Without it, organizations celebrate activity instead of achievement. The mature leader understands that alignment is not consensus—it is coherence, the ability to move as one without losing diversity of thought.

MOMENTUM

Drive the Speed

The fourth rung is **Momentum**, where the organization begins to *Drive the Speed*. Leadership here moves toward scaling impact with predictable velocity. By **Removing the Drag** of legacy thinking and institutional inertia, leaders accelerate the collective intelligence of the enterprise. Momentum maturity is the physics of transformation—it measures how efficiently energy converts into progress. At this stage, the organization's rhythm changes; it begins to operate with self-propelling force. The leader's role shifts from directing individuals to empowering systems, ensuring that innovation scales faster than resistance. Momentum is not chaos—it is controlled acceleration, the moment when transformation gains its gravity.

OPTIMIZATION

Refine the Flow

Refinement occurs at **Optimization**, characterized by *The Flow*. Here, leadership pauses to **Refine the Flow** by **Locking the Code** of influence through reflection and data-driven inquiry. Immature organizations measure what is easy; mature ones measure what matters. Optimization maturity is the discipline of precision—removing friction without losing flexibility. Leaders refine decisions still in motion, adjusting with humility and surgical accuracy. Ego and bias are replaced by evidence and iteration. The organization becomes self-healing, capable of continuous improvement without external intervention. Optimization is not about perfection—it is about adaptability, the ability to evolve faster than the environment demands.

AUTHORITY

Own the Market

The summit is reached at **Authority**, where leaders finally *Own the Market*. This stage manifests as *The Benchmark*—a legacy of endurance that can withstand external shocks and volatility. By **Fueling the Future**, leaders ensure that transformation is not a temporary win but a permanent standard. Authority maturity is the discipline of stewardship: sustaining excellence while preparing for reinvention. Immature organizations avoid this stage because it exposes uncomfortable truths; mature ones embrace it as the crucible of resilience. At this level, leadership becomes institutional—it transcends personality and becomes culture. The organization no longer reacts to the market; it defines it.

Business Maturity is the scaffolding that holds the weight of innovation. It ensures that speed does not collapse under fragility and that ambition does not outpace capability. Each level builds upon the last, forming a continuum from clarity to mastery. Foundation defines intent, Initiation activates leadership, Cohesion integrates systems, Momentum accelerates impact, Optimization refines intelligence, and Authority sustains legacy. In the AI era, maturity is not a luxury—it is survival. As algorithms amplify both strengths and weaknesses, leaders must design organizations that can learn, adapt, and endure. Business Maturity is therefore not a checklist but a **leadership discipline**—a way of thinking that transforms ambition into architecture. When leaders embrace this full sequence, they discover that maturity is not the end of transformation but its engine—the invisible structure that makes both explosive speed and long-term continuity possible.

BUSINESS MATURITY ANALOGY – CLIMBING MOUNT EVEREST

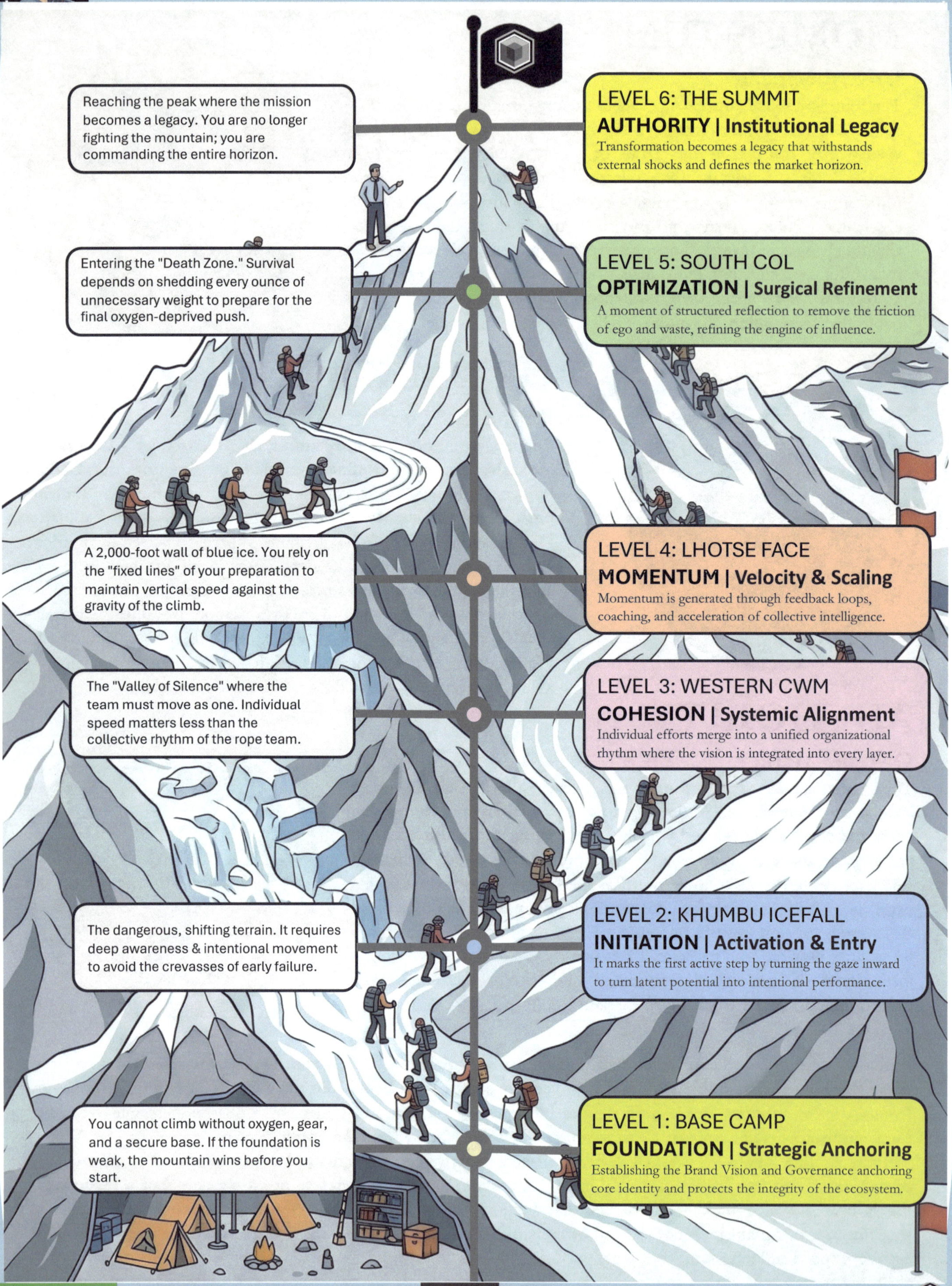

Ai6x6™ Transformation Blueprint

FOUNDATION | Data Governance

"The goal is to turn data into information, and information into insight."
- Carly Fiorina.

Former CEO of Hewlett-Packard (HP), the first woman to lead a Fortune 20 company. She was a prominent and often controversial figure in the technology industry, known for her leadership and strategic initiatives at HP, marking a significant milestone for women in corporate leadership.

The Foundation level is your Data Governance, defining the secure identity and market presence of your information through Zero-Trust standards. These rules set the ethical and structural boundaries that protect the integrity of your entire digital ecosystem.

Establishing robust policies, clear processes, and defined responsibilities for the meticulous management of an organization's data assets throughout their entire lifecycle, from creation to archival. This foundational element ensures data quality, regulatory compliance, and stringent security.

"Data Governance" forms the bedrock for trustworthy, reliable, and impactful data-driven decisions across all levels of the enterprise, minimizing risks, enhancing data usability, and building confidence in analytical outcomes, crucial for any data-led transformation initiative.

For business strategy maturity, the "**Data Governance**" chapter is foundational because it establishes the overall strategic framework for managing an organization's most critical asset: data. A mature business strategy defines clear policies, standards, and procedures for ensuring data quality, security, and accessibility. This governance ensures data integrity, compliance, and trustworthiness, which are non-negotiable for informed decision-making and strategic execution. It's about building a reliable data foundation that supports all business operations, enabling accurate insights and confident strategic moves. Here is how we approach our learning:

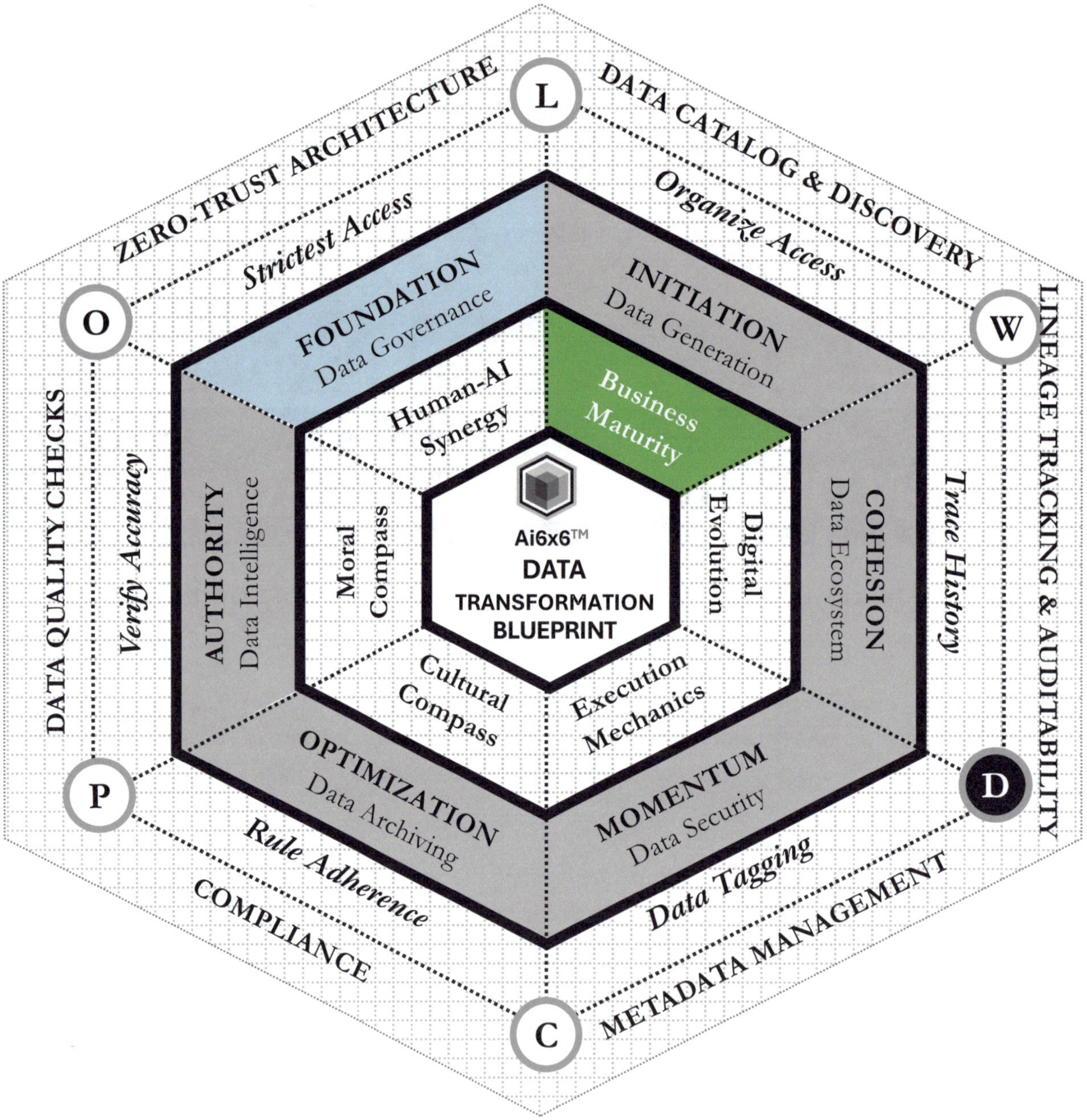

Ai6x6 BUSINESS MATURITY FOR GOVERNANCE
DATA TRANSFORMATION | DATA GOVERNANCE

For a business leader, implementing a Zero-Trust Architecture is no longer an advanced security concept but a fundamental imperative for protecting your organization's most valuable asset: its data. In an era of escalating cyber threats and complex attack vectors, the traditional perimeter-based security model is insufficient. Zero-trust dictates that no user, device, or application is inherently trustworthy, regardless of its location relative to the corporate network; every access request must be strictly verified. Central to this approach is the principle of least privilege access. As a leader, you must ensure that users are granted only the absolute minimum permissions necessary for their specific role and tasks, dramatically reducing the potential attack surface if an account is compromised. This granular control is vital.

Furthermore, micro-segmentation divides networks into isolated segments, limiting the lateral movement of threats within your infrastructure. If one segment is breached, the attack is contained, preventing widespread damage and protecting critical data stores. This containment strategy is a cornerstone of robust security. Crucially, continuous verification requires authentication at every stage of access and interaction, rather than assuming trust once a user is inside the network. This constant re-authentication based on context (e.g., device, location, behavior) adds multiple layers of defense. Neglecting a Zero-Trust Architecture can lead to catastrophic data breaches, significant financial losses, reputational damage, and severe regulatory penalties. By championing this proactive and rigorous security paradigm, you empower your organization to defend against sophisticated attacks, ensure data integrity, and build a resilient digital infrastructure that safeguards your business's future and customer trust.

WHAT?	WHY?	HOW?	BEWARE!
LEAST PRIVILEGE ACCESS	**REDUCED ATTACK SURFACE**	**ROLE-BASED ACCESS CONTROL (RBAC)**	**COMPLEXITY**
Granting minimum necessary permissions.	Minimizes potential damage from compromised accounts.	Define roles and assign granular permissions.	Managing fine-grained permissions across large organizations.
MICRO-SEGMENTATION	**THREAT CONTAINMENT**	**NETWORK ISOLATION**	**IMPLEMENTATION OVERHEAD**
Dividing networks into isolated segments.	Limits lateral movement of threats within the network.	Create small, isolated network segments with strict controls.	Significant architectural and operational complexity.
CONTINUOUS VERIFICATION	**ENHANCED AUTHENTICATION**	**MULTI-FACTOR AUTHENTICATION (MFA)**	**USER FRICTION**
Requiring authentication at every stage.	Provides multi-layered security, preventing unauthorized access.	Implement MFA and context-aware authentication.	Potentially increased login friction for users.

For a business leader, establishing a comprehensive data catalog and robust discovery capabilities is no longer merely a technical convenience but a strategic imperative for maximizing the value of your data assets. In an age where data is the new oil, the ability to quickly find, understand, and utilize relevant information is critical for agile decision-making and competitive advantage. Central to this is creating a centralized data catalog. This searchable repository for both structured and unstructured data sources acts as a single source of truth, allowing employees across the organization to easily locate the data they need, regardless of where it resides.

Furthermore, leveraging automated discovery tools is crucial. By using AI to classify, tag, and organize data across diverse platforms—from databases to cloud storage and legacy systems—you streamline the process of making data discoverable and understandable, reducing manual effort and human error. Crucially, the ultimate aim is data democratization. This involves empowering business users, not just data scientists, to access and utilize relevant data effectively. A well-organized data catalog facilitates self-service analytics, fostering a data-driven culture across all departments. Neglecting a robust data catalog can lead to data silos, duplicated efforts, inconsistent reporting, and a significant lag in decision-making, as employees struggle to find or trust the information they need. By championing data cataloging and discovery, you empower your organization to unlock the full potential of its data assets, accelerate insights, and foster a truly data-driven culture that drives innovation and strategic success. This transparency in data availability is key to a modern business.

WHAT?	WHY?	HOW?	BEWARE!
CENTRALIZED DATA CATALOG	**DATA ACCESSIBILITY**	**METADATA COLLECTION**	**INTEGRATION COMPLEXITY**
Creating a searchable repository.	Enables easy discovery and understanding of data assets.	Implement a tool to collect and organize metadata from all data sources.	Integrating disparate data sources into a single catalog.
AUTOMATED DISCOVERY TOOLS	**EFFICIENCY & ACCURACY**	**AI-POWERED SCANNING**	**AI LIMITATIONS**
Using AI to classify, tag, and organize data.	Streamlines data classification and reduces manual effort.	Deploy AI/ML tools to automatically scan, classify, and tag data.	AI misclassifications or inability to understand unique business contexts.
DATA DEMOCRATIZATION	**DATA-DRIVEN CULTURE**	**SELF-SERVICE TOOLS**	**DATA LITERACY**
Enabling business users to access and utilize data.	Fosters a culture where all employees can leverage data.	Provide user-friendly interfaces for data access and analysis.	Lack of data literacy among business users, leading to misinterpretation.

3. LINEAGE TRACKING AND AUDITABILITY

For a business leader, establishing robust data lineage tracking and ensuring comprehensive auditability are non-negotiable imperatives for maintaining transparency, fostering accountability, and building trust in your data assets. In an age of increasing regulatory scrutiny and data complexity, knowing exactly where your data comes from, how it's transformed, and who accesses it is critical for compliance, data quality, and confident decision-making. Central to this is data flow visualization. By tracking how data moves between systems, applications, and various stakeholders, you gain a clear, end-to-end understanding of your data pipelines. This visual representation reveals dependencies, identifies bottlenecks, and enhances problem-solving when issues arise.

Furthermore, maintaining meticulous audit trails and historical tracking is crucial. This involves capturing detailed logs of all modifications, access events, and usage history for every piece of data. Such immutable records are essential for forensic analysis in case of a breach, troubleshooting data discrepancies, and demonstrating compliance. Crucially, these capabilities directly support regulatory reporting. Providing structured, comprehensive audit records is vital for meeting the stringent requirements of industry standards and data privacy regulations (e.g., GDPR, HIPAA). This ensures that your organization can confidently prove its adherence to legal obligations. Neglecting data lineage and auditability can lead to compliance failures, an inability to trace data errors to their source, and a lack of trust in your data's accuracy, ultimately undermining strategic initiatives. By championing these practices, you empower your organization with transparency, accountability, and a high degree of confidence in its data, safeguarding both operational integrity and reputation. This is about building a trustworthy data foundation.

WHAT?	WHY?	HOW?	BEWARE!
DATA FLOW VISUALIZATION	**TRANSPARENCY & DEBUGGING**	**MAPPING TOOLS**	**COMPLEXITY**
Tracking data movement between systems.	Provides clear understanding of data movement, aids troubleshooting.	Implement tools to map data origins, transformations, and destinations.	High complexity in mapping data flows across vast, interconnected systems.
AUDIT TRAILS AND HISTORICAL TRACKING	**ACCOUNTABILITY & FORENSICS**	**LOGGING SYSTEMS**	**DATA VOLUME**
Maintaining logs of modifications and access.	Ensures accountability and supports security investigations.	Implement comprehensive logging for all data access and modifications.	Managing and storing large volumes of audit log data efficiently.
REGULATORY REPORTING	**COMPLIANCE ASSURANCE**	**REPORTING FRAMEWORKS**	**REGULATORY CHANGES**
Providing structured audit records for compliance.	Facilitates adherence to legal and industry regulations.	Develop automated reports aligned with regulatory requirements.	Keeping up with constant changes in data compliance regulations.

For a business leader, effective metadata management is a fundamental, yet often overlooked, strategic imperative for truly unlocking the value of your data assets. Metadata—data about data—provides the essential context and structure that transforms raw information into usable, understandable, and interoperable insights. Central to this is comprehensive metadata classification. As a leader, you must ensure that clear attributes like source, purpose, format, and ownership are defined for all data assets. This standardized categorization enhances efficient utilization, allowing users to quickly find and understand the relevance of specific datasets.

Furthermore, achieving semantic consistency is crucial. Standardizing metadata across all systems and databases improves interoperability and searchability, ensuring that terms and definitions are uniformly understood throughout the organization. This prevents misinterpretations and fosters accurate data analysis. Crucially, leveraging automated metadata extraction tools is vital for efficiency. By using AI tools to generate metadata directly from raw data sources, you reduce manual effort, minimize human error, and ensure that metadata is consistently updated and accurate, even as data volumes grow. Neglecting robust metadata management can lead to data silos, misinterpretations, duplicated data efforts, and significant inefficiencies in data discovery and analysis. By championing a systematic approach to structuring data descriptions, you empower your organization with cleaner, more understandable, and highly interoperable data, accelerating insights, improving data quality, and fostering a truly data-driven culture that directly supports strategic decision-making and innovation. This foundational work transforms your data from a liability into a powerful asset.

WHAT?	WHY?	HOW?	BEWARE!
METADATA CLASSIFICATION	**EFFICIENT UTILIZATION**	**ATTRIBUTE DEFINITION**	**CONSISTENCY**
Defining attributes for efficient utilization.	Improves data discoverability and understanding.	Establish clear attributes (source, purpose, sensitivity) for all data.	Ensuring consistent classification across diverse data types and systems.
SEMANTIC CONSISTENCY	**INTEROPERABILITY**	**GLOSSARY & STANDARDS**	**LEGACY SYSTEMS**
Standardizing metadata across systems.	Enhances data interoperability and searchability across platforms.	Develop a common business glossary and metadata standards.	Integrating and standardizing metadata from older, disparate systems.
AUTOMATED METADATA EXTRACTION	**ACCURACY & EFFICIENCY**	**AI-POWERED TOOLS**	**AI RELIANCE**
Using AI tools to generate metadata.	Reduces manual effort and improves metadata accuracy.	Implement AI/ML tools to automatically extract and populate metadata.	Over-reliance on automation leading to missed nuances or errors.

For a business leader, ensuring strict compliance in data governance is not merely a legal obligation but a strategic imperative that protects your organization from significant financial penalties, reputational damage, and loss of customer trust. In an increasingly regulated global economy, aligning your data practices with a myriad of legal and industry regulations is paramount. Central to this is ensuring rigorous regulatory adherence. As a leader, you must oversee compliance with critical data privacy frameworks such as GDPR, CCPA, HIPAA, and industry-specific mandates. This involves implementing policies and processes that meet or exceed these legal requirements across all data operations.

Furthermore, developing and enforcing clear data retention policies is crucial. This involves implementing structured guidelines for how long data should be stored, when it should be archived, and when it must be securely purged. Proper retention management balances legal obligations with efficient storage, reducing unnecessary risk. Crucially, establishing continuous risk assessment and mitigation practices is vital. This means continuously monitoring for potential regulatory breaches or violations, proactively identifying gaps in compliance, and implementing corrective measures before they escalate into costly incidents. Neglecting data compliance can lead to severe fines, legal challenges, mandatory reporting of breaches, and a significant erosion of customer confidence. By championing a proactive and comprehensive compliance strategy, you not only protect your business from legal and financial risks but also build a reputation as a trustworthy and responsible data steward, ultimately safeguarding your business's long-term viability and competitive standing in a highly regulated digital world.

WHAT?	WHY?	HOW?	BEWARE!
REGULATORY ADHERENCE	**LEGAL PROTECTION**	**POLICY IMPLEMENTATION**	**EVOLVING REGULATIONS**
Ensuring compliance with GDPR, CCPA, HIPAA.	Minimizes legal risks and avoids hefty fines.	Develop and enforce policies aligned with all relevant regulations.	Keeping up with constant changes in global data privacy laws.
DATA RETENTION POLICIES	**RISK REDUCTION**	**LIFECYCLE MANAGEMENT**	**BALANCING NEEDS**
Implementing guidelines for storing and purging data.	Reduces storage costs and compliance risks.	Define and automate rules for data storage, archival, and deletion.	Balancing legal retention needs with storage efficiency and privacy.
RISK ASSESSMENT AND MITIGATION	**PROACTIVE PROTECTION**	**AUDITS & MONITORING**	**COMPLEXITY**
Continuously monitoring for breaches.	Identifies and mitigates potential regulatory violations.	Conduct regular compliance audits and real-time monitoring for anomalies.	Identifying all potential vulnerabilities in a complex data ecosystem.

For a business leader, implementing rigorous data quality checks is a fundamental strategic imperative for maintaining the accuracy, reliability, and consistency across all your datasets. In today's data-driven world, flawed data leads to flawed insights, undermining decision-making, eroding customer trust, and ultimately impacting your business's bottom line. Central to this is applying robust data validation techniques. This involves implementing automated rules, formatting checks, and integrity constraints to ensure the correctness and completeness of data as it enters your systems, preventing errors at the source.

Furthermore, systematic cleansing and deduplication processes are crucial. This means actively identifying and removing inconsistencies, correcting errors, and eliminating redundant data entries across your datasets, ensuring that your information is clean, standardized, and free from inaccuracies that can skew analysis. Crucially, establishing ongoing monitoring mechanisms ensures continuous quality assurance. By implementing real-time detection systems and regular audits, you can proactively detect anomalies, identify inaccuracies, and address data quality issues as they emerge, rather than reactively fixing problems after they've caused damage. Neglecting data quality can lead to misinformed strategic decisions, inefficient operations, customer dissatisfaction due to incorrect information, and a loss of trust in your analytical capabilities. By championing a comprehensive approach to data quality, you empower your organization with trustworthy, reliable data, accelerating insights, enhancing operational efficiency, and building a foundation for confident, impactful decision-making that drives sustained business success. This commitment to data integrity is non-negotiable for modern enterprises.

WHAT?	WHY?	HOW?	BEWARE!
DATA VALIDATION TECHNIQUES	**DATA INTEGRITY**	**RULE IMPLEMENTATION**	**EVOLVING DATA**
Ensuring correctness through automated rules.	Guarantees data accuracy and completeness at the point of entry.	Define and apply automated validation rules for data inputs.	Adapting validation rules to new data types and sources.
CLEANSING AND DEDUPLICATION	**DATA RELIABILITY**	**DATA PROFILING**	**MANUAL EFFORT**
Removing inconsistencies and redundant entries.	Improves reliability and consistency of datasets.	Use tools to identify and correct errors, and merge duplicate records.	Large datasets requiring significant manual effort for cleansing.
ONGOING MONITORING	**PROACTIVE DETECTION**	**AUTOMATED ALERTS**	**FALSE POSITIVES**
Establishing continuous quality assurance.	Identifies anomalies and inaccuracies in real-time.	Implement dashboards and alerts for data quality issues.	Over-alerting or difficulty distinguishing true issues from noise.

Ai6x6™ Transformation Blueprint

INITIATION | Data Generation

"Data is the exhaust of the digital economy." - Michael Dell.

Founder, chairman, and CEO of Dell Technologies. He started Dell from his dorm room, pioneering direct-to-consumer computer sales and building a global powerhouse. His vision and adaptability made him a tech industry titan and a model of entrepreneurial success.

The Initiation level is your Data Generation, the act of turning raw observations into usable digital assets through real-time capture. By implementing modeling and streaming, the organization initiates the flow of information that feeds the AI engine.

The fundamental act of creating, collecting, and capturing new data from a multitude of diverse sources, ranging from customer interactions to operational sensors, web analytics, and IoT devices. This is the absolute starting point of all data value; without robust and accurate data generation, there's no fuel for meaningful analysis. There is no basis for insightful conclusions, and thus, a significant hindrance to any data-led transformation. High-quality data generation ensures a rich, reliable foundation for all subsequent data activities.

The **"Data Generation"** chapter is crucial for business strategy maturity as it mobilizes the raw data essential for strategic insights. A mature business strategy recognizes the importance of efficiently gathering and storing data from diverse sources like sensors, transactions, user inputs, or external databases. This systematic data generation ensures a comprehensive and timely influx of information, which is the lifeblood of data-driven decision-making. It's about establishing the pipelines that feed the strategic intelligence engine, enabling the business to capture and leverage every relevant data point for competitive advantage. Here is how we approach our learning:

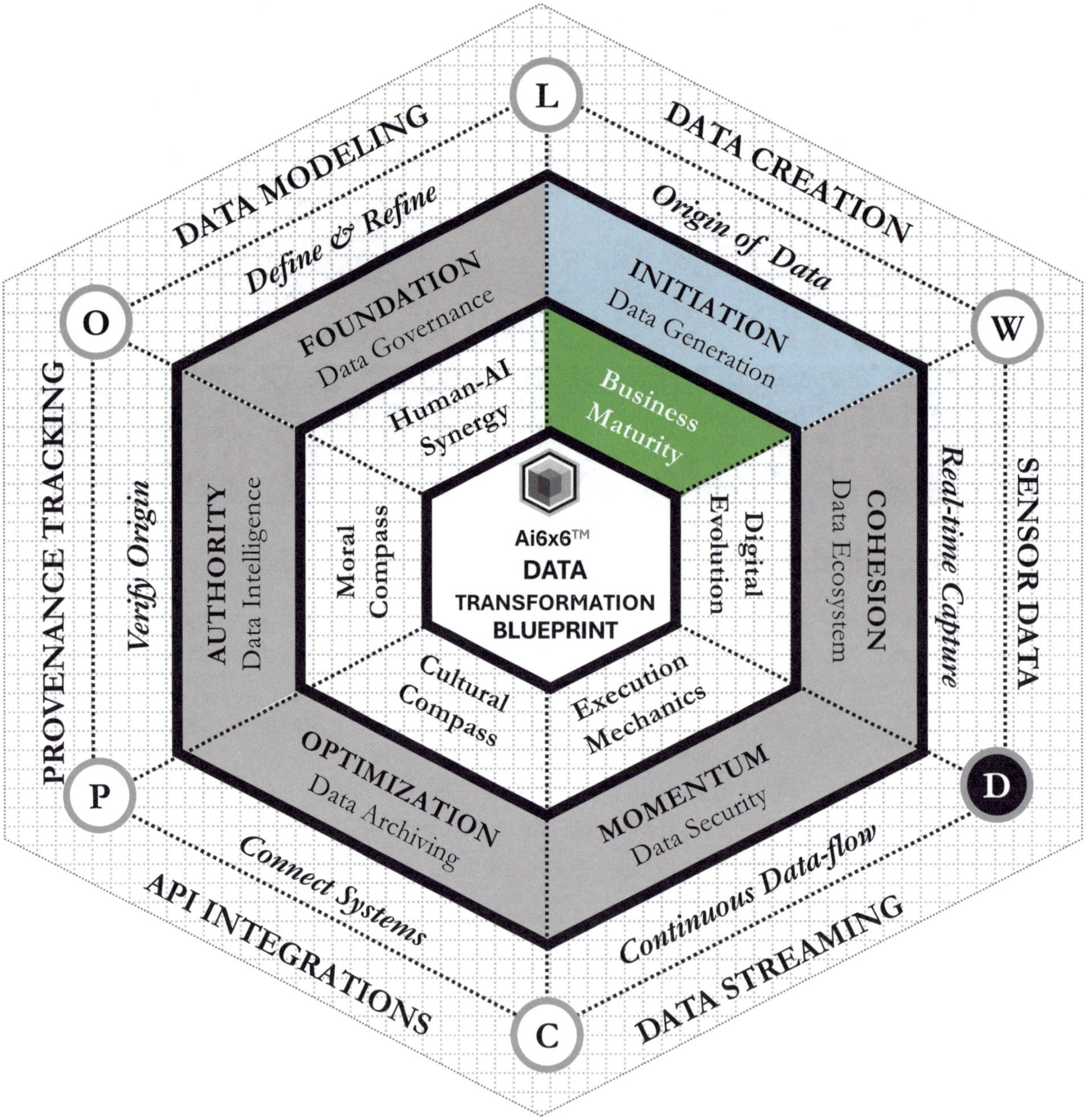

Ai6x6 BUSINESS MATURITY FRAMEWORK
DATA TRANSFORMATION | DATA GENERATION

For a business leader, effective data modeling is a foundational strategic imperative for unlocking the full potential of your data assets, particularly for analytical and AI-driven applications. It's about meticulously structuring and refining raw data into a coherent and optimized format that facilitates robust analysis, powerful machine learning, and scalable insights. Central to this is the potential for synthetic data generation. Creating artificial datasets that mirror the statistical properties of real data can significantly improve AI training and model accuracy, especially when real data is scarce, sensitive, or challenging to acquire. This innovation accelerates development without compromising privacy.

Furthermore, meticulous schema design and optimization are crucial. As a leader, you must ensure that structured data is organized logically for better performance, scalability, and ease of use. A well-designed schema is the backbone of efficient data retrieval and processing, enabling faster insights. Crucially, sophisticated data modeling directly supports advanced machine learning models. By refining AI-driven insights through well-organized data frameworks, you ensure that your algorithms operate on clean, relevant, and properly structured information, leading to more accurate predictions and actionable recommendations. Neglecting data modeling can lead to data silos, inconsistent data interpretation, inefficient analytical processes, and suboptimal performance of AI applications. By championing a disciplined approach to data modeling, you empower your organization to transform raw information into a powerful strategic asset, accelerating data-driven innovation, enhancing analytical capabilities, and ultimately driving a competitive advantage through more intelligent decision-making.

WHAT?	WHY?	HOW?	BEWARE!
SYNTHETIC DATA GENERATION	**AI TRAINING ENHANCEMENT**	**MODELING & SIMULATION**	**REALISM**
Creating artificial datasets for AI training.	Improves AI training and model accuracy, addresses data scarcity.	Use algorithms to generate artificial data mimicking real-world patterns.	Ensuring synthetic data accurately reflects real-world complexities.
SCHEMA DESIGN AND OPTIMIZATION	**PERFORMANCE & SCALABILITY**	**DATABASE DESIGN**	**EVOLUTION**
Organizing structured data for performance.	Enhances data retrieval, storage, and application performance.	Design logical and physical schemas, considering data types and relationships.	Adapting schemas to rapidly changing business requirements.
ADVANCED MACHINE LEARNING MODELS	**ACCURATE INSIGHTS**	**DATA PREPROCESSING**	**DATA COMPLEXITY**
Refining AI-driven insights.	Ensures AI models are trained on optimal data for precision.	Clean, transform, and normalize data for ML algorithm input.	Managing the complexity of diverse data types for ML.

For a business leader, understanding and optimizing data creation—the collection and integration of data from multiple disparate sources—is a fundamental strategic imperative for building a comprehensive and insightful view of your operations and customers. In today's interconnected business world, data originates from countless touchpoints, and your ability to bring it all together is key to unlocking its full value. Central to this is leveraging Customer Data Platforms (CDPs). These platforms consolidate user data from various channels (web, mobile, CRM, social) into a single, unified customer profile, enabling personalized insights, targeted analytics, and a 360-degree view of your customer base. This holistic perspective drives better customer experiences.

Furthermore, embracing automated capture technologies is crucial for efficiency and scalability. Using AI-driven methods to collect data from diverse sources—such as web scraping, IoT sensors, or public APIs—without manual input significantly reduces operational costs, minimizes human error, and ensures continuous data flow. Crucially, exploring decentralized platforms like blockchain and distributed systems offers novel approaches for secure, independent data collection. These technologies can enhance data integrity, privacy, and transparency, particularly in collaborative or highly regulated environments. Neglecting a cohesive data creation strategy can lead to fragmented customer views, data silos, inefficiencies in data collection, and an inability to perform comprehensive analytics. By championing advanced data creation methodologies, you empower your organization to build a rich, integrated data foundation, enabling deeper insights, personalized customer experiences, and ultimately driving a significant competitive advantage through a truly holistic understanding of your business landscape.

WHAT?	WHY?	HOW?	BEWARE!
CUSTOMER DATA PLATFORMS (CDPs)	**HOLISTIC CUSTOMER VIEW**	**DATA AGGREGATION**	**DATA CONSISTENCY**
Consolidating user data for insights.	Creates unified customer profiles for personalized insights.	Implement CDPs to collect and unify data from all customer touchpoints.	Ensuring data consistency and avoiding duplication across various sources.
AUTOMATED CAPTURE	**EFFICIENCY & SCALABILITY**	**AI INTEGRATION**	**DATA ACCURACY**
Using AI-driven methods to collect data without manual input.	Reduces manual effort and increases data collection speed.	Deploy AI tools for web scraping, sensor data collection, and API integrations.	Ensuring the accuracy and reliability of automatically captured data.
DECENTRALIZED PLATFORMS	**SECURITY & TRANSPARENCY**	**BLOCKCHAIN IMPLEMENTATION**	**COMPLEXITY**
Leveraging blockchain for secure data collection.	Enhances data security, privacy, and auditability.	Explore DLTs for immutable and distributed data collection.	High technical complexity and scalability challenges with decentralized systems.

For a business leader, understanding how to harness the immense potential of sensor data—real-time information flowing from connected devices and systems—is a strategic imperative for driving operational efficiency, enabling predictive capabilities, and unlocking new business models. The proliferation of IoT devices means a constant stream of granular data that, when properly utilized, can revolutionize decision-making. Central to this is integrating IoT sensors into your operations. These devices can gather a vast array of environmental, operational, and behavioral data, providing unprecedented visibility into everything from machinery performance and supply chain logistics to customer movements and resource consumption. This raw data is a goldmine for optimization.

Furthermore, implementing smart infrastructure monitoring is crucial, particularly in industrial and urban applications. By collecting sensor data from critical infrastructure (e.g., factories, utilities, buildings), you can enable predictive maintenance, identify anomalies before they become critical failures, and optimize resource utilization, leading to significant cost savings and improved reliability. Using streaming technologies and edge computing to process and act upon sensor-generated data instantaneously allows for immediate operational adjustments, personalized customer experiences, and rapid response to critical events. Neglecting sensor data integration means missing out on competitive advantages, operating with less visibility, and failing to capitalize on opportunities for automation and optimization. By championing the strategic use of sensor data, you empower your organization to gain unprecedented operational intelligence, drive proactive decision-making, and create new efficiencies that directly contribute to your business's agility, resilience, and long-term success in the digital age.

WHAT?	WHY?	HOW?	BEWARE!
IOT SENSORS	**REAL-TIME VISIBILITY**	**SENSOR DEPLOYMENT**	**DATA VOLUME**
Gathering environmental, operational, and behavioral data.	Provides rich, granular insights into physical environments and operations.	Integrate IoT sensors into assets, environments, and processes.	Managing and processing massive volumes of continuous sensor data.
SMART INFRASTRUCTURE MONITORING	**PROACTIVE OPTIMIZATION**	**SENSOR NETWORKS**	**INTEGRATION**
Collecting data for predictive maintenance.	Enables predictive maintenance and operational efficiency.	Implement sensor networks to monitor infrastructure and assets.	Integrating sensor data from diverse proprietary systems.
REAL-TIME SENSOR ANALYTICS	**INSTANT ACTION**	**STREAMING PROCESSING**	**LATENCY**
Using streaming technologies to process data.	Facilitates immediate operational adjustments and rapid response.	Implement real-time data streaming and analytics platforms.	Ensuring low latency for processing and acting on real-time data.

For a business leader, embracing data streaming capabilities is a strategic imperative that enables continuous, real-time data flow, unlocking instant processing and analysis for critical business operations. In an increasingly dynamic and event-driven world, relying on batch processing for data can lead to delayed insights and missed opportunities. Your ability to act on data as it's generated is a significant competitive differentiator. Central to this is building event-driven architectures. By designing systems that capture and respond to real-time triggers from system interactions, user behavior, and external events, you enable dynamic applications that can adapt and react instantaneously to changing conditions, powering personalized experiences and immediate operational adjustments.

Furthermore, leveraging edge computing is crucial. Processing data closer to its source—at the 'edge' of the network—reduces latency, minimizes bandwidth requirements, and enables faster insights, particularly for IoT devices and remote operations. This distributed processing model enhances efficiency and responsiveness. This involves designing robust infrastructure capable of handling rapid data inputs from diverse sources like social media feeds, financial transactions, and live monitoring systems, ensuring data flows smoothly without bottlenecks or data loss. Neglecting data streaming means operating with outdated information, reacting to events rather than anticipating them, and failing to capitalize on the immediacy demanded by modern consumers and business processes. By championing continuous data flow, you empower your organization to gain real-time intelligence, drive immediate action, and build highly responsive applications, ultimately securing a significant competitive advantage and transforming how your business interacts with its environment and customers.

WHAT?	WHY?	HOW?	BEWARE!
EVENT-DRIVEN ARCHITECTURES	**DYNAMIC APPLICATIONS**	**MESSAGE QUEUES**	**COMPLEXITY**
Capturing and responding to real-time triggers.	Enables immediate responsiveness to real-time events.	Design systems that use message queues for asynchronous event processing.	Building and managing complex event-driven microservices.
EDGE COMPUTING	**FASTER INSIGHTS**	**DISTRIBUTED INFRASTRUCTURE**	**SECURITY & MANAGEMENT**
Processing data closer to its source.	Reduces latency and bandwidth, enabling quicker insights.	Deploy processing capabilities at the data source (edge devices).	Securing and managing a vast, distributed edge infrastructure.
HIGH-FREQUENCY DATA PIPELINES	**SCALABLE INGESTION**	**STREAMING PLATFORMS**	**DATA INTEGRITY**
Managing rapid data inputs.	Handles massive volumes of data from real-time sources efficiently.	Implement high-throughput data streaming platforms (e.g., Kafka).	Ensuring data integrity and order in high-volume, real-time streams.

For a business leader, mastering API integrations is a strategic imperative that unlocks seamless interoperability, automates workflows, and accelerates data exchange across your entire technological ecosystem. In today's interconnected digital landscape, where data resides in countless applications and services, the ability to connect these sources efficiently is paramount to operational agility and informed decision-making. Central to this is leveraging integration hubs. These centralized platforms provide a unified environment for managing multi-source data exchange, simplifying the complexity of connecting disparate systems and enabling smoother data flow.

Furthermore, ensuring robust interoperability between systems is crucial. This involves designing and implementing solutions that allow smooth communication and data sharing across different technologies, databases, and legacy systems, breaking down data silos and creating a unified view of your operations. Crucially, adopting a Service-Oriented Architecture (SOA) or microservices approach structures your integrations for modularity and scalability. This design principle allows for independent development, deployment, and management of services, making your data-driven applications more resilient, adaptable, and easier to maintain as your business evolves. Neglecting effective API integrations can lead to fragmented data, manual inefficiencies, delayed data access, and a significant hindrance to automation efforts, ultimately impacting productivity and the speed of insights. By championing comprehensive API integration strategies, you empower your organization to build a highly connected, automated, and intelligent ecosystem, accelerating data utilization, enhancing operational efficiency, and driving innovation across all functions for sustained business success.

WHAT?	WHY?	HOW?	BEWARE!
INTEGRATION HUBS	**SIMPLIFIED CONNECTIVITY**	**PLATFORM DEPLOYMENT**	**VENDOR LOCK-IN**
Centralized platforms for data exchange.	Streamlines managing multi-source data exchange.	Implement an API management platform or integration hub.	Dependence on a single integration platform.
INTEROPERABILITY BETWEEN SYSTEMS	**SEAMLESS DATA FLOW**	**STANDARDIZED APIS**	**LEGACY SYSTEMS**
Ensuring smooth communication.	Enables smooth communication and data sharing across technologies.	Develop and use standardized APIs for all system interactions.	Challenges integrating with older systems lacking modern API support.
SERVICE-ORIENTED ARCHITECTURE (SOA)	**MODULAR SCALABILITY**	**MICROSERVICES ADOPTION**	**COMPLEX DEPLOYMENT**
Structuring integrations for scalability.	Creates flexible and scalable data-driven applications.	Design systems using modular, independent services.	Increased complexity in deploying and managing distributed services.

For a business leader, establishing robust provenance tracking is a non-negotiable strategic imperative for ensuring transparency and authenticity in your data's origins and transformations. In an era of increasing data complexity and regulatory demands, knowing the full history of your data—where it came from, who touched it, and how it was changed—is critical for trust, compliance, and accurate analysis. Central to this is detailed data lineage visualization. By mapping the entire journey of data from its initial collection point through every transformation, system, and usage, you gain a clear, auditable trail.

Furthermore, maintaining meticulous audit trails and security logs is crucial. Documenting every access, modification, and integrity measure for your data assets provides an immutable record of activity. Such logs are essential for security forensics, demonstrating accountability, and ensuring compliance with data governance policies. Crucially, rigorous regulatory compliance in provenance is vital. Aligning your data sourcing, processing, and storage practices with ethical and legal standards (e.g., GDPR, industry-specific regulations) protects your organization from fines and reputational damage. This proactive approach ensures data accountability. Neglecting provenance tracking can lead to distrust in your data, difficulties in meeting compliance mandates, an inability to trace errors, and ultimately, flawed decision-making based on unverifiable information. By championing comprehensive provenance tracking, you empower your organization with irrefutable data authenticity, transparency, and accountability, safeguarding both operational integrity and the trust of your stakeholders. This commitment to data history is fundamental for data-driven credibility.

WHAT?	WHY?	HOW?	BEWARE!
DATA LINEAGE VISUALIZATION	**TRANSPARENCY & DEBUGGING**	**LINEAGE TOOLS**	**COMPLEXITY**
Mapping the journey of data.	Provides clear, auditable map of data transformations.	Implement tools to automatically map data flow and transformations.	Mapping lineage across complex, distributed data systems.
AUDIT TRAILS AND SECURITY LOGS	**ACCOUNTABILITY & SECURITY**	**COMPREHENSIVE LOGGING**	**STORAGE & ANALYSIS**
Documenting access and modifications.	Ensures accountability and aids in security investigations.	Implement robust logging for all data interactions and changes.	Managing and analyzing vast volumes of log data efficiently.
REGULATORY COMPLIANCE IN PROVENANCE	**LEGAL ADHERENCE**	**POLICY IMPLEMENTATION**	**EVOLVING REGULATIONS**
Aligning data sourcing with standards.	Ensures data origins and transformations meet legal and ethical standards.	Develop and enforce policies for data sourcing and usage.	Keeping up with changing compliance requirements.

Ai6x6™ Transformation Blueprint

COHESION | Data Ecosystem

"The ability to collect, analyze, and act on data is the new competitive advantage." - Satya Nadella.

CEO of Microsoft, credited with profoundly revitalizing the company by shifting its focus to cloud computing and AI. Under his leadership, Microsoft's market value significantly increased, as he steered the tech giant into new growth areas.

The Cohesion level is your Data Ecosystem, occurring when disparate sources are harmonized into a single, interoperable landscape. This ecosystem orchestration ensures that information flows across silos to support unified strategic decisions.

The intricate and interconnected environment comprising all data sources, processing tools, analytical platforms, storage solutions, and human users within an organization. A well-integrated and optimized data ecosystem enables seamless data flow, fosters efficient collaboration, and facilitates comprehensive insights.

"Data Ecosystem" is very critical for achieving holistic and impactful organizational transformation. It ensures data accessibility, interoperability, and scalability, allowing for agile responses to evolving business needs and maximizing data's strategic value.

"**Data Ecosystem**" is vital for business strategy maturity because it represents the core process of actively working with and deriving value from data. A mature business strategy leverages a robust data ecosystem to transform raw data into meaningful insights through advanced algorithms, AI, or statistical methods. This acceleration enables faster, more accurate strategic decisions, optimizes operational efficiency, and uncovers new opportunities. It's about creating a dynamic environment where data is continuously processed and analyzed, empowering the business to quickly adapt to market changes and execute its strategy with precision. Here is how we approach our learning:

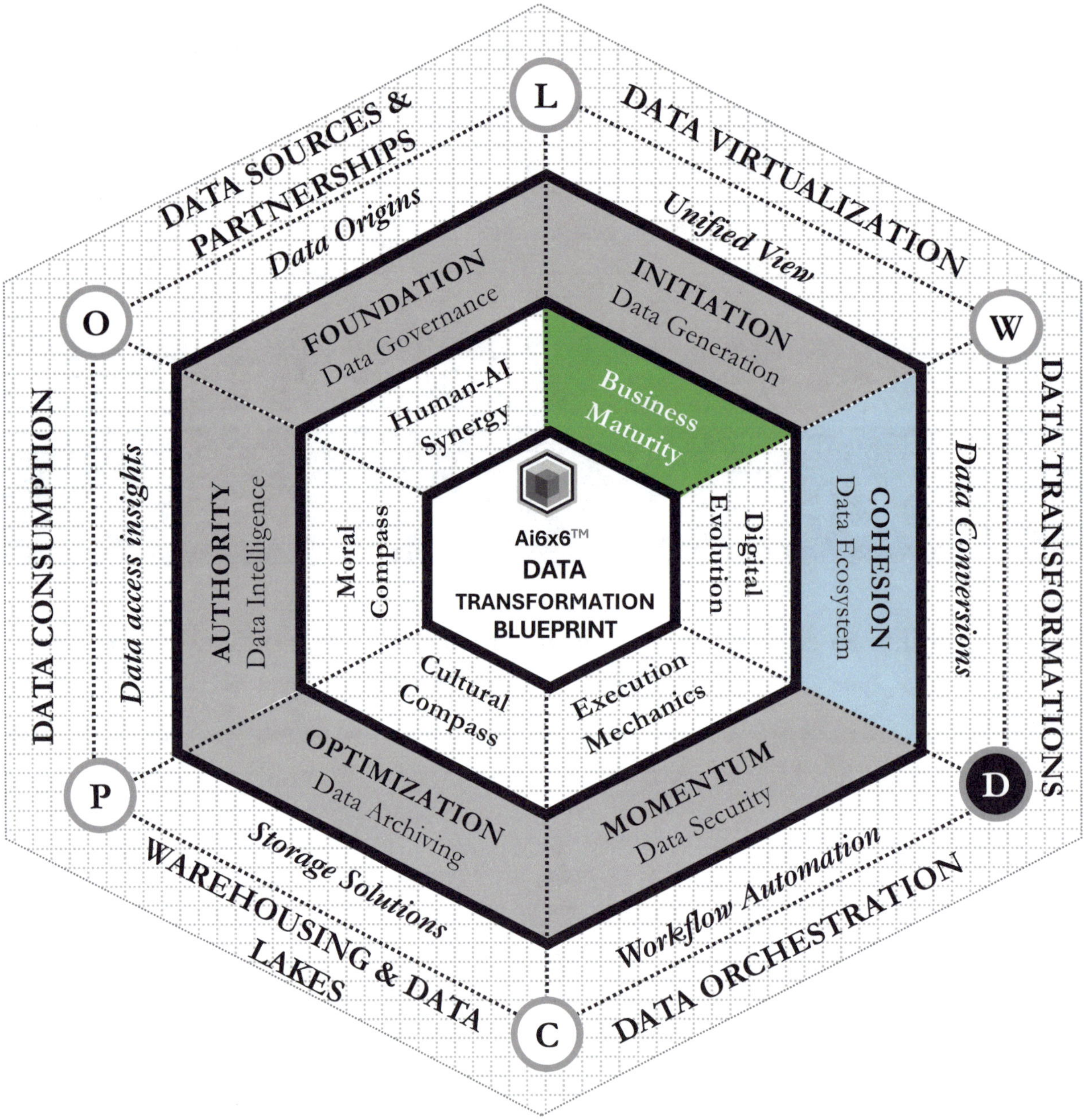

Ai6x6 BUSINESS MATURITY FRAMEWORK
DATA TRANSFORMATION | DATA ECOSYSTEM

For a business leader, strategically managing data sources and cultivating effective partnerships are fundamental to building a robust data ecosystem capable of driving actionable insights and competitive advantage. In today's data-rich world, limiting your data inputs means limiting your intelligence. Your ability to expand data access through diverse origins and strategic collaborations is paramount. Central to this is mastering multi-source data aggregation. This involves integrating structured, semi-structured, and unstructured data from a wide variety of origins—internal systems, external APIs, social media, IoT devices, and more—to create a comprehensive and holistic view of your operations and customers.

Furthermore, actively pursuing external data partnerships is crucial. Establishing secure and mutually beneficial agreements with third-party providers can significantly enhance your insights, providing access to market trends, demographic data, or specialized industry information that you couldn't generate internally. These collaborations broaden your analytical capabilities. Crucially, leveraging open data initiatives can unlock broader analytical capabilities. Tapping into publicly available datasets—from government statistics to scientific research—provides additional context and enriches your internal data, enabling more comprehensive analysis and innovative solutions. Neglecting diverse data sources and strategic partnerships can lead to narrow insights, incomplete market understanding, and a significant competitive disadvantage. By championing a proactive approach to data acquisition, you empower your organization to build a richer, more diverse data foundation, driving deeper intelligence, accelerating innovation, and ultimately securing your business's long-term success.

WHAT?	WHY?	HOW?	BEWARE!
MULTI-SOURCE DATA AGGREGATION	**HOLISTIC VIEW**	**ETL/ELT PIPELINES**	**DATA CONSISTENCY**
Integrating structured, semi-structured, unstructured data.	Creates a comprehensive and richer understanding of business.	Build robust pipelines to ingest and unify data from various formats.	Ensuring consistency and quality across disparate data types.
EXTERNAL DATA PARTNERSHIPS	**ENHANCED INSIGHTS**	**LEGAL & TECHNICAL AGREEMENTS**	**DATA GOVERNANCE**
Establishing secure agreements with third-party providers.	Provides access to specialized and broader market intelligence.	Negotiate secure data sharing agreements and API access.	Managing data privacy, security, and compliance with external data.
OPEN DATA INITIATIVES	**BROADER ANALYTICS**	**PUBLIC DATA INTEGRATION**	**RELIABILITY & QUALITY**
Leveraging publicly available datasets.	Enriches internal data, enabling more comprehensive analysis.	Identify relevant open datasets and integrate into your analytics.	Ensuring the reliability and quality of public data sources.

2. DATA VIRTUALIZATION

For a business leader, embracing data virtualization is a strategic imperative that allows for seamless, real-time access to disparate datasets without the costly and complex duplication of data. In a world where data often resides in numerous siloed systems, data virtualization acts as a unifying layer, enabling agile analytics and efficient data consumption. Central to this is enabling federated queries. This technology allows users to access and combine data from multiple sources in real-time, without physically moving or copying the data. This means analysts can get immediate answers by querying virtualized views, rather than waiting for lengthy ETL processes.

Furthermore, utilizing metadata-driven accessibility is crucial. By structuring virtualized views with rich metadata, you simplify complex data landscapes for business users, making it easier to discover, understand, and utilize the data they need without deep technical knowledge of underlying systems. This promotes data democratization. Crucially, ensuring cross-platform interoperability is vital. Data virtualization seamlessly connects data residing in various environments, including cloud, on-premise, and hybrid setups, providing a unified logical view. This flexibility supports complex enterprise architectures and migrations. Neglecting data virtualization can lead to data silos persisting, increased storage costs due to redundant data copies, slower time-to-insight, and a fragmented view of your business. By championing data virtualization, you empower your organization with immediate, flexible access to its entire data landscape, accelerate decision-making, and reduce the operational overhead associated with data integration, ultimately driving greater agility and competitive advantage through a truly unified data experience.

WHAT?	WHY?	HOW?	BEWARE!
FEDERATED QUERIES	**REAL-TIME ACCESS**	**VIRTUALIZATION PLATFORM**	**PERFORMANCE**
Enabling real-time access to multiple sources.	Provides immediate insights without data replication.	Implement a data virtualization platform to create logical data views.	Query performance issues across very disparate or slow data sources.
METADATA-DRIVEN ACCESSIBILITY	**SIMPLIFIED DISCOVERY**	**METADATA INTEGRATION**	**METADATA MANAGEMENT**
Structuring virtualized views.	Makes complex data landscapes easier to understand for business users.	Ensure rich metadata is incorporated into virtualized layers.	Keeping metadata accurate and up-to-date across all sources.
CROSS-PLATFORM INTEROPERABILITY	**FLEXIBLE ACCESS**	**CONNECTIVITY**	**COMPLEX ENVIRONMENT**
Ensuring compatibility across environments.	Provides unified access across cloud, on-premise, and hybrid environments.	Establish robust connectors for various database types and cloud services.	Managing connectivity and security in highly distributed environments.

For a business leader, mastering data transformations is a fundamental strategic imperative for converting raw, often messy, information into structured, usable insights that truly drive analytical value and informed decision-making. Data in its original state is rarely ready for direct analysis; it needs refinement to unlock its full potential. Central to this is implementing efficient ETL (Extract, Transform, Load) processes. This involves standardizing, cleansing, and structuring data from various sources so it's ready for analytics and warehousing. Robust ETL pipelines ensure data quality and consistency, which are foundational for accurate reporting and machine learning.

Furthermore, embracing schema evolution strategies is crucial. In a dynamic business environment, data structures (schemas) need to adapt dynamically to fit changing business needs and new data sources. Your leadership must ensure flexibility in how data is modeled and transformed to avoid rigid bottlenecks that hinder agility. Crucially, leveraging AI-enhanced transformations is revolutionizing this process. By utilizing machine learning for automated classification, enrichment, and anomaly detection during transformation, you can significantly reduce manual effort, improve accuracy, and accelerate the data preparation phase, making your data pipelines more intelligent and efficient. Neglecting data transformations can lead to inaccurate insights, wasted analytical effort, frustrated data users, and an inability to support advanced analytics or AI initiatives. By championing a disciplined and technologically advanced approach to data transformation, you empower your organization to convert raw data into a reliable, actionable asset, accelerating insights, improving data quality, and ultimately driving a competitive advantage through superior data utilization. This foundational step unlocks true data intelligence.

WHAT?	WHY?	HOW?	BEWARE!
ETL (EXTRACT, TRANSFORM, LOAD) PROCESSES	**ANALYTICS READINESS**	**PIPELINE DEVELOPMENT**	**COMPLEXITY**
Standardizing and structuring data.	Prepares data for accurate analytics and warehousing.	Build robust ETL/ELT pipelines for data ingestion and transformation.	Managing complex transformations for diverse and large datasets.
SCHEMA EVOLUTION STRATEGIES	**AGILITY & FLEXIBILITY**	**FLEXIBLE SCHEMA DESIGN**	**DATA MIGRATION**
Adapting datasets dynamically.	Ensures data models adapt to changing business requirements.	Adopt schema-on-read principles and version control for schemas.	Migrating existing data to new schema versions without disruption.
AI-ENHANCED TRANSFORMATIONS	**EFFICIENCY & ACCURACY**	**ML INTEGRATION**	**AI VALIDATION**
Leveraging machine learning for automated enrichment.	Automates data classification, cleaning, and enrichment.	Integrate machine learning algorithms into transformation pipelines.	Ensuring the accuracy and reliability of AI-driven transformations.

For a business leader, mastering data orchestration is a strategic imperative that automates and streamlines complex data integration and processing workflows, ensuring efficiency, reliability, and agility across your entire data ecosystem. In today's landscape of massive data volumes and diverse sources, manual management of data pipelines is unsustainable. Central to this is implementing event-driven data pipelines. By designing systems that trigger updates and processes based on real-time system interactions and user behavior, you create a highly responsive and automated data flow, enabling immediate insights and proactive actions rather than batch-dependent delays.

Furthermore, leveraging robust workflow automation frameworks is crucial. These frameworks ensure optimized sequencing for efficient data handling, managing dependencies between different data tasks, and automating scheduling. This minimizes manual intervention, reduces human error, and ensures that data processing occurs smoothly and consistently. Crucially, adopting scalable execution models is vital for managing distributed workloads dynamically and ensuring peak performance. As data volumes fluctuate and processing demands increase, your orchestration layer must intelligently allocate resources and scale up or down as needed, maintaining efficiency without manual oversight. Neglecting data orchestration can lead to inefficient data pipelines, increased operational costs, delayed insights, and a reactive approach to data management. By championing comprehensive data orchestration, you empower your organization to automate its data workflows, accelerate data availability for analytics and AI, and build a highly efficient and resilient data ecosystem that directly contributes to operational excellence and strategic agility. This automation is key to unlocking the full potential of your data.

WHAT?	WHY?	HOW?	BEWARE!
EVENT-DRIVEN DATA PIPELINES	**REAL-TIME PROCESSING**	**STREAMING ARCHITECTURES**	**COMPLEX DEPENDENCIES**
Triggering updates based on system interactions.	Enables immediate processing and responsiveness to data changes.	Implement message queues and event stream processing.	Managing complex dependencies in real-time, event-driven flows.
WORKFLOW AUTOMATION FRAMEWORKS	**EFFICIENCY & RELIABILITY**	**ORCHESTRATION TOOLS**	**DEBUGGING**
Ensuring optimized sequencing.	Streamlines data handling and reduces manual errors.	Use tools like Apache Airflow, Prefect, or Dagster.	Debugging and troubleshooting complex automated workflows.
SCALABLE EXECUTION MODELS	**PERFORMANCE & FLEXIBILITY**	**CLOUD-NATIVE SOLUTIONS**	**COST OPTIMIZATION**
Managing distributed workloads dynamically.	Ensures peak performance under varying data loads.	Utilize serverless computing or Kubernetes for dynamic scaling.	Balancing performance with cost efficiency in scalable cloud environments.

For a business leader, strategically structuring your data storage solutions through a thoughtful combination of data warehousing and data lakes is paramount for efficient scalability and accessibility of your valuable information. This hybrid approach allows you to optimize for both structured, high-performance analytics and the flexible storage of diverse, raw data, catering to all your analytical needs. Central to this is leveraging cloud-native data warehouses. These modern solutions provide highly scalable and high-performance storage for structured, processed data, optimized for complex analytical queries and business intelligence reporting.

Furthermore, implementing flexible data lake architectures is crucial for handling massive-scale retention of multi-format datasets. Data lakes are ideal for storing raw, unprocessed, and diverse data—structured, semi-structured, and unstructured—allowing for future analysis without predefined schemas. This flexibility supports advanced analytics and machine learning initiatives. Crucially, optimizing for hybrid storage combines the strengths of both warehouses and lakes. By strategically determining which data resides in which system, you can balance immediate processing needs with long-term retention and cost efficiency, creating a seamless data flow between the two environments. Neglecting a well-architected storage strategy can lead to data silos, inefficient data retrieval, escalating storage costs, and an inability to support both routine business intelligence and advanced analytical use cases. By championing this dual approach, you empower your organization with robust, scalable, and highly accessible data infrastructure, ensuring that all forms of data are readily available for driving insights, fostering innovation, and supporting strategic decision-making across the enterprise.

WHAT?	WHY?	HOW?	BEWARE!
CLOUD-NATIVE DATA WAREHOUSES	**OPTIMIZED ANALYTICS**	**CLOUD MIGRATION**	**DATA GOVERNANCE**
Supporting structured, high-performance storage.	Provides high-performance storage for structured business intelligence.	Migrate or build warehouses on cloud platforms (e.g., Snowflake, BigQuery).	Ensuring data governance and security in cloud environments.
FLEXIBLE DATA LAKE ARCHITECTURES	**RAW DATA STORAGE**	**SCALABLE STORAGE**	**DATA SWAMPS**
Enabling large-scale retention of multi-format datasets.	Allows retention of diverse, raw data for future analytics.	Implement cloud-based object storage (e.g., S3, ADLS) for data lakes.	Data lakes becoming unmanageable "data swamps" without proper governance.
HYBRID STORAGE OPTIMIZATION	**BALANCED PERFORMANCE**	**STRATEGIC PLACEMENT**	**COMPLEX MANAGEMENT**
Combining warehouse and lake functionalities.	Balances immediate processing needs with long-term, cost-effective storage.	Define clear rules for data movement between lake and warehouse.	Managing data consistency and movement across hybrid architectures.

For a business leader, optimizing data consumption is the ultimate strategic imperative, translating your meticulously managed data assets into actionable insights that directly drive decision-making and business growth. The value of data lies in its utilization, and your ability to enable efficient access and intelligent consumption is paramount. Central to this is providing robust business intelligence and analytics tools. This involves creating intuitive, real-time dashboards and visualization tools that transform complex data into easily digestible formats, allowing leaders and teams to monitor performance, identify trends, and make informed decisions swiftly.

Furthermore, leveraging AI-powered recommendations significantly enhances decision-making. By implementing predictive modeling and machine learning algorithms, you can generate proactive suggestions, anticipate customer needs, optimize operations, and identify emerging opportunities, moving beyond descriptive analytics to prescriptive guidance. Crucially, offering API-based data delivery provides structured access for seamless consumption across various systems and applications. This programmatic access enables automation, integration with third-party tools, and ensures that data flows effortlessly to where it's needed, powering dynamic insights and new digital products. Neglecting efficient data consumption can lead to data hoarding, missed opportunities, reactive decision-making, and a failure to capitalize on your significant data investments. By championing comprehensive data consumption strategies, you empower your organization with direct, insightful access to its intelligence, fostering a truly data-driven culture that accelerates strategic choices, enhances operational efficiency, and ultimately secures your business's competitive edge and long-term success.

WHAT?	WHY?	HOW?	BEWARE!
BUSINESS INTELLIGENCE AND ANALYTICS	**INFORMED DECISIONS**	**DASHBOARD DEVELOPMENT**	**DATA LITERACY**
Providing real-time dashboards.	Transforms data into actionable insights for business growth.	Implement BI tools (e.g., Tableau, Power BI) for dashboards.	Ensuring all users can effectively interpret and act on BI dashboards.
AI-POWERED RECOMMENDATIONS	**PROACTIVE GUIDANCE**	**ML MODEL DEPLOYMENT**	**MODEL BIAS**
Enhancing decision-making with predictive modeling.	Provides forward-looking insights for strategic decision-making.	Develop and deploy machine learning models for predictions.	Risk of bias in AI models leading to inaccurate or unfair recommendations.
API-BASED DATA DELIVERY	**SEAMLESS INTEGRATION**	**API DEVELOPMENT**	**API SECURITY**
Offering structured access for consumption.	Enables easy, programmatic access for external systems and apps.	Create well-documented APIs for internal and external data access.	Ensuring robust security for data exposed via APIs.

Ai6x6™ Transformation Blueprint

MOMENTUM | Data Security

"Security is not a feature, it is a foundation." - Tim Cook.

CEO of Apple Inc., succeeding Steve Jobs. He has overseen Apple's significant growth in services and wearables, maintaining its position as one of the world's most valuable companies. Cook is known for his operational expertise and strong advocacy for user privacy.

The Momentum level is your Data Security, sustained when the speed of data use is matched by the strength of its protection. Advanced encryption and threat detection allow the organization to scale rapidly without compromising trust.

Implementing stringent measures and protocols to protect data from any unauthorized access, accidental corruption, deliberate manipulation, or catastrophic loss, ensuring its confidentiality, integrity, and availability. This is absolutely essential for maintaining customer trust, ensuring strict regulatory compliance (e.g., GDPR, HIPAA), and safeguarding sensitive information.

"Data Security" thereby enables the confident and secure utilization of data across all business functions. Robust security prevents breaches, preserves reputation, and ensures business continuity in an increasingly digital landscape.

The "**Data Security**" chapter is paramount for business strategy maturity as it focuses on continuously optimizing protection measures and adapting to new threats. A mature business strategy prioritizes safeguarding data from breaches, maintaining privacy, and rigorously following regulatory guidelines. This proactive approach to data security builds and preserves stakeholder trust, mitigates significant financial and reputational risks, and ensures business continuity. It's about embedding security into the strategic fabric, ensuring that the organization's data assets remain protected and reliable, which is essential for long-term strategic viability and compliance. Here is how we approach our learning:

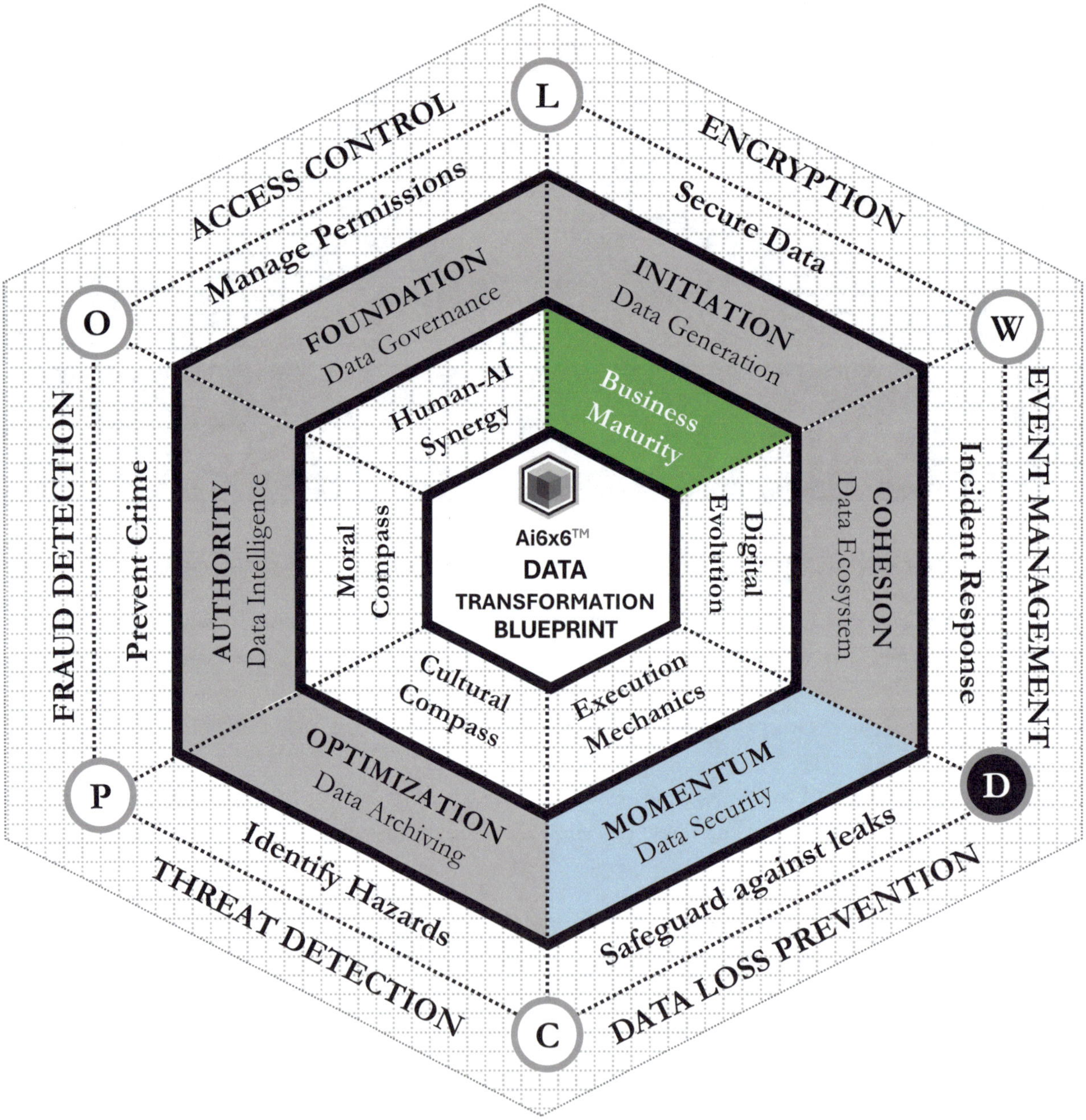

Ai6x6 BUSINESS MATURITY FRAMEWORK
DATA TRANSFORMATION | DATA SECURITY

For a business leader, establishing robust access control mechanisms is a foundational pillar of data security, paramount for protecting sensitive information, maintaining privacy, and ensuring compliance with regulatory guidelines. In an environment where data breaches are a constant threat, precisely managing who can access what data is critical to minimizing risk. Central to this is implementing role-based security. By assigning data access levels based on predefined organizational roles and responsibilities, you ensure that employees only have permissions relevant to their job functions, dramatically reducing the attack surface and potential for unauthorized access. This granular control enhances efficiency and security.

Furthermore, integrating decentralized identity verification, potentially using blockchain or advanced identity-based authentication, offers enhanced security. These methods create more resilient and tamper-proof identity systems, making it harder for unauthorized entities to impersonate legitimate users and gain access. Crucially, strictly adhering to the least privilege principle is vital. This security best practice dictates that every user, system, or application should be granted only the absolute minimum access rights necessary to perform its intended function, minimizing the impact of a compromise. Neglecting strong access control can lead to unauthorized data exposure, insider threats, compliance violations, and a significant erosion of customer trust. By championing a meticulous and dynamic approach to access management, you empower your organization to protect its valuable data assets, ensure regulatory compliance, and build a resilient security posture that safeguards your business's long-term integrity and reputation in an increasingly data-dependent world.

WHAT?	WHY?	HOW?	BEWARE!
ROLE-BASED SECURITY	**GRANULAR CONTROL**	**RBAC IMPLEMENTATION**	**COMPLEXITY**
Assigning data access levels based on roles.	Restricts access to sensitive data, minimizing exposure.	Define roles and map them to specific data access permissions.	Managing roles and permissions across large, complex organizations.
DECENTRALIZED IDENTITY VERIFICATION	**ENHANCED SECURITY**	**DLT/BIOMETRIC INTEGRATION**	**ADOPTION & INTEROPERABILITY**
Using blockchain or identity-based authentication.	Provides more robust and tamper-proof identity verification.	Explore blockchain-based identity or advanced biometrics.	Challenges in broad adoption and integration with existing systems.
LEAST PRIVILEGE PRINCIPLE	**RISK MINIMIZATION**	**AUDIT & ADJUSTMENT**	**USER FRICTION**
Restricting access to only necessary data.	Limits potential damage if an account or system is compromised.	Regularly audit and adjust permissions to ensure least privilege is maintained.	Balancing security with user convenience and productivity.

2. ENCRYPTION

For a business leader, deploying robust encryption strategies is a fundamental and non-negotiable imperative for safeguarding your organization's data throughout its entire lifecycle, preventing unauthorized access, and maintaining privacy. In a world where data breaches are a constant threat, encryption provides a critical layer of defense that renders sensitive information unreadable to malicious actors. Central to this is understanding and strategically applying symmetric and asymmetric encryption methods. These techniques are used to secure data both when it's at rest (stored in databases or on servers) and in transit (as it moves across networks), ensuring protection.

Furthermore, implementing end-to-end encryption is crucial. This ensures that data remains protected from its point of origin to its final destination, with only the intended recipient able to decrypt and access the information. This method is particularly vital for secure communication channels and sensitive data transfers. Crucially, forward-thinking leaders are now exploring quantum encryption. Leveraging quantum cryptography offers next-generation security against evolving threats, including potential attacks from quantum computers, future-proofing your data protection strategies. Neglecting comprehensive encryption can lead to devastating data breaches, regulatory non-compliance, severe financial penalties, and a complete erosion of customer and stakeholder trust. By championing a multi-layered and advanced approach to encryption, you empower your organization to protect its sensitive information proactively, ensure data confidentiality, and build a resilient security posture that directly safeguards your business's reputation, operational integrity, and long-term viability in the digital age. This cryptographic defense is essential for peace of mind.

WHAT?	WHY?	HOW?	BEWARE!
SYMMETRIC & ASYMMETRIC ENCRYPTION	**COMPREHENSIVE PROTECTION**	**KEY MANAGEMENT**	**KEY MANAGEMENT COMPLEXITY**
Securing stored and transmitted data.	Secures data both at rest and in transit.	Implement robust key management systems for encryption/decryption.	Managing and protecting encryption keys securely.
END-TO-END ENCRYPTION	**DATA CONFIDENTIALITY**	**PROTOCOL INTEGRATION**	**INTEROPERABILITY**
Ensuring data remains protected throughout its lifecycle.	Guarantees data privacy from origin to destination.	Implement E2EE protocols for communication and data transfer.	Challenges with E2EE across disparate systems and third-party integrations.
QUANTUM ENCRYPTION	**FUTURE-PROOFING**	**RESEARCH & PILOTS**	**MATURITY & ADOPTION**
Leveraging quantum cryptography for next-generation security.	Protects against advanced future threats, including quantum computing.	Invest in research and pilot programs for quantum-safe algorithms.	Technology is nascent, with high costs and low widespread adoption.

3. EVENT MANAGEMENT

For a business leader, establishing robust security event management capabilities is a critical imperative for proactively monitoring and swiftly responding to security incidents in real-time, minimizing their impact on your organization. In today's landscape of persistent cyber threats, a reactive approach is simply inadequate; rapid detection and response are paramount. Central to this is comprehensive security event logging. This involves systematically capturing detailed records of all system activities, network traffic, and user interactions across your IT infrastructure. These logs are essential for forensic analysis in the event of a breach, providing the necessary evidence to understand the attack vector and mitigate damage.

Furthermore, developing clear incident response workflows is crucial. As a leader, you must ensure that well-defined protocols are in place for detecting, investigating, and mitigating threats as they emerge. This includes identifying responsible teams, establishing communication plans, and outlining steps for containment and recovery. Crucially, leveraging automated threat intelligence is revolutionizing incident response. By using AI-driven Security Operations Centers (SOCs) and advanced analytics, you can automatically detect anomalies, identify suspicious patterns, and flag potential threats with greater speed and accuracy than human analysis alone, enabling immediate action. Neglecting effective security event management can lead to delayed breach detection, extended downtime, increased data loss, and significant reputational damage. By championing a proactive, real-time approach to monitoring and response, you empower your organization to defend against cyberattacks, ensure business continuity, and build a resilient security posture that directly safeguards your business's long-term integrity and customer trust.

WHAT?	WHY?	HOW?	BEWARE!
SECURITY EVENT LOGGING	**FORENSIC ANALYSIS**	**LOGGING INFRASTRUCTURE**	**LOG VOLUME**
Capturing detailed records of system activities.	Provides crucial evidence for breach investigations.	Implement centralized logging systems (SIEM) across all systems.	Managing and effectively analyzing massive volumes of log data.
INCIDENT RESPONSE WORKFLOWS	**RAPID MITIGATION**	**RESPONSE PLAN DEVELOPMENT**	**TESTING**
Establishing protocols for detecting and mitigating threats.	Ensures swift and coordinated response to security incidents.	Create clear, tested protocols for incident detection, analysis, and recovery.	Conducting regular, realistic incident response drills.
AUTOMATED THREAT INTELLIGENCE	**PROACTIVE DETECTION**	**AI-POWERED SOC**	**FALSE POSITIVES**
Using AI-driven security operations centers.	Detects anomalies and potential threats with greater speed.	Implement AI-driven security information and event management (SIEM) systems.	High volume of false positives requiring manual investigation.

For a business leader, implementing robust data loss prevention (DLP) strategies is a critical imperative for safeguarding sensitive information from leaks, unauthorized exfiltration, and breaches, protecting your organization from devastating financial and reputational consequences. In an era of increasing insider threats and stringent data privacy regulations, proactively controlling where sensitive data resides and how it moves is paramount. Central to this is meticulous classification and tagging. As a leader, you must ensure that all confidential and sensitive data is accurately labeled and categorized (e.g., PII, financial, intellectual property). This tagging enables proper access restrictions and ensures that DLP policies can be effectively applied across your systems.

Furthermore, comprehensive endpoint protection is crucial. This involves securing all devices (laptops, mobile phones, USB drives) and network egress points against unauthorized data exfiltration. DLP solutions monitor and block attempts to copy, print, email, or transfer sensitive data outside approved channels, preventing accidental or malicious leaks. Crucially, developing compliance-driven DLP policies is vital. Enforcing regulatory measures aligned with frameworks like GDPR, HIPAA, and CCPA ensures that your DLP efforts not only protect data but also meet legal obligations, minimizing data exposure risks and avoiding costly penalties. Neglecting robust DLP can lead to severe data breaches, non-compliance fines, loss of intellectual property, and significant damage to customer trust. By championing a comprehensive and proactive DLP strategy, you empower your organization to control its sensitive data effectively, ensure regulatory adherence, and build a resilient security posture that directly safeguards your business's long-term integrity and competitive advantage in a data-driven world.

WHAT?	WHY?	HOW?	BEWARE!
CLASSIFICATION AND TAGGING	**POLICY ENFORCEMENT**	**DATA INVENTORY & CLASSIFICATION**	**ACCURACY**
Labeling confidential data.	Enables effective application of DLP policies.	Discover and label sensitive data across all systems.	Ensuring accurate and consistent classification of data.
ENDPOINT PROTECTION	**DATA EXFILTRATION PREVENTION**	**DLP SOFTWARE DEPLOYMENT**	**USER FRICTION**
Securing devices and networks.	Blocks unauthorized transfer of sensitive data from devices.	Deploy DLP agents on endpoints and network gateways.	Overly restrictive policies hindering legitimate user workflows.
COMPLIANCE-DRIVEN DLP POLICIES	**REGULATORY ADHERENCE**	**POLICY DEFINITION**	**FALSE POSITIVES**
Enforcing regulatory measures.	Ensures DLP efforts align with legal privacy requirements.	Define DLP rules based on GDPR, HIPAA, CCPA, etc.	DLP systems flagging legitimate activities as potential breaches.

For a business leader, cultivating robust threat detection capabilities is a non-negotiable strategic imperative for identifying vulnerabilities and suspicious activities within your networks and databases, proactively defending against evolving cyberattacks. In an era of sophisticated and persistent threats, reactive security is insufficient; rapid identification is key to minimizing damage. Central to this is leveraging behavioral analytics. By using AI and machine learning to analyze abnormal patterns in system activity—such as unusual login times, data access, or network traffic—you can detect anomalies that indicate insider threats or external compromises.

Furthermore, implementing intrusion detection systems (IDS) is crucial for real-time monitoring. These systems continuously monitor network traffic for signs of cyber threats, unauthorized access attempts, and malicious activity, providing immediate alerts to security teams. This constant vigilance is vital for early warning. Crucially, establishing continuous vulnerability assessment programs ensures proactive identification of security weaknesses. Regularly performing scans, penetration tests, and security audits helps to uncover exploitable flaws in your systems and applications before attackers can exploit them, allowing for timely remediation. Neglecting strong threat detection can lead to undetected breaches, prolonged exposure to vulnerabilities, significant data loss, and severe reputational damage. By championing a multi-layered and intelligent approach to threat detection, you empower your organization to identify and respond to cyber threats with speed and precision, ensuring a resilient security posture that directly safeguards your business's long-term integrity, operational continuity, and customer trust in an increasingly hostile digital landscape.

WHAT?	WHY?	HOW?	BEWARE!
BEHAVIORAL ANALYTICS	**INSIDER THREATS**	**ML ALGORITHMS**	**FALSE POSITIVES**
Using AI and machine learning to detect abnormal patterns.	Identifies unusual behavior indicating potential breaches.	Deploy AI/ML solutions to analyze user and system behavior patterns.	Generating too many false positives, leading to alert fatigue.
INTRUSION DETECTION SYSTEMS (IDS)	**EARLY WARNING**	**IDS DEPLOYMENT**	**DATA VOLUME**
Monitoring network traffic for cyber threats.	Provides real-time alerts for suspicious network activity.	Implement IDS sensors at network entry/exit points and critical segments.	Processing and analyzing vast amounts of network traffic data.
CONTINUOUS VULNERABILITY ASSESSMENT	**PROACTIVE REMEDIATION**	**REGULAR SCANNING**	**COMPLEXITY**
Performing proactive scans.	Identifies and fixes security weaknesses before exploitation.	Schedule automated vulnerability scans and manual penetration tests.	Managing and prioritizing remediation efforts for numerous vulnerabilities.

For a business leader, investing in advanced fraud detection systems is a critical strategic imperative for preventing financial and identity-related cybercrimes, protecting your organization's assets, and maintaining customer trust. In an age of sophisticated fraud schemes, traditional detection methods are often insufficient; real-time, intelligent monitoring is paramount. Central to this is robust transaction monitoring. By using predictive analytics and rule-based systems to analyze financial operations in real-time, you can detect fraudulent behavior, identify suspicious transactions, and flag anomalies that deviate from normal patterns, preventing financial losses proactively.

Furthermore, incorporating biometric authentication adds a powerful layer of security for identities. Utilizing fingerprint, facial recognition, or behavioral biometrics enhances user verification, making it significantly harder for fraudsters to gain unauthorized access to accounts or systems. This multi-factor approach strengthens identity protection. Crucially, leveraging AI-driven fraud prevention is revolutionizing the fight against financial crime. By applying machine learning algorithms to detect irregular patterns and anomalies across vast datasets, AI can identify emerging fraud schemes and flag suspicious activities with greater speed and accuracy than human analysis alone, providing continuous, adaptive defense. Neglecting advanced fraud detection can lead to significant financial losses, reputational damage, loss of customer confidence, and potential legal liabilities. By championing comprehensive and intelligent fraud prevention strategies, you empower your organization to protect its financial integrity, secure customer data, and maintain a trustworthy reputation, ultimately safeguarding your business's long-term viability and success in a digitally connected world.

WHAT?	WHY?	HOW?	BEWARE!
TRANSACTION MONITORING	**FINANCIAL PROTECTION**	**PREDICTIVE ANALYTICS**	**FALSE POSITIVES**
Detecting fraudulent behavior in financial operations.	Prevents financial losses from fraudulent transactions.	Implement real-time monitoring and rule-based systems for transactions.	High volume of legitimate transactions flagged as suspicious.
BIOMETRIC AUTHENTICATION	**IDENTITY SECURITY**	**BIOMETRIC INTEGRATION**	**PRIVACY CONCERNS**
Using fingerprint, facial recognition, and behavioral biometrics.	Enhances user verification and secures identities.	Integrate fingerprint, facial, or behavioral biometric systems.	User privacy concerns and data storage implications of biometrics.
AI-DRIVEN FRAUD PREVENTION	**ADAPTIVE DEFENSE**	**ML ALGORITHM DEPLOYMENT**	**ADVERSARIAL ATTACKS**
Applying machine learning algorithms to detect patterns.	Detects new and evolving fraud schemes efficiently.	Train and deploy ML models to identify irregular patterns.	Sophisticated fraudsters attempting to bypass AI models.

Ai6x6™ Transformation Blueprint

OPTIMIZATION | Data Archiving

"Data is the lifeblood of innovation, and its preservation ensures future breakthroughs." - Sheryl Sandberg.

Former COO of Meta Platforms (Facebook), known for scaling Facebook's advertising business. She is a prominent advocate for women in leadership and author of "Lean In," influencing discussions on gender equality. Her business acumen is widely recognized.

The Optimization level is your Data Archiving, the art of knowing what to keep and what to discard to maintain a lean system. Precise retention and disposal ensure the "oiled machine" only carries the high-value data needed for Nirvana-level insights.

Systematically storing inactive or less frequently accessed data securely and cost-effectively for long-term retention, historical analysis, and compliance with regulatory requirements. This practice preserves invaluable historical context and significantly reduces immediate operational storage costs by moving data off expensive active systems.

"Data Archiving" ensures that critical data remains available for future analysis, auditing, or strategic insights. It balances accessibility with cost-efficiency, ensuring legal and historical data integrity while optimizing current system performance.

"Data Archiving" is critical for business strategy maturity because it focuses on managing the transition of data from active to historical states, evolving its storage and accessibility. A mature business strategy includes systematic processes for storing historical data for future reference while meticulously managing access policies, retention, and disposal. This ensures compliance with regulations, optimizes storage costs, and preserves valuable historical context for strategic analysis. It's about efficiently adapting data management practices to its lifecycle, ensuring data remains accessible for long-term strategic insights without compromising current operational efficiency. Here is how we approach our learning:

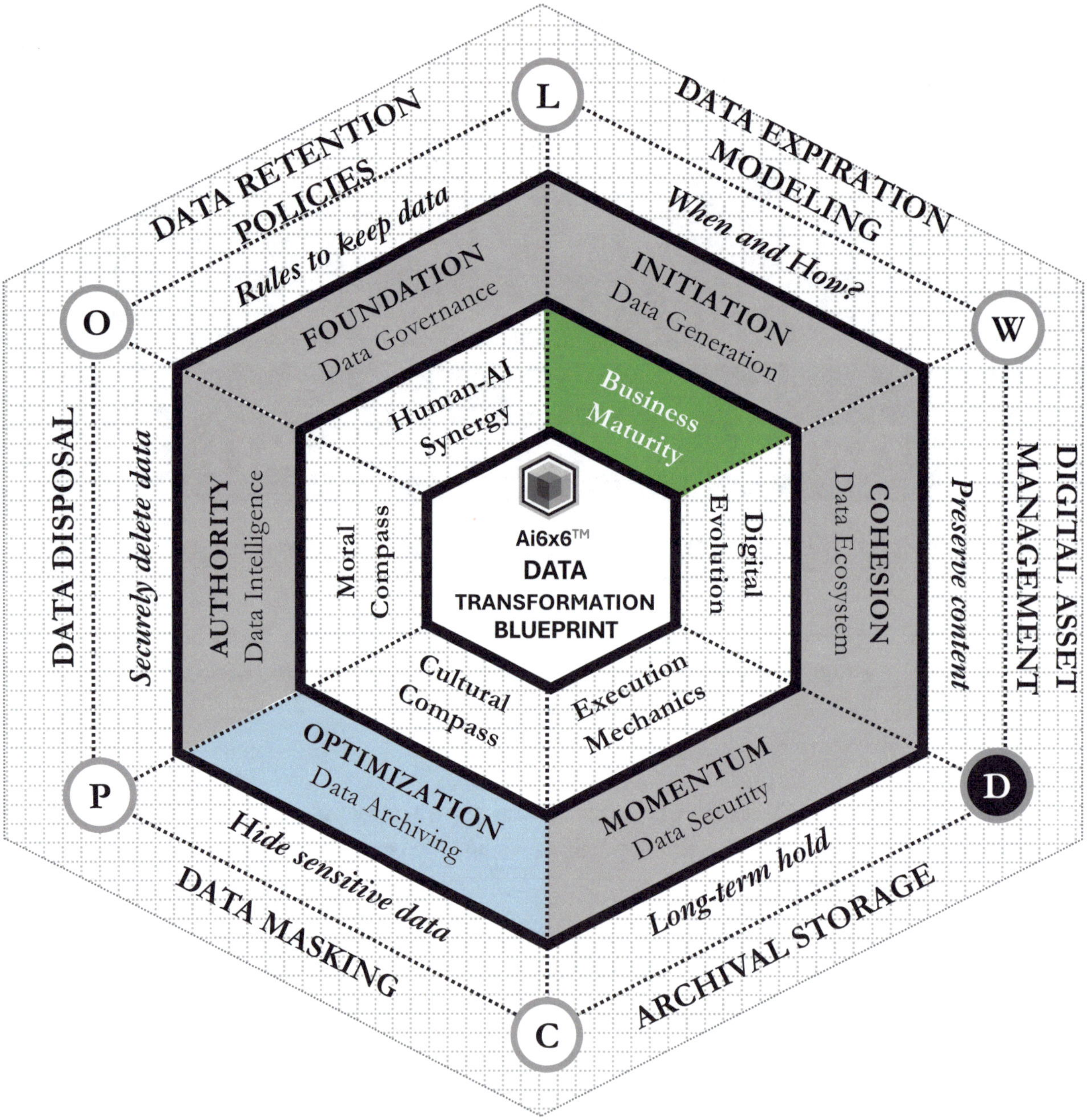

Ai6x6 BUSINESS MATURITY FRAMEWORK
DATA TRANSFORMATION | DATA ARCHIVING

For a business leader, establishing clear data retention policies is a fundamental strategic imperative for managing information lifecycle, ensuring compliance, and optimizing storage costs. These policies define precisely how long data should be stored before it is either archived for long-term reference or securely deleted, balancing legal obligations with operational efficiency. Central to this is adhering to industry-specific retention guidelines. As a leader, you must ensure your policies comply with stringent regulations such as GDPR, HIPAA, financial reporting requirements, and other sector-specific mandates, avoiding costly legal penalties and reputational damage for non-compliance.

Furthermore, meticulously balancing retention vs. accessibility is crucial. Determining which data should remain readily available for immediate operational use versus data that can be moved to lower-cost, long-term archives for future reference or audit purposes optimizes storage resources and access speeds. This strategic decision impacts efficiency. Crucially, leveraging automated policy enforcement streamlines the process. By using AI-driven solutions to implement retention rules based on predefined business needs, data type, and regulatory requirements, you reduce manual effort, minimize human error, and ensure consistent application of policies across vast datasets. Neglecting clear data retention policies can lead to excessive storage costs, legal liabilities due to prolonged data keeping or premature deletion, and a disorganized data landscape. By championing a systematic and automated approach to data retention, you empower your organization with compliant, efficient, and well-governed data, safeguarding your business's integrity and optimizing its data storage strategy for the long term.

WHAT?	WHY?	HOW?	BEWARE!
INDUSTRY-SPECIFIC RETENTION GUIDELINES	**REGULATORY COMPLIANCE**	**LEGAL RESEARCH**	**EVOLVING REGULATIONS**
Ensuring compliance with regulations.	Minimizes legal risks and ensures adherence to data laws.	Research and implement policies based on relevant industry and data privacy laws.	Keeping pace with frequent changes in legal and industry retention requirements.
RETENTION VS. ACCESSIBILITY BALANCE	**OPTIMIZED STORAGE**	**DATA CLASSIFICATION**	**BUSINESS NEEDS**
Determining data availability.	Balances cost-effective long-term storage with data accessibility needs.	Categorize data by sensitivity and access frequency to guide storage tiers.	Difficulty in defining clear business needs for historical data access.
AUTOMATED POLICY ENFORCEMENT	**EFFICIENCY & CONSISTENCY**	**AUTOMATION TOOLS**	**FALSE POSITIVES**
Using AI-driven solutions.	Reduces manual effort and ensures consistent policy application.	Deploy AI/ML-driven tools to automatically apply retention rules.	Automated systems misclassifying data or applying incorrect retention rules.

2. DATA EXPIRATION MODELING

For a business leader, implementing data expiration modeling is a strategic imperative for proactively predicting and efficiently managing data lifecycle phases, enabling optimized archiving and disposal processes. In a world of ever-growing data volumes, intelligently determining when data loses its immediate value is crucial for cost control, performance, and compliance. Central to this is comprehensive lifecycle classification. This involves categorizing data based on its usage frequency, relevance, and criticality over time. By understanding the "aging" of data (e.g., active, near-line, archive, delete), you can make informed decisions about its appropriate storage tier and retention period.

Furthermore, leveraging predictive expiration algorithms is crucial. By employing AI and analytics, you can forecast optimal archiving and disposal timelines, anticipating when specific datasets will no longer be actively needed for operations but might still require long-term retention for compliance or historical analysis. This foresight optimizes storage costs. This intelligent automation ensures that data is moved to the most cost-effective storage tier at the right time, freeing up high-performance storage and reducing manual effort. Neglecting data expiration modeling can lead to unnecessary storage costs, compliance risks from retaining data too long, or premature deletion of valuable information. By championing a proactive and intelligent approach to managing data lifecycle phases, you empower your organization to optimize its data infrastructure, reduce operational overhead, and ensure that data is available when needed and disposed of securely when no longer required, ultimately contributing to a more efficient and compliant data ecosystem.

WHAT?	WHY?	HOW?	BEWARE!
LIFECYCLE CLASSIFICATION	**INFORMED STORAGE**	**DATA PROFILING**	**COMPLEXITY**
Categorizing data based on usage and relevance.	Guides efficient storage tiering and retention decisions.	Analyze data usage patterns and define lifecycle stages (hot, warm, cold).	Defining clear classification criteria for diverse data types.
PREDICTIVE EXPIRATION ALGORITHMS	**COST OPTIMIZATION**	**ML MODELING**	**ACCURACY**
Forecasting optimal archiving and disposal timelines.	Minimizes storage costs by predicting ideal archival/disposal times.	Develop AI/ML models to forecast data relevance and expiration.	Ensuring predictive models accurately reflect future data utility.
ADAPTIVE STORAGE MANAGEMENT	**EFFICIENCY & AUTOMATION**	**AUTOMATED WORKFLOWS**	**INFLEXIBILITY**
Automating transitions based on data aging.	Streamlines data movement across storage tiers.	Implement automated policies for data tiering and deletion.	Automated rules being too rigid for evolving business needs.

For a business leader, implementing robust digital asset management (DAM) is a strategic imperative for organizing, preserving, and ensuring the long-term usability of valuable digital content. In today's media-rich environment, managing vast libraries of images, videos, documents, and creative files efficiently is crucial for brand consistency, operational efficiency, and legal compliance. Central to this is meticulous metadata tagging and indexing. By enhancing searchability through comprehensive categorization of archived digital assets with relevant metadata (e.g., keywords, creators, dates, usage rights), you ensure that users can quickly find and retrieve the exact content they need, preventing wasted time and duplicated efforts.

Furthermore, integrating strong version control and historical tracking is crucial. Maintaining access to previous iterations of digital files provides an invaluable audit trail, supports compliance requirements, and enables teams to revert to earlier versions if needed, protecting creative and historical assets. Crucially, implementing robust access control for archived assets is vital. This ensures that security measures are in place to protect historical data from unauthorized modifications, accidental deletion, or intellectual property theft, safeguarding your valuable digital content even in long-term storage. Neglecting digital asset management can lead to disorganized content libraries, difficulty in locating approved brand assets, potential copyright infringements, and inefficiencies in creative workflows. By championing a comprehensive DAM strategy, you empower your organization to leverage its digital content effectively, maintain brand consistency, and protect valuable intellectual property, ultimately contributing to a more efficient, compliant, and creatively agile business.

WHAT?	WHY?	HOW?	BEWARE!
METADATA TAGGING AND INDEXING	**ASSET DISCOVERY**	**CLASSIFICATION**	**MANUAL EFFORT**
Enhancing searchability of assets.	Improves ease of finding and reusing digital content.	Apply consistent metadata tags and keywords to all digital assets.	Large volumes of assets requiring significant manual tagging.
VERSION CONTROL AND HISTORICAL TRACKING	**AUDITABILITY & RECOVERY**	**DAM SYSTEM IMPLEMENTATION**	**STORAGE OVERHEAD**
Maintaining access to previous iterations.	Ensures historical integrity and enables reversion to previous versions.	Use a DAM system with built-in version control.	Increased storage requirements due to multiple versions.
ACCESS CONTROL FOR ARCHIVED ASSETS	**ASSET PROTECTION**	**PERMISSION MANAGEMENT**	**COMPLEXITY**
Implementing security measures.	Safeguards valuable digital content from unauthorized access.	Define and enforce role-based access for archived assets.	Managing granular access permissions across diverse asset types.

For a business leader, implementing scalable and efficient archival storage solutions is a critical strategic imperative for preserving historical data, balancing long-term retention needs with cost-effectiveness. As data volumes explode, intelligently structuring your archives is crucial for compliance, future analysis, and efficient resource management. Central to this is distinguishing between cold vs. hot storage. Understanding the difference between low-cost, long-term archival storage (cold) and readily accessible, frequently queried archives (hot) allows for strategic data placement, optimizing both access speeds and cost efficiency based on usage patterns.

Furthermore, meticulously comparing tape storage and cloud archiving solutions is crucial for informed decision-making. While traditional tape storage offers a highly cost-effective and secure offline option for vast cold archives, modern cloud-based archival solutions provide immense scalability, flexibility, and easier access, each with its own benefits for specific data types and retrieval needs. Crucially, ensuring redundancy and disaster recovery capabilities for archived data is paramount. Implementing robust backup strategies, geographic replication, and disaster recovery plans guarantees that your valuable historical data remains accessible and protected against loss, even in the event of unforeseen catastrophic events. Neglecting a well-thought-out archival storage strategy can lead to escalating costs, difficulties in retrieving historical data for audits or analysis, and potential data loss in the event of system failures. By championing a strategic approach to archival storage, you empower your organization to cost-effectively preserve its historical data, ensure compliance, and maintain access to valuable long-term insights, ultimately safeguarding your business's integrity and future analytical capabilities.

WHAT?	WHY?	HOW?	BEWARE!
COLD VS. HOT STORAGE	**COST OPTIMIZATION**	**TIERING STRATEGY**	**DATA CLASSIFICATION**
Differentiating storage tiers.	Balances immediate access needs with cost-effective long-term storage.	Define rules for moving data between hot, warm, and cold storage tiers.	Accurately classifying data based on access frequency.
TAPE STORAGE AND CLOUD ARCHIVING	**FLEXIBLE OPTIONS**	**EVALUATION & SELECTION**	**VENDOR LOCK-IN**
Comparing physical and cloud solutions.	Offers diverse solutions for long-term data preservation.	Assess cost, security, and retrieval times for each method.	Potential vendor lock-in with specific cloud archiving solutions.
REDUNDANCY AND DISASTER RECOVERY	**DATA AVAILABILITY**	**BACKUP & REPLICATION**	**COST & COMPLEXITY**
Ensuring data remains accessible and protected.	Protects archived data against loss and ensures accessibility during disasters.	Implement geographic replication and redundant storage solutions.	High costs and complexity of maintaining robust DR for archives.

For a business leader, implementing robust data masking techniques is a critical strategic imperative for protecting sensitive information within archived datasets, ensuring privacy, and maintaining compliance without compromising data usability for analytical or testing purposes. In an age of stringent data privacy regulations, simply storing data is not enough; it must be protected at a granular level. Central to this is utilizing anonymization techniques. This involves removing personally identifiable information (PII) from archived data while preserving its statistical properties and usability for analysis, testing, or development environments. This allows for data utilization without exposing sensitive details.

Furthermore, comprehensive encryption for archived data is crucial. Securing stored records through cryptographic techniques ensures that even if unauthorized access occurs, the confidential information remains unreadable and protected, adding a vital layer of defense to your long-term storage. Crucially, establishing controlled access policies ensures that masked data is retrievable by authorized personnel without exposing confidential information. This means implementing strict role-based access controls and authentication mechanisms that grant access only to the sanitized versions of the data, while highly sensitive original data remains restricted. Neglecting data masking can lead to severe data breaches, non-compliance with privacy regulations, significant financial penalties, and a complete erosion of customer trust if sensitive data in archives is exposed. By championing a proactive approach to data masking, you empower your organization to utilize its historical data safely, ensure privacy, and maintain regulatory adherence, ultimately safeguarding your business's reputation and long-term viability in a data-driven world.

WHAT?	WHY?	HOW?	BEWARE!
ANONYMIZATION TECHNIQUES	**PRIVACY PRESERVATION**	**PSEUDONYMIZATION**	**REVERSIBILITY**
Removing personally identifiable information.	Protects sensitive data while preserving its usability.	Implement techniques like tokenization, shuffling, or redaction.	Ensuring that anonymized data cannot be re-identified.
ENCRYPTION FOR ARCHIVED DATA	**DATA SECURITY**	**CRYPTOGRAPHIC ALGORITHMS**	**KEY MANAGEMENT**
Securing stored records.	Protects archived data from unauthorized access.	Apply strong encryption to data before archiving.	Securely managing encryption keys for long-term archived data.
CONTROLLED ACCESS POLICIES	**SECURE UTILIZATION**	**ROLE-BASED ACCESS (RBAC)**	**POLICY COMPLEXITY**
Ensuring masked data is retrievable by authorized personnel.	Allows safe access to data for authorized purposes.	Implement strict RBAC for masked and unmasked data.	Managing granular access policies for different masked versions.

For a business leader, implementing robust data disposal protocols is a critical strategic imperative for securely removing obsolete data, minimizing storage costs, and significantly reducing compliance risks. In an age of data sprawl and stringent privacy regulations, simply deleting files is insufficient; permanent, verifiable erasure is paramount. Central to this is adhering to secure deletion protocols. This involves using permanent erasure methods—such as cryptographic erasure or physical destruction for storage media—to ensure that obsolete data cannot be recovered by unauthorized parties, preventing potential breaches and data leakage from discarded assets.

Furthermore, compliance-driven disposal is crucial. As a leader, you must ensure that all data purging activities strictly adhere to regulatory requirements like GDPR, HIPAA, and industry-specific mandates. This prevents legal liabilities and ensures that data is not retained beyond its legal purpose, or disposed of incorrectly. Crucially, considering environmental impact considerations is becoming increasingly important. Implementing eco-friendly disposal methods for both physical storage media (e.g., hard drives) and digital data (e.g., cloud data centers) aligns your business with sustainability goals and enhances your corporate social responsibility. Neglecting secure data disposal can lead to data breaches from forgotten or improperly deleted data, severe regulatory fines, and escalating storage costs. By championing a systematic and environmentally conscious approach to data disposal, you empower your organization to reduce its data footprint, mitigate legal and security risks, and uphold its commitment to privacy, ultimately safeguarding your business's long-term integrity and responsible data stewardship.

WHAT?	WHY?	HOW?	BEWARE!
SECURE DELETION PROTOCOLS	**DATA PROTECTION**	**CRYPTOGRAPHIC ERASURE**	**VERIFICATION**
Using permanent erasure methods.	Ensures data cannot be recovered after deletion.	Implement secure wipe tools or physical destruction for media.	Confirming complete and irreversible data deletion.
COMPLIANCE-DRIVEN DISPOSAL	**LEGAL COMPLIANCE**	**RETENTION POLICY INTEGRATION**	**EVOLVING REGULATIONS**
Ensuring regulatory adherence when purging information.	Avoids regulatory fines and legal liabilities.	Automate disposal based on defined retention policies.	Keeping up with changing legal requirements for data disposal.
ENVIRONMENTAL IMPACT CONSIDERATIONS	**SUSTAINABILITY**	**RECYCLING & ENERGY EFFICIENCY**	**COST**
Implementing eco-friendly disposal methods.	Aligns disposal practices with environmental responsibility goals.	Partner with certified recyclers; optimize cloud resource usage.	Eco-friendly disposal methods sometimes incurring higher costs.

Ai6x6™ Transformation Blueprint

AUTHORITY | Data Intelligence

"In God we trust. All others must bring data." - Jim Barksdale.

Jim Barksdale: Former CEO of Netscape Communications, a pioneering internet company. He led Netscape through the early "browser wars," playing a significant role in shaping the web's development and popularizing internet browsing for the masses.

The Authority level is your Data Intelligence, achieved when information stops being a record and starts being a predictor of future behavior. Utilizing pattern analytics grants the organization the wisdom to lead the market rather than react to it.

Fostering an organizational mindset and pervasive environment where decisions at all levels are primarily based on rigorous data insights and empirical evidence, rather than relying solely on intuition, anecdote, or personal bias. This fundamentally transforms operations by promoting objectivity, enabling continuous improvement, and establishing a significant competitive advantage in the marketplace.

"Data Intelligence" empowers employees with facts, encourages experimentation, and leads to more informed, strategic, and ultimately more successful outcomes across the entire organization.

The "**Data Intelligence**" chapter is fundamental for business strategy maturity as it focuses on sustaining value from data and innovating how it's used for future insights. A mature business strategy ensures that data is effectively shared across applications, reports, or dashboards for informed decision-making. This amplification democratizes access to insights, fosters a data-driven culture, and continuously enhances the organization's ability to leverage its data assets. It's about maximizing the strategic impact of data, ensuring it consistently drives growth, efficiency, and competitive advantage across all business functions. Here is how we approach our learning:

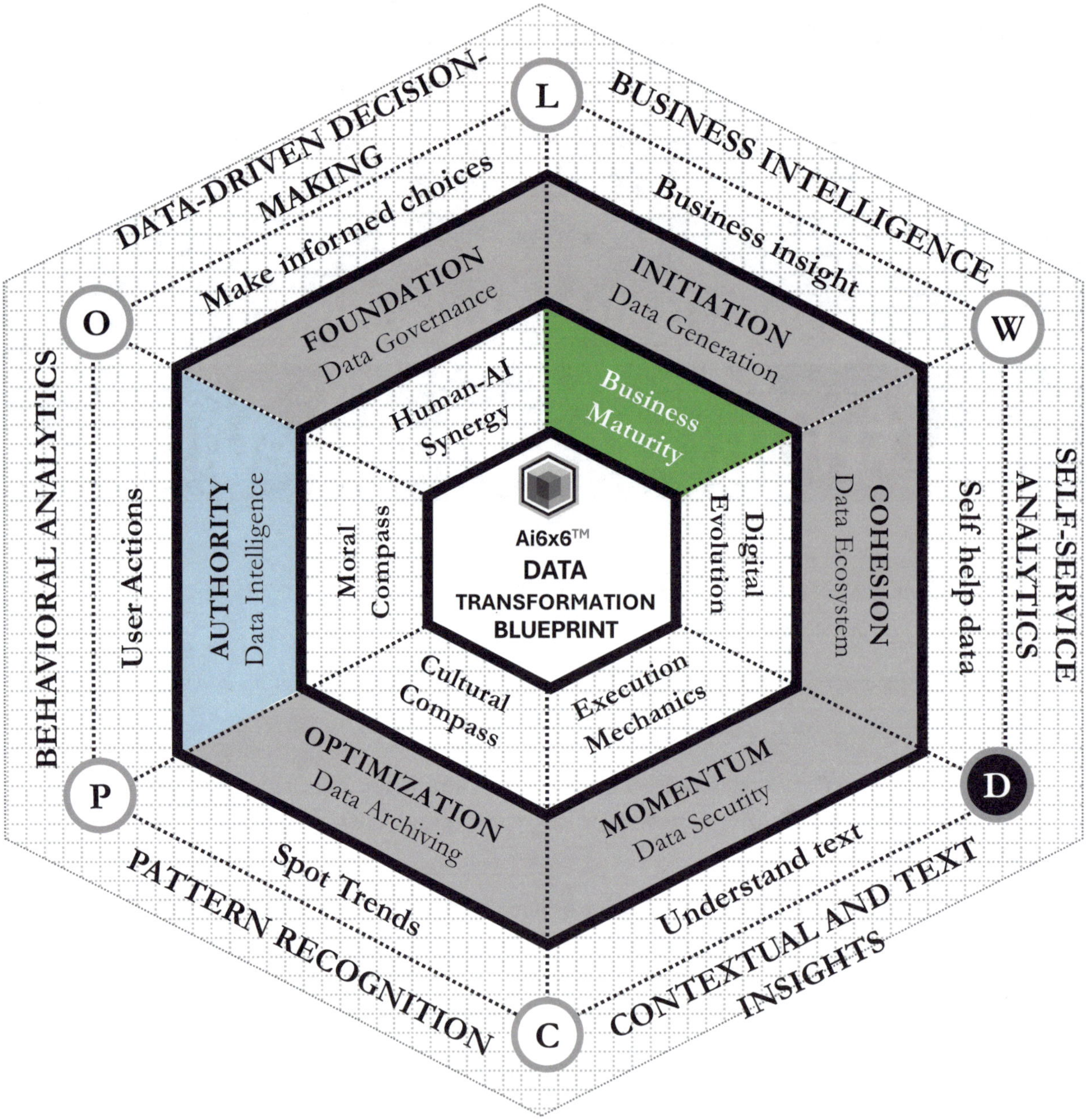

Ai6x6 BUSINESS MATURITY FOR INTELLIGENCE
DATA TRANSFORMATION | DATA ANALYTICS

For a business leader, mastering data-driven decision-making is a strategic imperative that leverages analytics to guide all strategic business choices, fundamentally transforming how your organization operates and competes. In today's complex and competitive landscape, relying on intuition alone is a recipe for stagnation; evidence-based strategies are paramount for success. Central to this is adopting an evidence-based strategy. By systematically using data insights, you can minimize uncertainty, reduce guesswork, and significantly improve decision outcomes, ensuring that every strategic move is backed by quantifiable facts rather than assumptions.

Furthermore, implementing real-time decision support systems is crucial. By integrating AI-driven solutions for instant analysis and recommendations directly into your operational workflows, you empower leaders and teams to make agile, informed choices at the speed of business, capitalizing on fleeting opportunities and responding swiftly to challenges. Crucially, robustly measuring impact is vital for continuous improvement. Tracking key performance indicators (KPIs) that directly evaluate the effectiveness of your decisions allows you to learn from outcomes, iterate on strategies, and continuously refine your approach for greater future success. This closes the feedback loop. Neglecting data-driven decision-making can lead to misinformed strategies, wasted resources, missed market opportunities, and a reactive stance that hinders growth. By championing a culture where data intelligence informs every critical choice, you empower your organization to operate with greater precision, accelerate innovation, and gain a significant competitive advantage, ultimately driving sustained business success in a data-centric world.

WHAT?	WHY?	HOW?	BEWARE!
EVIDENCE-BASED STRATEGY	**IMPROVED OUTCOMES**	**DATA INTEGRATION**	**DATA OVERLOAD**
Using data insights to minimize uncertainty.	Reduces guesswork and enhances decision accuracy.	Integrate data from all relevant sources into analytical platforms.	Too much data leading to analysis paralysis.
REAL-TIME DECISION SUPPORT	**INSTANT ANALYSIS**	**AI/ML DEPLOYMENT**	**TRUST IN AI**
Implementing AI-driven solutions.	Provides immediate recommendations for agile decision-making.	Deploy AI/ML models that offer real-time insights and predictions.	Building trust in AI-generated recommendations among decision-makers.
MEASURING IMPACT	**ACCOUNTABILITY & LEARNING**	**KPI DEFINITION**	**ATTRIBUTION**
Tracking KPIs to evaluate decision effectiveness.	Ensures accountability and facilitates continuous learning.	Define clear KPIs for each strategic decision and track performance.	Difficulty attributing specific business outcomes directly to certain decisions.

For a business leader, leveraging robust business intelligence (BI) is a strategic imperative that transforms raw data into actionable insights, fueling informed decision-making and driving sustainable business growth. BI is the lens through which you understand your current performance and identify future opportunities, moving beyond mere data collection to true strategic understanding. Central to this is creating intuitive dashboards and reporting systems. By developing visual tools that monitor performance, track key trends, and present complex data in easily digestible formats, you empower leaders and teams across the organization to quickly grasp critical insights and respond effectively.

Furthermore, harnessing predictive analytics is crucial. By using historical data to forecast future trends, anticipate market shifts, and identify emerging opportunities, you move from reactive problem-solving to proactive strategic planning, positioning your business for future success. Crucially, ensuring enterprise data integration connects multiple data sources across your organization. This provides a comprehensive, unified view of operations—from sales and marketing to finance and supply chain—eliminating data silos and enabling holistic analysis that drives better cross-functional decision-making. Neglecting robust business intelligence can lead to fragmented insights, reactive strategies, missed market opportunities, and a struggle to understand the true drivers of performance. By championing BI capabilities, you empower your organization with clarity, foresight, and a powerful analytical engine, accelerating decision-making, improving operational efficiency, and ultimately securing a significant competitive advantage in the data-driven economy. This is about converting information into intelligence.

WHAT?	WHY?	HOW?	BEWARE!
DASHBOARDS AND REPORTING	**PERFORMANCE MONITORING**	**BI PLATFORM DEPLOYMENT**	**DASHBOARD FATIGUE**
Creating visual tools to monitor performance.	Provides clear, real-time insights into business performance.	Implement BI tools (e.g., Tableau, Power BI) for dashboards.	Too many dashboards leading to information overload and disuse.
PREDICTIVE ANALYTICS	**FUTURE FORESIGHT**	**ML MODEL DEVELOPMENT**	**ACCURACY**
Using historical data to forecast future trends.	Enables proactive strategic planning and opportunity identification.	Build and deploy machine learning models for forecasting.	Ensuring the accuracy and reliability of predictive forecasts.
ENTERPRISE DATA INTEGRATION	**HOLISTIC VIEW**	**ETL/ELT PIPELINES**	**DATA SILOS**
Connecting multiple data sources.	Provides a comprehensive, unified view of business operations.	Integrate data from all enterprise systems into a central data warehouse/lake.	Breaking down entrenched data silos across departments.

For a business leader, empowering your employees with self-service analytics is a strategic imperative that democratizes data access, accelerates insights, and fosters a truly data-driven culture throughout your organization. In an agile environment, relying solely on data analysts to generate every report creates bottlenecks; enabling independent data exploration is crucial. Central to this is providing no-code analytics platforms. These user-friendly tools empower employees across various departments—even those without technical expertise—to generate their own insights, ask their own questions of the data, and make data-informed decisions swiftly, reducing dependency on IT or data teams.

Furthermore, developing interactive visualizations is crucial for simplifying complex data. By presenting information through intuitive charts, graphs, and dynamic dashboards, you make data more accessible and understandable, enabling users to explore trends and patterns easily without needing deep analytical skills. Crucially, leveraging AI-assisted query tools is revolutionizing self-service capabilities. By incorporating natural language processing (NLP), employees can simply type their questions in plain language and receive relevant data insights, making data exploration as intuitive as a search engine query. Neglecting self-service analytics can lead to delayed insights, frustrated business users, and a bottleneck in data consumption that hinders agility and innovation. By championing tools and training that empower independent data access and analysis, you foster a culture of curiosity, accelerate decision-making at all levels, and unlock the collective intelligence of your workforce, ultimately driving greater efficiency and competitive advantage for your business.

WHAT?	WHY?	HOW?	BEWARE!
NO-CODE ANALYTICS PLATFORMS	**DATA DEMOCRATIZATION**	**PLATFORM DEPLOYMENT**	**DATA LITERACY**
Enabling employees to generate insights.	Empowers non-technical users to access and analyze data.	Implement user-friendly, no-code/low-code BI tools.	Ensuring users have sufficient data literacy to interpret results correctly.
INTERACTIVE VISUALIZATIONS	**EASY UNDERSTANDING**	**DASHBOARD DESIGN**	**OVER-SIMPLIFICATION**
Simplifying complex data with charts and graphs.	Makes complex data digestible and actionable for a wider audience.	Create intuitive and interactive dashboards with drill-down capabilities.	Risk of over-simplifying data, leading to misinterpretation.
AI-ASSISTED QUERY TOOLS	**INTUITIVE EXPLORATION**	**NLP INTEGRATION**	**ACCURACY**
Using natural language processing (NLP) for exploration.	Enables data exploration through natural language.	Integrate NLP capabilities into self-service analytics platforms.	NLP models misinterpreting complex or ambiguous user queries.

For a business leader, extracting meaning from unstructured data through contextual and text insights is a strategic imperative for gaining deeper analytical understanding and unlocking hidden value. In an era where a vast amount of critical information resides in text (e.g., customer reviews, social media, emails, documents), your ability to derive intelligence from this data source is crucial for comprehensive market understanding and informed decision-making. Central to this is conducting robust sentiment analysis. By understanding customer emotions, opinions, and feedback trends from text data, you can gauge public perception, identify areas of dissatisfaction or delight, and proactively respond to market sentiment, directly impacting brand reputation and customer loyalty.

Furthermore, leveraging natural language processing (NLP) is crucial. NLP transforms raw text-based data into structured insights, extracting key entities, relationships, and themes. This enables scalable analysis of vast amounts of textual information that would be impossible to process manually, opening new avenues for intelligence. By automatically categorizing and summarizing recurring subjects, you can quickly grasp the most important conversations, emerging issues, or popular discussions relevant to your business, informing product development and strategic communications. Neglecting contextual and text insights means missing out on a wealth of qualitative data, leading to incomplete market understanding, reactive decision-making, and a struggle to truly understand the voice of the customer. By championing these advanced analytical techniques, you empower your organization to gain richer, more nuanced intelligence, accelerate innovation, and respond more effectively to market dynamics, ultimately securing a significant competitive advantage through comprehensive data understanding.

WHAT?	WHY?	HOW?	BEWARE!
SENTIMENT ANALYSIS	**CUSTOMER INSIGHTS**	**NLP & ML DEPLOYMENT**	**NUANCE**
Understanding customer emotions and feedback trends.	Gauges public perception and identifies areas for improvement.	Use NLP and machine learning models to analyze text for sentiment.	Difficulty in accurately interpreting sarcasm, irony, or complex emotional nuances.
NATURAL LANGUAGE PROCESSING (NLP)	**TEXT DATA UTILIZATION**	**NLP MODEL TRAINING**	**LANGUAGE DIVERSITY**
Transforming text-based data into structured insights.	Unlocks insights from unstructured text data at scale.	Train NLP models for entity recognition, part-of-speech tagging, etc.	Handling multiple languages and highly domain-specific jargon.
TOPIC MODELING	**HIGH-LEVEL UNDERSTANDING**	**ALGORITHM APPLICATION**	**INTERPRETABILITY**
Identifying themes and trends from large text datasets.	Summarizes key discussions and emerging subjects.	Apply algorithms like Latent Dirichlet Allocation (LDA) to text corpora.	Interpreting the relevance and actionable insights from identified topics.

For a business leader, mastering pattern recognition is a strategic imperative for detecting recurring trends and critical anomalies within your data, enabling powerful predictive insights that drive proactive decision-making and competitive advantage. In today's data-intensive environment, buried patterns hold the key to foresight. Central to this is leveraging sophisticated machine learning models. By training algorithms to recognize intricate data patterns across vast datasets, you can develop robust forecasting capabilities, predict future outcomes, and anticipate market shifts with greater accuracy than traditional methods.

Furthermore, implementing robust anomaly detection systems is crucial. These systems continuously spot irregularities that deviate significantly from normal patterns, indicating potential fraud, operational errors, security breaches, or even emerging market opportunities that require immediate attention. This early warning capability is invaluable. Crucially, applying time-series analysis allows you to identify cyclical trends and seasonality in data. Understanding how data evolves over time, and recognizing periodic fluctuations, enables better resource planning, inventory management, and marketing campaign timing. Neglecting advanced pattern recognition can lead to missed opportunities, reactive problem-solving, susceptibility to fraud, and a failure to capitalize on subtle market signals. By championing the integration of machine learning, anomaly detection, and time-series analysis, you empower your organization to transform raw data into predictive intelligence, enabling proactive strategies, mitigating risks, and ultimately securing a significant competitive advantage through superior foresight. This analytical capability is a cornerstone of intelligent business operations.

WHAT?	WHY?	HOW?	BEWARE!
MACHINE LEARNING MODELS	**PREDICTIVE FORECASTING**	**MODEL DEVELOPMENT**	**DATA QUALITY**
Training algorithms to recognize data patterns.	Enables accurate predictions and anticipation of future events.	Build and train ML algorithms (e.g., regression, classification) on historical data.	ML model performance heavily reliant on clean, relevant data.
ANOMALY DETECTION	**EARLY WARNING**	**ALGORITHM APPLICATION**	**FALSE POSITIVES**
Spotting irregularities indicating issues or opportunities.	Identifies critical deviations for proactive intervention.	Deploy algorithms to monitor data streams for unusual patterns.	Distinguishing genuine anomalies from normal variations or noise.
TIME-SERIES ANALYSIS	**RESOURCE OPTIMIZATION**	**STATISTICAL MODELING**	**COMPLEX PATTERNS**
Identifying cyclical trends and seasonality.	Improves forecasting for planning and resource allocation.	Apply statistical models to analyze data points collected over time.	Identifying non-linear or multi-seasonal patterns in complex data.

For a business leader, embracing behavioral analytics is a strategic imperative for deeply understanding user actions, optimizing engagement strategies, and refining overall business approaches. In today's digital economy, every click, interaction, and path taken by a user provides invaluable insights into their preferences, motivations, and pain points. Central to this is meticulous customer journey mapping. By tracking user interactions across all digital platforms—websites, mobile apps, social media, and customer service channels—you can visualize their complete path, identify key touchpoints, and pinpoint areas of friction or opportunity, enabling a more seamless customer experience.

Furthermore, conducting robust clickstream analysis is crucial. By analyzing browsing behavior—what users click on, how long they stay on a page, and their navigation paths—you can improve personalization, optimize website design, and enhance conversion rates, directly impacting sales and user satisfaction. Crucially, leveraging user segmentation allows you to group audiences based on their behavior, preferences, and engagement history. This enables highly targeted marketing campaigns, personalized product recommendations, and tailored communication strategies that resonate more deeply with specific customer segments, maximizing impact and ROI. Neglecting behavioral analytics can lead to generic customer experiences, inefficient marketing spend, high churn rates, and a fundamental misunderstanding of your customer base. By championing these analytical techniques, you empower your organization to truly understand its users, optimize their experiences, and build strategies that drive deeper engagement, higher conversion rates, and ultimately, sustained business growth through a profound customer-centric approach.

WHAT?	WHY?	HOW?	BEWARE!
CUSTOMER JOURNEY MAPPING	**EXPERIENCE OPTIMIZATION**	**DATA INTEGRATION**	**COMPLEXITY**
Tracking user interactions across platforms.	Identifies pain points and opportunities to improve customer experience.	Integrate data from all customer touchpoints to map journeys.	Mapping and visualizing intricate, multi-channel customer journeys.
CLICKSTREAM ANALYSIS	**PERSONALIZATION & CONVERSION**	**BEHAVIORAL TRACKING**	**DATA VOLUME**
Analyzing browsing behavior.	Improves personalization and enhances conversion rates.	Implement tools to track user clicks, navigation, and time on page.	Processing and analyzing massive volumes of clickstream data.
USER SEGMENTATION	**TARGETED STRATEGIES**	**CLUSTERING ALGORITHMS**	**SEGMENT DEFINITION**
Grouping audiences based on behavior and preferences.	Enables highly personalized marketing and product offerings.	Apply machine learning algorithms to group users by behavior.	Defining meaningful and actionable user segments.

BLUEPRINT SERIES

SECTION D

Digital Evolution

Ai6x6™ Transformation Blueprint

Story 1 — The Birth of COBOL and Business Computing (1959–1965): Early enterprises struggled to translate business logic into machine-readable form, limiting the usefulness of emerging computers. COBOL introduced a standardized language that mirrored human reasoning, enabling organizations to encode processes consistently. This shift allowed systems to scale beyond isolated experiments. Human analysts defined rules while machines executed them with precision. The partnership created the first bridge between business intent and digital execution.

Story 2 — ARPANET and Packet Switching (1969–1972): Communication networks were fragile, centralized, and prone to failure. Packet switching introduced a decentralized architecture that allowed information to flow even when parts of the network failed. Humans designed protocols while machines routed data autonomously. This resilience transformed digital communication from a technical novelty into a strategic infrastructure. ARPANET demonstrated how distributed systems could scale without collapsing. It laid the foundation for the modern internet.

Story 3 — Xerox PARC and the Graphical User Interface (1973–1983): Computers were powerful but inaccessible to most users, requiring specialized knowledge to operate. Xerox PARC introduced the graphical user interface, enabling humans to interact with machines visually rather than through code. This shift democratized computing and redefined usability. Machines handled complexity behind the scenes while humans engaged through intuitive metaphors. The interface became a catalyst for mass digital adoption. It showed that digital evolution accelerates when technology adapts to human cognition.

Story 4 — The Rise of Enterprise Resource Planning (1980s–1990s): Organizations operated in silos, with fragmented systems and inconsistent data. ERP platforms unified finance, operations, and supply chains into integrated digital environments. Machines enforced standardization while humans redesigned processes around shared information. This integration improved visibility, coordination, and control. ERP became the backbone of global enterprises. It demonstrated how digital evolution transforms not just tools but entire operating models.

Story 5 — The Commercial Internet and E-Commerce (1994–2005): The internet shifted from academic infrastructure to commercial ecosystem, enabling new forms of interaction, distribution, and value creation. Websites became digital storefronts, and payment systems matured into secure transactional platforms. Machines automated logistics while humans reimagined customer experience. The digital marketplace expanded global reach and lowered barriers to entry. It marked a turning point where digital capability became essential for competitiveness.

Story 6 — Cloud Computing and Elastic Infrastructure (2006–2015): Organizations struggled with rigid, capital-intensive infrastructure that limited experimentation and scale. Cloud computing introduced elastic resources that could expand or contract instantly. Machines automated provisioning while humans focused on architecture and innovation. This shift accelerated deployment cycles, reduced cost, and enabled global reach. Cloud became the foundation for modern digital ecosystems. It demonstrated how digital evolution thrives when infrastructure becomes fluid, on-demand, and programmable.

LESSONS FROM A DATA TRANSFORMATION PERSPECTIVE

Digital evolution transforms data from isolated records into interconnected, real-time intelligence. Early systems captured structured information, while later architectures enabled distributed communication, integrated platforms, and elastic storage. As digital capability expanded, data became more accessible, more contextual, and more actionable. Machines provided scale and automation, but humans shaped meaning, governance, and strategic interpretation. Data maturity emerges when organizations treat information as a shared asset that flows across functions and systems. Digital evolution ultimately turns data into the connective tissue of the enterprise—supporting foresight, coordination, and continuous improvement through disciplined human–machine collaboration.

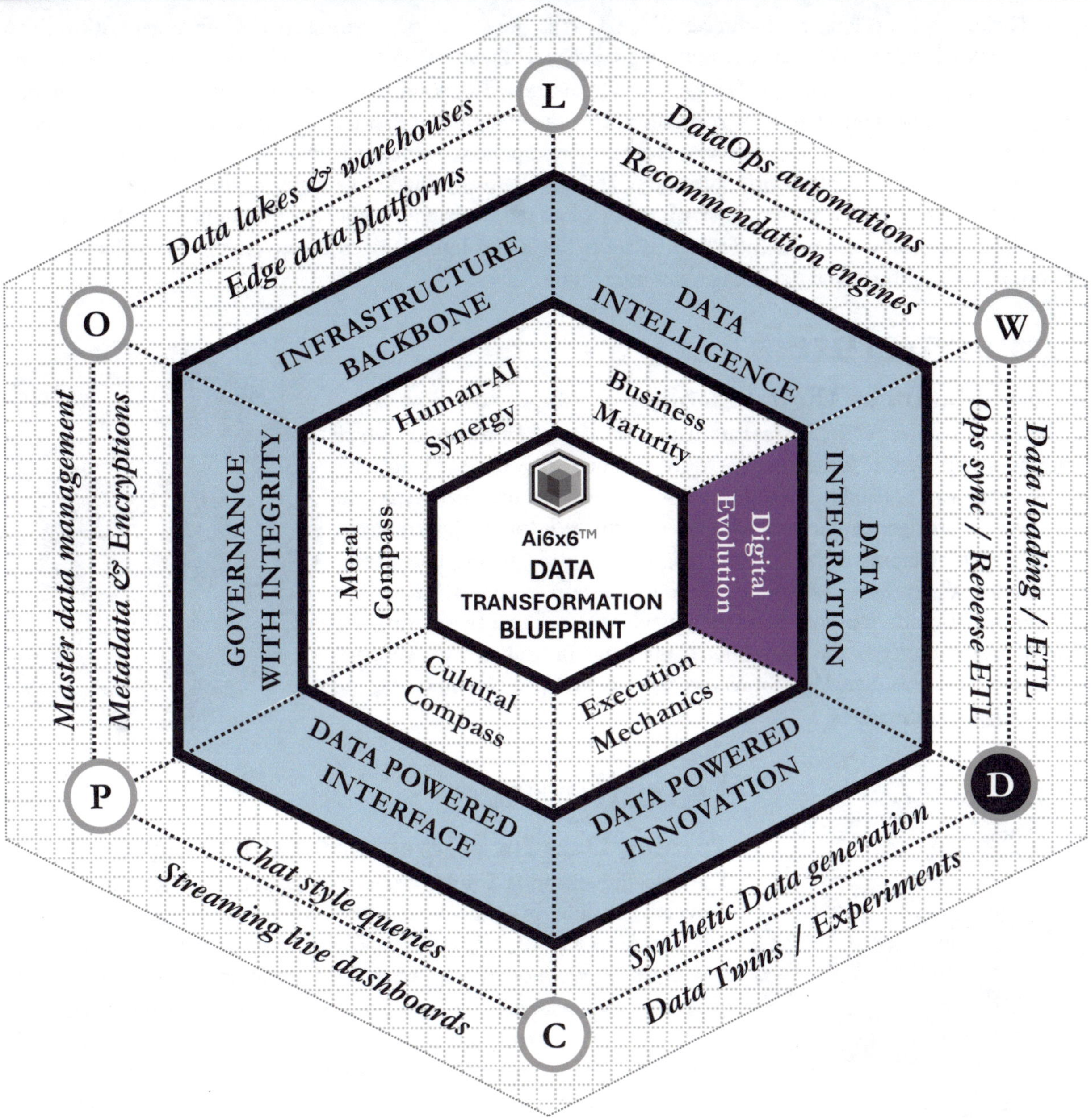

The **Ai6x6 Research Lab** chose this lens to solve the "complexity gap" that often leaves organizations drowning in disconnected information. We believe that for data to be truly transformative, it must move from being a visible problem to becoming the "Invisible Backbone" of the enterprise. This isn't about storage capacity; it is about the mental shift toward "Thinking in Data" and creating a culture where information flows with natural ease. It is vital because an organization cannot innovate or "Invent New Realities" if its data is trapped in silos or lacks a human-centric interface. This lens ensures that the digital evolution is guided by the **Guardians of Information**.

By mastering this evolution, the enterprise moves from data fragmentation to absolute **Connected Contexts**. This lens ensures that the infrastructure is built on a foundation of integrity, where data is not just processed, but humanized to drive decision-making. When a leader understands the "Interface" of their information, they gain the freedom to lead with an undistracted focus on the future. This lens provides the psychological peace of mind that comes from knowing the organization's intelligence is secure, reliable, and designed for human synergy. It is about the power that comes from clarity, ensuring that the transition from raw bits to meaningful insights is a seamless and authoritative evolution of the business soul.

Digital Evolution is no longer a technical initiative. It is the structural rewiring of how an organization thinks, operates, and competes. For years, leaders treated digital transformation as a project—something with a start date, a roadmap, and a finish line. But the world has moved beyond that framing. Digital Evolution is not a destination; it is a permanent state of adaptation. It is the recognition that technology is no longer an enabler of strategy but a **shaper** of it. The organizations that thrive are those that treat digital capability as a living system—continuously upgraded, continuously integrated, and continuously aligned with the realities of the market. This lens reframes Digital Evolution as a leadership discipline rather than a technology program. It challenges leaders to see digital capability not as a collection of tools but as an interconnected ecosystem that influences every decision, every workflow, and every customer interaction.

Infrastructure

The Foundation of Digital Stability

The journey begins with **Infrastructure**, the foundation upon which everything else depends. Infrastructure is not just hardware or cloud capacity; it is the stability, scalability, and reliability of the systems that carry the organization's intelligence. Leaders who underestimate infrastructure discover that innovation collapses under its own weight. Without a strong foundation, every new capability becomes fragile, every integration becomes brittle, and every attempt at modernization becomes a patch rather than a transformation. Infrastructure maturity is the discipline of building for endurance—ensuring that the organization's digital backbone can support the speed, complexity, and volatility of modern markets.

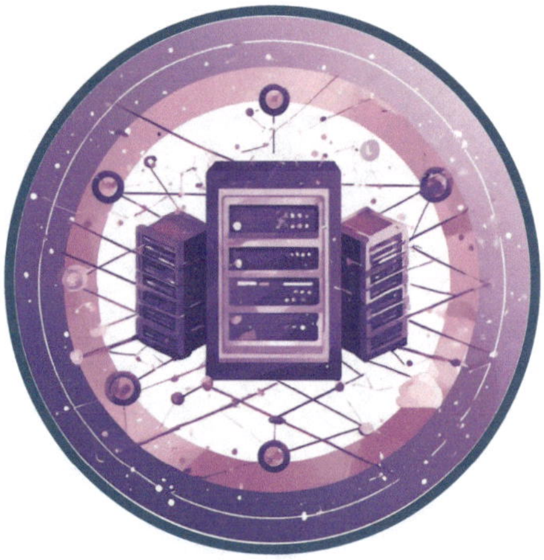

Intelligence

Turning Data Into Organizational Foresight

The next dimension is **Intelligence**, representing the organization's ability to convert data into insight and insight into action. Intelligence is not about dashboards or reports; it is about building a system that learns. It requires data that flows freely, models that are trained responsibly, and decisions that are informed by evidence rather than assumption. Leaders who embrace intelligence understand that it is not a technical asset but a strategic one. Intelligence shapes priorities, exposes risks, and reveals opportunities that would otherwise remain invisible. It is the cognitive engine of Digital Evolution—the mechanism through which the organization perceives, interprets, and responds to its environment.

Integration

Making the Enterprise Whole

Integration is where Digital Evolution becomes real. Organizations often accumulate technologies faster than they can connect them. Systems become islands, data becomes siloed, and teams become fragmented. Integration is the discipline that ensures the enterprise operates as a coherent whole. It is the work of connecting platforms, harmonizing data, and designing workflows that move across boundaries. Without integration, digital investments become isolated experiments rather than enterprise capabilities. Integration maturity is the art of coherence—ensuring that the organization's digital landscape functions as a single, unified system rather than a patchwork of disconnected tools.

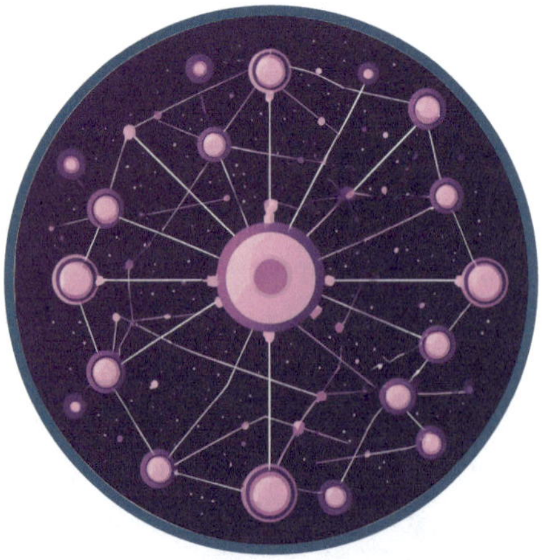

Innovation

Expanding the Field of Possibility

The fourth dimension, **Innovation**, expands the field of possibility. Innovation is not about chasing trends or adopting technology for its own sake. It is about using digital capability to reimagine how value is created, delivered, and experienced. Innovation demands that leaders challenge assumptions, explore alternatives, and test ideas quickly. It requires a culture where experimentation is encouraged, failure is reframed as learning, and progress is measured by insight rather than perfection. Innovation maturity is the discipline of curiosity—using digital tools not just to optimize the present but to invent the future.

Interface

Where Technology Meets Human Behavior

Interface represents the human experience of Digital Evolution. It is the point where technology meets behavior, where systems meet expectations, and where design meets trust. Leaders who ignore interface discover that even the most powerful systems fail if people cannot or will not use them. Interface is not cosmetic; it is strategic. It shapes adoption, accelerates learning, and determines whether digital capability becomes embedded or abandoned. Interface maturity is the discipline of empathy—designing digital experiences that are intuitive, accessible, and aligned with human needs.

Integrity

The Guardrail of Responsible Evolution

The final dimension, **Integrity**, grounds Digital Evolution in responsibility. Integrity ensures that systems are secure, data is protected, and decisions are ethical. It prevents speed from becoming recklessness and innovation from becoming harm. Leaders who prioritize integrity build organizations that are trusted, resilient, and prepared for the scrutiny that accompanies digital capability. Integrity maturity is the discipline of stewardship—ensuring that digital evolution strengthens the organization's moral and operational foundations rather than compromising them.

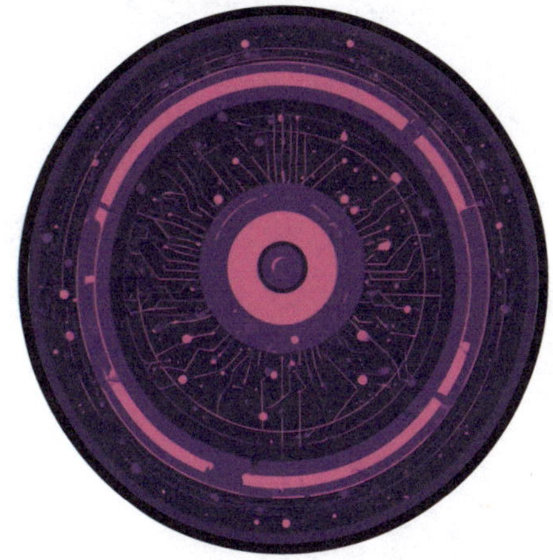

Digital Evolution is not a technology story. It is a leadership story. It is the recognition that the future belongs to organizations that treat digital capability as a core competency, not a side project. It is the shift from building tools to building systems, from adopting technology to evolving with it, and from reacting to change to shaping it. In this future, Digital Evolution is not a department, a project, or a budget line. It is the operating philosophy of the enterprise. It shapes how decisions are made, how systems are built, how teams collaborate, and how value is created. It turns technology from a source of disruption into a source of stability. It transforms the organization from a follower of change into a shaper of it.

DIGITAL EVOLUTION ANALOGY – KEEPING OURSELVES HEALTHY

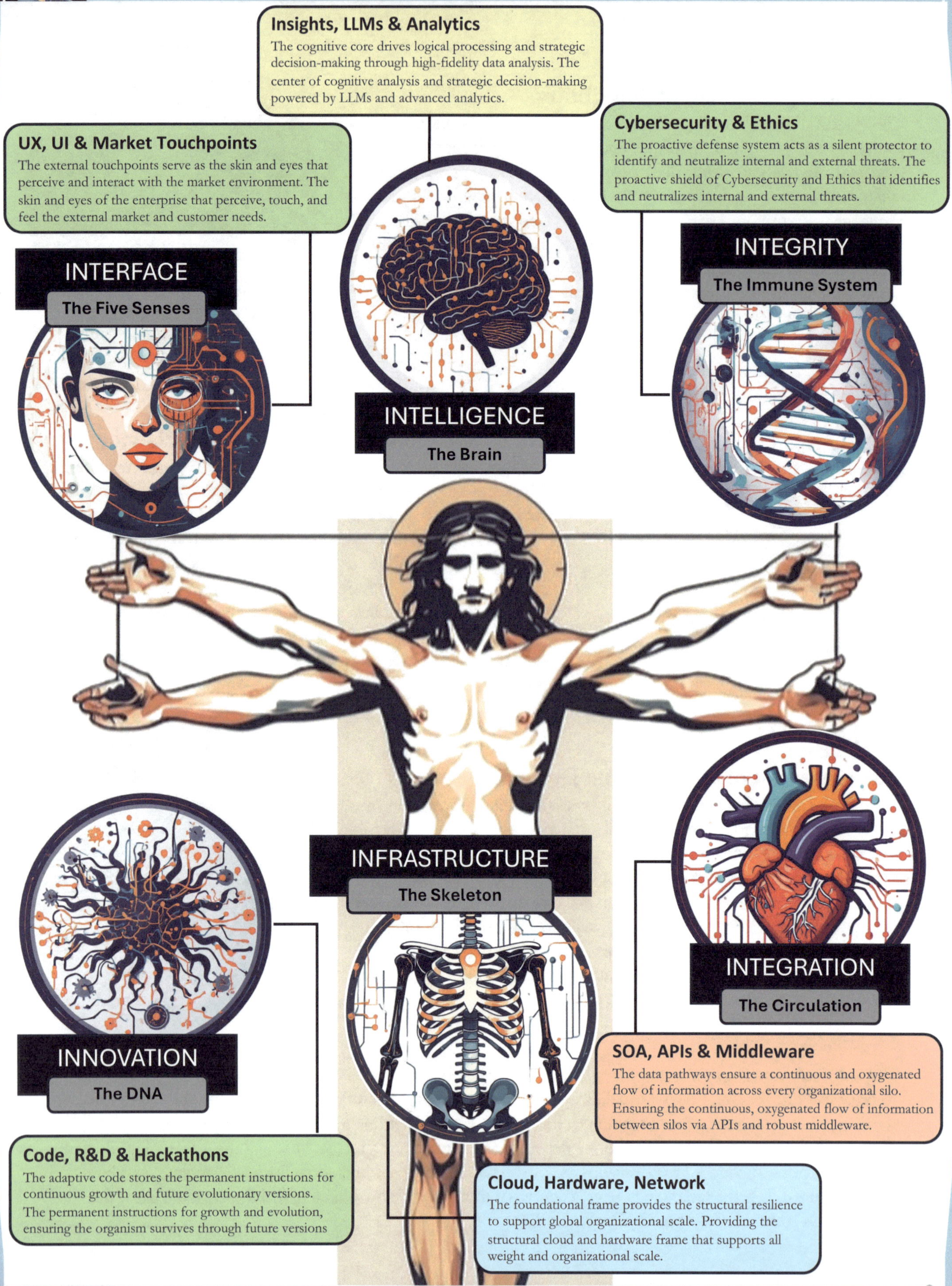

Ai6x6™ Transformation Blueprint

'Infrastructure' Backbone

"Data is the new oil—but only if it's refined." — Clive Humby

British mathematician and data science pioneer; coined one of the most referenced metaphors in modern analytics.

Infrastructure isn't just storage—it's strategic formation. Data lakes, warehouses, and in-memory grids aren't visible to customers, but they shape every insight, every product decision, every market move. "Invisible backbone" means leadership isn't focused on the vault—it's shaping the vault to scale. Mature transformation uses architecture not just to hold data, but to accelerate access, lineage, and reusability. This backbone must flex across real-time demand, decentralized environments, and hybrid cloud ecosystems. If information is capital, then infrastructure is creditworthiness. The stronger the spine, the louder the intelligence can speak.

For data transformation, the "Infrastructure Backbone" chapter is critical for digital maturity because it lays the essential groundwork for an organization's ability to collect, store, and process data at scale. A digitally mature enterprise understands that robust data infrastructure, encompassing cloud data lakes, data warehouses, and big data platforms, is not just a technical necessity but a strategic asset. This backbone enables the organization to handle the immense volume and velocity of modern data, ensuring reliable access, high-speed processing, and foundational support for all subsequent data-driven initiatives and insights across the business.

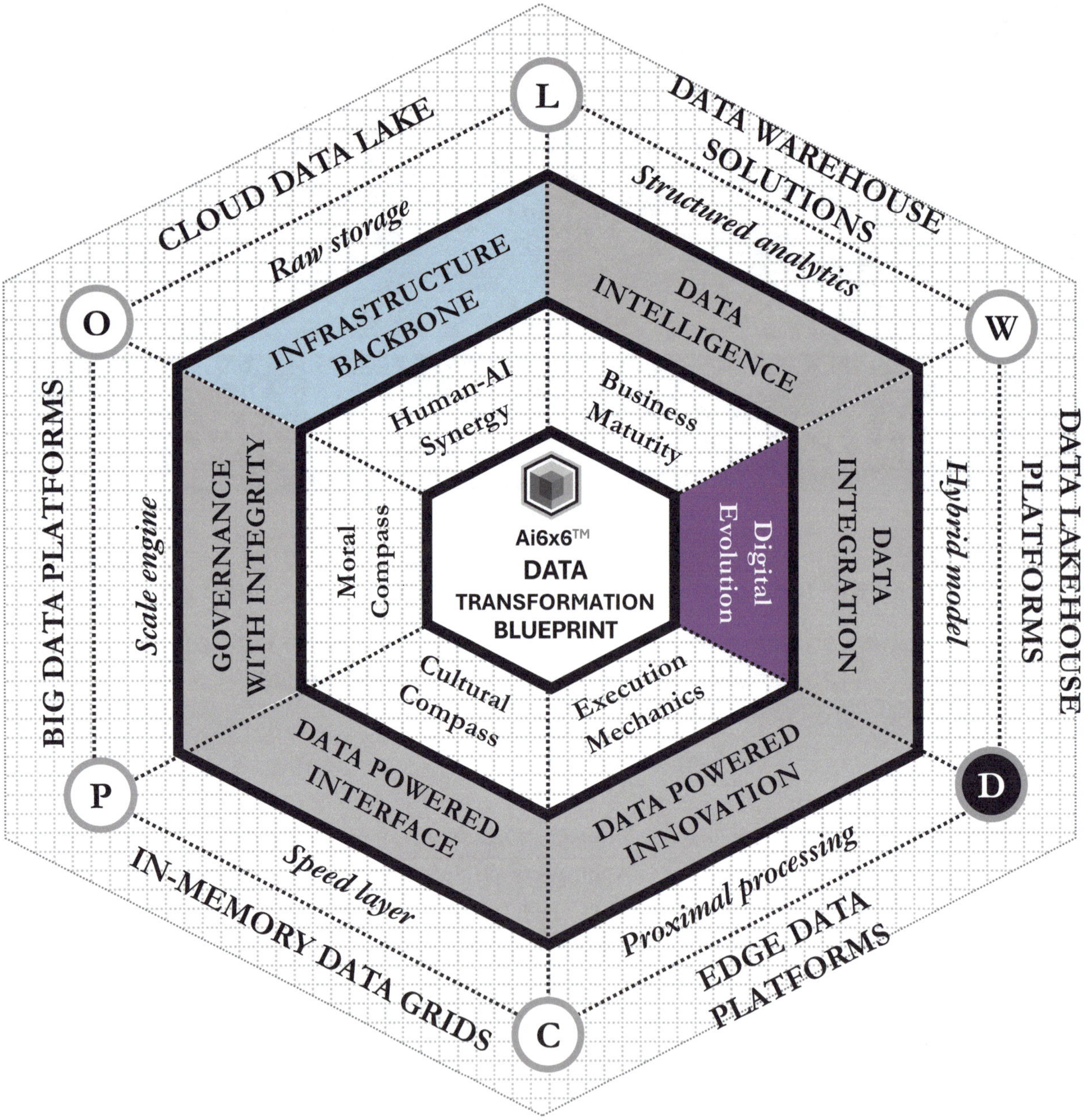

Ai6x6 DIGITAL EVOLUTION FOR **DATA LIFECYCLE | INFRASTRUCTURE**

1. CLOUD DATA LAKE

A Cloud Data Lake is a centralized repository built on cloud infrastructure that allows organizations to store vast amounts of raw data in its native format, without requiring a predefined schema, from diverse sources such as operational databases, IoT devices, social media, and streaming applications. Unlike traditional data warehouses, data lakes are designed for scalability, flexibility, and cost-effectiveness, enabling the storage of structured, semi-structured, and unstructured data. This architecture supports advanced analytics, machine learning, and AI initiatives by providing a single source for all enterprise data, making it readily available for various analytical workloads and empowering organizations to derive deeper insights and innovate faster.

Without a Cloud Data Lake, organizations often face significant challenges in storing, processing, and analyzing the rapidly growing volumes and varieties of data generated today, particularly unstructured and semi-structured data from new sources. Relying on traditional data warehouses or disparate data silos leads to limitations in scalability, high costs for storing raw data, and an inability to perform advanced analytics like machine learning due to rigid schemas. The absence of a centralized, flexible data lake results in missed opportunities for deep insights, hindered innovation, and a fragmented view of enterprise data, ultimately limiting an organization's capacity to become truly data-driven and agile in a competitive market.

Implementation	Challenges
* Define data storage strategy & governance * Select cloud provider (AWS, Azure, GCP) * Design data ingestion pipelines (batch/stream) * Implement data catalog & metadata management * Establish security, access controls & encryption * Plan data lifecycle management & cost ops	* Data governance & quality management * Ensuring data security & compliance * Cost optimization for storage & compute * Managing diverse data formats & schemas * Building robust data ingestion pipelines * Data discovery & usability for analysts

Products	Platform Cost	Project Cost	Duration
Amazon S3 (Data Lake Storage)	$5-30K	$30-150K	2-7 m
Azure Data Lake Storage Gen2	$5-25K	$20-120K	2-6 m
Google Cloud Storage (Data Lake)	$5-20K	$20-100K	2-6 m
Databricks (Lakehouse layer)	$10-60K	$50-300K	3-9 m
Snowflake (External Stage)	$10-50K	$40-250K	3-9 m
Cloudera Data Platform (Cloud)	$20-80K	$100-400K	4-12 m

Data Warehouse Solutions are specialized relational databases designed for analytical processing, aggregating and structuring data from multiple operational systems into a single, consistent repository to support business intelligence, reporting, and strategic decision-making. Unlike transactional databases, data warehouses are optimized for complex query performance over large datasets, employing schemas (like star or snowflake) that facilitate efficient data retrieval and analysis. By providing a centralized, clean, and historical view of an organization's data, these solutions enable accurate trend analysis, performance monitoring, and forecasting, empowering business users with reliable insights to optimize operations and drive informed strategic choices.

Without robust Data Warehouse Solutions, organizations often struggle with fragmented data across disparate operational systems, leading to inconsistent reporting, unreliable analytics, and significant manual effort for data consolidation. This lack of a centralized, structured repository for historical data hinders the ability to perform accurate trend analysis, measure performance effectively, or generate reliable insights for strategic decision-making. The absence of a dedicated data warehouse results in "data silos," delayed or inaccurate reporting, and an inability to fully leverage historical information, ultimately impacting an organization's capacity to make informed business decisions and maintain a competitive edge.

Implementation	Challenges
* Define business requirements & KPIs * Select data warehouse platform * Design data models (star schema, etc.) * Develop ETL/ELT pipelines for data loading * Implement data governance & quality controls * Integrate with BI tools & user training	* Complex data modeling & schema design * Ensuring data quality & consistency * High initial setup & ongoing maintenance costs * Performance optimization for large queries * Managing data governance & access control * Integrating disparate source systems

Products	Platform Cost	Project Cost	Duration
Snowflake	$20-100K	$100-500K	4-12 m
Google BigQuery	$15-80K	$80-400K	3-11 m
Amazon Redshift	$10-70K	$60-350K	3-10 m
Azure Synapse Analytics	$15-70K	$70-350K	3-10 m
Teradata Vantage	$30-150K	$200-800K	6-18 m
Oracle Autonomous Data Warehouse	$25-120K	$150-700K	5-16 m

Data Lakehouse Platforms represent a new architectural paradigm that combines the flexibility and cost-effectiveness of a data lake with the data management and performance capabilities of a data warehouse. They enable direct querying of data in its raw format while providing traditional data warehouse features such as ACID transactions, schema enforcement, data governance, and strong performance for BI and SQL workloads. By unifying data warehousing and data lake functionalities into a single platform, Data Lakehouses simplify data architectures, reduce data duplication, accelerate data access for diverse analytical use cases including AI/ML, and offer greater agility and efficiency for organizations seeking to maximize value from all their data.

Without leveraging Data Lakehouse Platforms, organizations often face architectural complexity and data fragmentation by operating separate data lakes for raw, unstructured data and data warehouses for structured, curated data. This dual architecture leads to data duplication, increased storage and processing costs, slower data movement between systems, and significant challenges in maintaining data consistency and governance across different environments. The absence of a unified data lakehouse results in delayed time-to-insight for analytics and AI, limited data agility, and higher operational overhead, ultimately hindering an organization's ability to fully leverage its diverse data assets for innovation and competitive advantage.

Implementation	Challenges
* Assess current data lake & DW architecture * Select a data lakehouse platform * Migrate data from existing systems * Design unified data models & governance * Implement data pipelines (streaming/batch) * Integrate with BI & ML tools	* Complex migration from existing data infrastructure * Ensuring data consistency & transactional integrity * Skillset requirements for new architecture * Managing data governance across raw & curated data * Performance optimization for diverse workloads * Cost management for unified platform

Products	Platform Cost	Project Cost	Duration
Databricks Lakehouse Platform	$20-100K	$150-700K	6-18 m
Apache Iceberg / Delta Lake / Hudi (Open Source)	N/A	$50-300K	3-9 m
Google Cloud Dataproc (for Open Source)	$10-60K	$60-300K	3-9 m
Amazon EMR (for Open Source)	$10-50K	$50-250K	3-9 m
Azure Synapse Analytics (Lakehouse capabilities)	$15-70K	$80-400K	4-12 m
Dremio (Lakehouse Query Engine)	$10-40K	$50-200K	3-8 m

4. EDGE DATA PLATFORMS

Edge Data Platforms are distributed computing and storage solutions designed to process and analyze data generated at or near the source (the "edge" of the network), rather than sending it all to a centralized cloud or data center. These platforms enable real-time data ingestion, local processing, and immediate insights from IoT devices, sensors, and remote locations. By minimizing latency, reducing bandwidth consumption, and enhancing data privacy, Edge Data Platforms support critical applications like autonomous vehicles, smart factories, and remote asset monitoring. This architecture empowers organizations to make faster, more localized decisions, improve operational efficiency, and drive innovation in environments where connectivity or real-time responsiveness is paramount.

Without Edge Data Platforms, organizations dealing with vast amounts of data generated at remote or geographically dispersed locations (e.g., IoT devices, manufacturing plants, retail stores) are forced to transmit all raw data to a centralized cloud or data center for processing. This creates significant challenges related to network latency, bandwidth costs, data privacy, and the inability to make real-time, localized decisions. The absence of edge processing capabilities results in slower responses to critical events, increased operational expenses due to data transfer, and missed opportunities for immediate insights, ultimately hindering the efficiency, responsiveness, and innovation potential of distributed operations.

Implementation	Challenges
* Identify edge data sources & processing needs * Select edge computing hardware & software * Design distributed data architecture & synchronization * Implement robust security for edge devices * Develop edge analytics models & applications * Plan remote management & monitoring	* Managing distributed infrastructure & devices * Ensuring security & data privacy at the edge * Limited compute/storage capacity at edge locations * Network connectivity & reliability issues * Deploying & updating software at scale * Integrating edge data with central platforms

Products	Platform Cost	Project Cost	Duration
AWS IoT Greengrass	$5-30K	$20-100K	2-6 m
Azure IoT Edge	$5-25K	$20-90K	2-5 m
Google Cloud IoT Core (Managed Service)	$5-20K	$20-80K	2-5 m
Balena.io	$1-10K	$5-30K	1-2 m
IBM Edge Application Manager	$10-40K	$50-200K	3-8 m
Cisco Kinetic for Cities/IoT	$10-50K	$60-250K	3-9 m

In-Memory Data Grids (IMDGs) are distributed systems that store vast amounts of data in RAM across a cluster of servers, providing extremely high-speed data access and processing capabilities for demanding, low-latency applications. Unlike traditional databases that rely on disk storage, IMDGs eliminate disk I/O bottlenecks, making them ideal for real-time analytics, transactional processing (IMDB), caching, and complex event processing. By enabling near-instantaneous data retrieval and computation, IMDGs significantly accelerate application performance, improve responsiveness for critical operations, and support high-throughput workloads, empowering organizations to handle massive data streams and deliver real-time insights for competitive advantage.

Without In-Memory Data Grids, organizations often struggle to achieve the ultra-low latency and high-throughput performance required for real-time applications, interactive analytics, or high-volume transactional workloads. Relying solely on disk-based databases creates performance bottlenecks due to slower data retrieval, leading to delayed responses, frustrated users, and missed opportunities for immediate action. The absence of in-memory processing capabilities results in a reactive rather than proactive operational posture, limits the ability to handle peak loads, and ultimately hinders the development and deployment of cutting-edge, real-time data-driven applications crucial for competitive advantage in areas like financial trading, fraud detection, and personalized customer experiences.

Implementation	Challenges
* Identify low-latency data access needs * Select appropriate IMDG platform * Design data partitioning & replication strategy * Integrate with application logic & data sources * Implement robust data persistence & recovery * Optimize memory usage & cluster scaling	* High memory requirements & cost * Complex cluster management & operations * Ensuring data consistency & durability * Managing data serialization & deserialization * Integrating with existing persistence layers * Scaling for massive data volumes & concurrent access

Products	Platform Cost	Project Cost	Duration
Apache Ignite (Open Source)	N/A	$30-150K	2-7 m
Hazelcast IMDG	$10-50K	$60-250K	3-9 m
GridGain (Enterprise Ignite)	$15-70K	$80-350K	3-10 m
Redis Enterprise	$10-60K	$50-300K	3-9 m
SAP HANA (In-Memory Database)	$30-150K	$200-800K	6-18 m
Apache Geode (Pivotal GemFire)	$15-80K	$80-400K	4-11 m

Big Data Platforms are comprehensive technology stacks designed to handle the immense volume, velocity, and variety of data that traditional databases cannot process efficiently. These platforms typically include distributed storage systems (like HDFS), processing frameworks (like Spark or Hadoop MapReduce), and specialized tools for data ingestion, querying, and analysis. They enable organizations to store and process petabytes of structured, semi-structured, and unstructured data, derive insights from massive datasets, and support advanced analytics, machine learning, and data science initiatives at scale. By providing the foundational infrastructure for large-scale data processing, Big Data Platforms empower organizations to uncover hidden patterns, develop predictive models, and drive significant business value from previously inaccessible data.

Without robust Big Data Platforms, organizations faced with exponential data growth struggle to store, process, and analyze the massive volumes and diverse formats of data generated from modern sources like IoT, web logs, and social media. Relying on traditional data systems leads to severe performance bottlenecks, prohibitive storage costs, and an inability to extract meaningful insights from unstructured or semi-structured data. The absence of a scalable, distributed big data infrastructure results in missed opportunities for advanced analytics, hindered innovation in areas like AI and machine learning, and a diminished capacity to leverage all available data for competitive advantage and informed decision-making.

Implementation	Challenges
* Define data volume, velocity & variety needs * Select platform components (storage, compute) * Design distributed architecture & data pipelines * Implement data governance & security at scale * Build data processing & analytics applications * Plan cluster management & resource optimization	* High complexity in setup & management * Significant cost for infrastructure & talent * Ensuring data quality & governance at scale * Skillset gap for distributed systems & data engineering * Performance tuning for diverse workloads * Data security & compliance in distributed environments

Products	Platform Cost	Project Cost	Duration
Cloudera Data Platform (CDP)	$20-100K	$150-700K	6-18 m
Hortonworks Data Platform (HDP) (Cloudera)	N/A	N/A	N/A
Apache Hadoop / Spark (Open Source)	N/A	$50-300K	3-9 m
Amazon EMR (Managed Hadoop/Spark)	$10-60K	$50-300K	3-9 m
Google Cloud Dataproc	$10-50K	$40-250K	3-9 m
Azure HDInsight	$10-50K	$40-250K	3-9 m

CHAPTER D2

Ai6x6™ Transformation Blueprint

Data 'Intelligence'

"The world is one big data problem." — Andrew McAfee , MIT researcher and co-author of "The Second Machine Age," known for his thinking on digital economics and machine intelligence.

Data Intelligence isn't extracted—it's constructed. From stream engines to AutoML and recommendation platforms, transformation hinges on framing problems in data-native terms. "Thinking in data" means shifting from storytelling to signal tracing. Leaders who mature digitally don't look at dashboards—they architect behaviors from them. Data doesn't give answers—it gives patterns, probabilities, and provocations. Transformation requires asking smarter questions and building smarter defaults. When analytics become ambient and search becomes cognitive, decision making isn't guided—it's reimagined. In this chapter, the enterprise no longer sees data—it starts thinking in it.

The "Data Intelligence" chapter is paramount for data transformation's digital maturity as it focuses on extracting meaningful insights and automating decision-making from raw data. A digitally mature organization moves beyond simple reporting to leverage advanced analytics, AI, and stream processing engines to uncover hidden patterns, predict trends, and recommend actions in real-time. This intelligence capability transforms data from a mere record into a strategic asset, empowering the organization to make smarter, faster, and more impactful decisions across all business functions, driving competitive advantage and operational excellence.

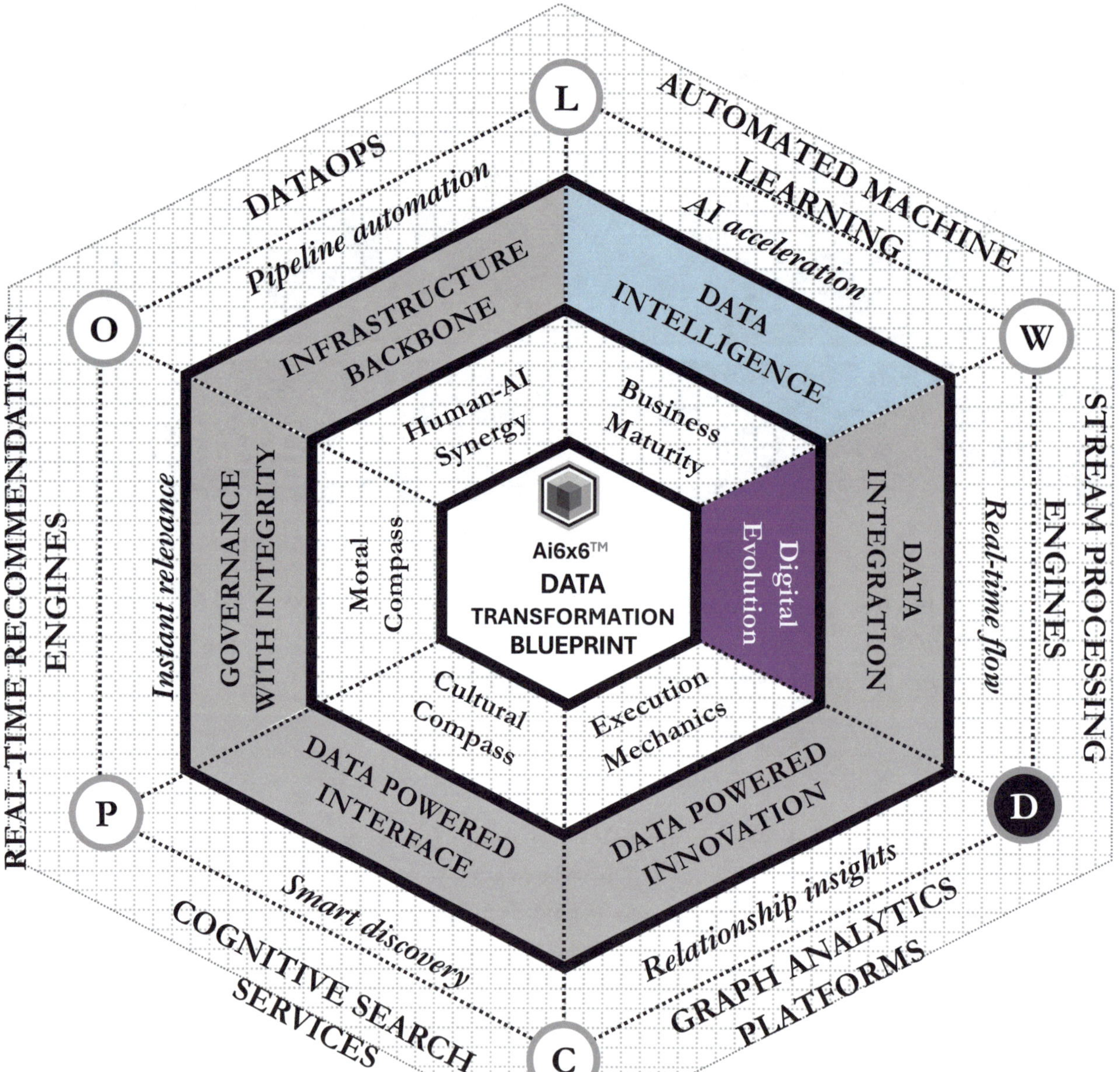

Ai6x6 DIGITAL EVOLUTION FOR **DATA LIFECYCLE | INTELLIGENCE**

DataOps is an automated, process-oriented methodology that integrates agile development, DevOps principles, and statistical process control to improve the quality, speed, and collaboration of data analytics. It focuses on automating the entire data lifecycle, from data ingestion and transformation to analysis and deployment of insights, treating data pipelines and analytics as a continuous delivery process. By fostering collaboration between data engineers, data scientists, and business analysts, DataOps reduces errors, accelerates time-to-insight, and ensures data quality and reliability. This enables organizations to deliver more accurate and timely data products, enhancing decision-making capabilities and maximizing the value derived from their data assets.

Without implementing DataOps, organizations often face fragmented data workflows, manual processes prone to errors, and significant delays in delivering reliable data insights to business stakeholders. This leads to inconsistent data quality, frustrated data teams struggling with repetitive tasks, and a slow pace of innovation in analytics and machine learning. The absence of an automated, collaborative, and quality-focused DataOps methodology results in untrustworthy data products, missed business opportunities due to delayed insights, and a higher cost of data management, ultimately hindering an organization's ability to leverage data effectively for competitive advantage and rapid decision-making.

Implementation	Challenges
* Assess current data lifecycle processes * Implement automation for data pipelines * Establish version control for data & code * Foster cross-functional collaboration * Integrate testing & monitoring into workflows * Adopt agile methodologies for data projects	* Cultural shift towards collaboration & automation * Integrating disparate tools across the data stack * Managing data governance & compliance in pipelines * Skillset gap for hybrid data/DevOps roles * Ensuring data quality and reliability at scale * Measuring clear ROI on DataOps initiatives

Products	Platform Cost	Project Cost	Duration
DataKitchen	$10-50K	$60-250K	3-9 m
Dataiku	$15-70K	$80-350K	3-10 m
Talend (Data Fabric)	$15-60K	$70-300K	3-9 m
Keboola	$5-30K	$20-100K	2-6 m
Apache Airflow (Open Source Orchestration)	N/A	$30-150K	2-7 m
Prefect / Dagster (Open Source)	N/A	$20-100K	2-6 m

Automated Machine Learning (AutoML) refers to the process of automating the end-to-end application of machine learning, from raw dataset to deployable ML model, with minimal human intervention. AutoML platforms streamline tasks such as data preprocessing, feature engineering, algorithm selection, hyperparameter tuning, and model deployment. By democratizing access to powerful ML capabilities, AutoML enables a wider range of users, including business analysts and domain experts, to build and deploy high-performing machine learning models without extensive data science expertise. This accelerates the development and deployment of AI solutions, reduces the time-to-value for machine learning projects, and empowers organizations to derive more predictive insights faster.

Without Automated Machine Learning (AutoML), organizations typically rely on highly specialized and scarce data scientists to manually perform every step of the machine learning lifecycle, from data preparation to model deployment. This leads to significant delays in developing and deploying ML solutions, higher costs due to resource intensity, and a bottleneck in leveraging predictive analytics across various business functions. The absence of AutoML results in slower experimentation, reduced ability to scale ML initiatives, and missed opportunities to infuse intelligence into operations, ultimately hindering an organization's capacity to derive timely and impactful insights from its data and maintain a competitive edge in AI adoption.

Implementation	Challenges
* Define business problem for ML application * Prepare & clean initial dataset * Select appropriate AutoML platform * Train & evaluate automated models * Interpret model results & explainability * Deploy & monitor models in production	* Ensuring data quality for automated processes * Understanding model explainability and bias * Trusting automated model selection * Integration with existing data infrastructure * Managing model drift & re-training * Cost optimization for compute resources

Products	Platform Cost	Project Cost	Duration
Google Cloud AutoML	$5-30K	$20-100K	2-6 m
Azure Machine Learning (AutoML)	$5-25K	$20-90K	2-5 m
Amazon SageMaker Autopilot	$5-20K	$20-80K	2-5 m
DataRobot	$20-90K	$100-450K	4-12 m
H2O.ai Driverless AI	$15-70K	$80-350K	3-10 m
Dataiku (AutoML features)	$15-70K	$80-350K	3-10 m

Stream Processing Engines are software systems designed to continuously ingest, process, and analyze data in real-time as it is generated, rather than storing it first and processing in batches. These engines handle high-velocity data streams from sources like IoT sensors, financial transactions, web clicks, and social media feeds, performing transformations, aggregations, and complex event detection with minimal latency. By enabling immediate insights and rapid responses to incoming data, Stream Processing Engines support critical applications such as fraud detection, personalized recommendations, real-time dashboards, and predictive maintenance. This capability empowers organizations to react instantly to dynamic events, optimize operations, and gain a significant competitive advantage in time-sensitive scenarios.

Without Stream Processing Engines, organizations are limited to batch processing of data, leading to significant delays in extracting insights from rapidly generated information and a reactive rather than proactive approach to critical events. This means they cannot respond instantly to anomalies, personalize experiences in real-time, or optimize operations based on immediate data flows. The absence of real-time stream processing capabilities results in missed opportunities for immediate fraud detection, slower responses to emerging market trends, and an inability to deliver up-to-the-minute dashboards, ultimately hindering an organization's agility and competitiveness in data-intensive, time-sensitive environments.

Implementation	Challenges
* Identify real-time data sources & needs * Select a suitable stream processing engine * Design robust, fault-tolerant data pipelines * Develop real-time analytics & applications * Implement monitoring & alerting for streams * Ensure data consistency & exactly-once processing	* Complex distributed system management * Ensuring fault tolerance and data durability * Handling high data volumes and velocity * Maintaining low latency & real-time guarantees * Skillset gap for stream processing expertise * Integrating with diverse upstream and downstream systems

Products	Platform Cost	Project Cost	Duration
Apache Kafka Streams (Open Source)	N/A	$30-150K	2-7 m
Apache Flink (Open Source)	N/A	$40-200K	2-8 m
Google Cloud Dataflow	$5-30K	$20-100K	2-6 m
Amazon Kinesis	$5-25K	$20-90K	2-5 m
Azure Stream Analytics	$5-20K	$20-80K	2-5 m
Confluent Platform (Kafka Enterprise)	$10-60K	$50-300K	3-9 m

4. GRAPH ANALYTICS PLATFORMS

Graph Analytics Platforms are specialized tools and databases designed to store, manage, and analyze highly interconnected data, where relationships between entities are as important as the entities themselves. Unlike traditional relational databases, graph databases explicitly model relationships as "edges" connecting "nodes," enabling efficient querying and analysis of complex networks, such as social connections, supply chains, fraud rings, or IT infrastructure dependencies. By revealing hidden patterns, influential nodes, and critical pathways, Graph Analytics Platforms support use cases like fraud detection, recommendation engines, network security, and customer 360-degree views, empowering organizations to uncover deeper insights from relational data that are difficult or impossible with conventional methods.

Without Graph Analytics Platforms, organizations struggle to effectively analyze highly interconnected data, where the relationships between data points are critical to understanding the full context and deriving meaningful insights. Relying on traditional relational databases to model complex relationships often leads to cumbersome queries, performance bottlenecks, and an inability to uncover hidden patterns or subtle connections within networks (e.g., fraud rings, social influence). The absence of specialized graph capabilities results in missed opportunities for advanced analysis in areas like cybersecurity, supply chain optimization, and personalized recommendations, ultimately hindering an organization's ability to gain a holistic and intelligent understanding from its relational data.

Implementation	Challenges
* Identify use cases with rich relationship data * Select appropriate graph database/platform * Design graph data model (nodes, edges, properties) * Ingest & integrate data into graph format * Develop graph algorithms & query logic * Integrate with existing BI & analytics tools	* Complex data modeling for relationships * Ingesting & transforming relational data to graph * Skillset gap for graph query languages (e.g., Cypher) * Performance optimization for large graphs * Integrating with existing enterprise data stack * Visualizing and interpreting complex graph results

Products	Platform Cost	Project Cost	Duration
Neo4j Enterprise	$15-70K	$80-350K	3-10 m
Amazon Neptune	$10-50K	$60-250K	3-9 m
TigerGraph	$20-90K	$100-450K	4-12 m
ArangoDB	$5-30K	$20-100K	2-6 m
Microsoft Azure Cosmos DB (Graph API)	$5-25K	$20-90K	2-5 m
Google Cloud Firestore (Graph-like queries)	$2-15K	$10-70K	1-4 m

Cognitive Search Services integrate AI capabilities, such as natural language processing (NLP), machine learning, and computer vision, into enterprise search engines to provide more intelligent, contextual, and relevant search results from various structured and unstructured data sources. Unlike traditional keyword-based search, cognitive search can understand intent, extract entities, identify relationships, and analyze content within documents, images, and audio, providing insights beyond simple keyword matches. By transforming fragmented enterprise data into an easily discoverable and actionable knowledge base, these services empower employees to find information faster, improve decision-making, enhance productivity, and unlock new insights from vast amounts of corporate data, democratizing access to critical knowledge.

Without Cognitive Search Services, organizations often struggle with employees spending excessive time searching for information across disparate systems, relying on outdated or keyword-based search that frequently delivers irrelevant results. This leads to frustrated users, redundant efforts, and a significant loss of productivity, as valuable insights remain buried within unstructured data. The absence of intelligent, context-aware search capabilities hinders effective knowledge sharing, slows down decision-making, and prevents organizations from fully leveraging their vast repositories of documents, images, and other unstructured data, ultimately impacting efficiency and innovation across the enterprise.

Implementation	Challenges
* Identify diverse data sources for indexing * Select a cognitive search platform * Configure AI models (NLP, CV) for content enrichment * Implement security & access controls * Design intuitive search interface & relevance tuning * Plan continuous content ingestion & model updates	* Integrating & indexing diverse unstructured data * Ensuring accuracy & relevance of search results * Managing data privacy & access control * Fine-tuning AI models for specific domains * Overcoming user adoption & trust challenges * Cost optimization for AI processing & storage

Products	Platform Cost	Project Cost	Duration
Azure Cognitive Search	$5-30K	$20-100K	2-6 m
Google Cloud Search	$5-25K	$20-90K	2-5 m
Amazon Kendra	$5-20K	$20-80K	2-5 m
Elastic Stack (Elasticsearch with NLP)	$2-15K	$10-70K	1-4 m
Sinequa	$15-70K	$80-350K	3-10 m
Coveo	$10-60K	$50-300K	3-9 m

Real-Time Recommendation Engines leverage machine learning and artificial intelligence to analyze user behavior, preferences, and contextual data in real-time, providing personalized suggestions for products, content, services, or actions at the moment of interaction. Unlike batch-processed recommendations, these engines continuously update their models and deliver dynamic, highly relevant suggestions with minimal latency, adapting instantly to changing user needs or market conditions. By enhancing user experience, increasing conversion rates, driving engagement, and optimizing discovery, Real-Time Recommendation Engines are crucial for e-commerce, media, content platforms, and internal enterprise applications, directly impacting revenue, customer satisfaction, and operational efficiency.

Without Real-Time Recommendation Engines, organizations often provide static or delayed suggestions to users, based on historical data that may not reflect current preferences or real-time context. This leads to irrelevant recommendations, lower user engagement, and missed opportunities for immediate upselling, cross-selling, or content discovery. The absence of dynamic, real-time personalization results in a less engaging user experience, reduced conversion rates, and a slower response to changing market dynamics, ultimately hindering an organization's ability to maximize revenue, improve customer satisfaction, and maintain a competitive edge in today's on-demand digital economy.

Implementation	Challenges
* Define recommendation goals & data sources * Select ML framework & real-time serving layer * Develop & train recommendation algorithms * Implement real-time data pipelines & feature stores * Design A/B testing & feedback loops * Deploy & monitor model performance in production	* Collecting & processing real-time user behavior data * Ensuring low latency for predictions at scale * Managing model complexity & cold start problem * Integrating with diverse data sources & applications * Addressing algorithmic bias & fairness * Measuring clear ROI on personalized recommendations

Products	Platform Cost	Project Cost	Duration
Amazon Personalize	$5-30K	$20-100K	2-6 m
Google Cloud Recommendations AI	$5-25K	$20-90K	2-5 m
Microsoft Azure Personalizer	$5-20K	$20-80K	2-5 m
Recombee	$1-10K	$5-30K	1-2 m
Braze / Amplitude (for engagement focus)	$10-50K	$60-250K	3-9 m
Custom ML Development (TensorFlow, PyTorch)	N/A	$100-500K	4-12 m

Ai6x6™ Transformation Blueprint

Data 'Integrations'

"Data really powers everything that we do." — Jeff Weiner

Former CEO of LinkedIn; made strategic use of data flow between talent, skills, and networks.

Disconnected data means disconnected decisions. Reverse ETL, unified metadata, and change data capture tools convert fractured platforms into continuous, living ecosystems. Mature transformation isn't about syncing rows—it's about synchronizing context. When every system speaks the same metadata, governance and personalization scale together. "Data integrations" implies that transformation is no longer vertical—it becomes relational. Enterprises don't just track data—they nurture ecosystems that link behavior, transaction, and identity. Intelligence flourishes when every connection is a conversation.

"Connected Contexts" is vital for data transformation's digital maturity because it addresses the critical challenge of unifying disparate data sources across an organization. A digitally mature enterprise ensures seamless data flow and interoperability between systems through ELT tools, reverse ETL platforms, and message queue systems. This breaks down data silos and creates a comprehensive, holistic view of information. This robust integration capability is essential for generating accurate insights, supporting cross-functional collaboration, and enabling the agility required to respond effectively to dynamic business needs and leverage data for strategic advantage.

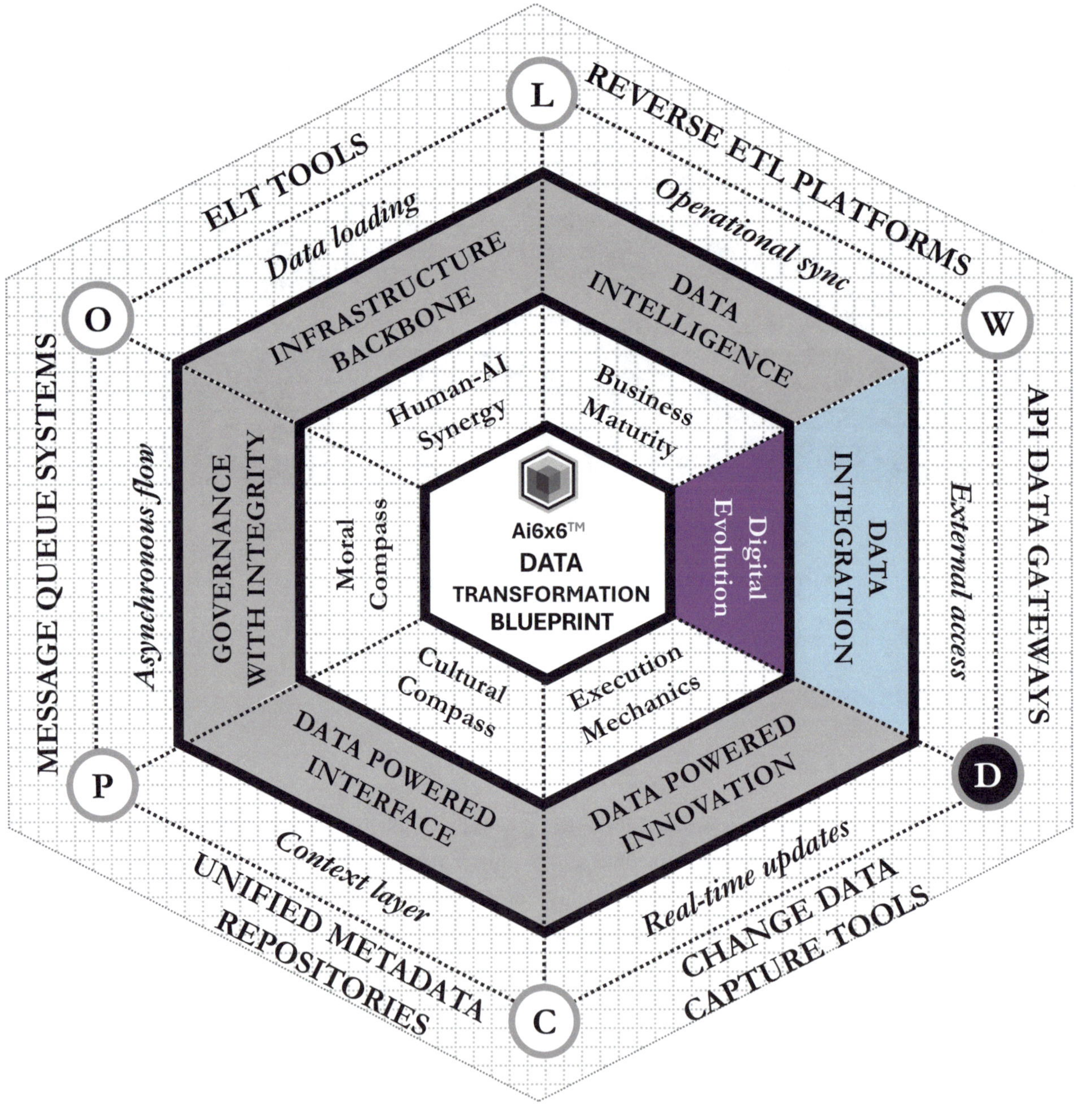

Ai6x6 DIGITAL EVOLUTION FOR **DATA LIFECYCLE | INTEGRATION**

ELT (Extract, Load, Transform) Tools are data integration solutions that first extract raw data from various source systems, then load it directly into a centralized data repository like a data lake or cloud data warehouse, and finally transform it within that powerful target system. Unlike traditional ETL (Extract, Transform, Load), ELT leverages the scalable compute and storage capabilities of modern data platforms to perform transformations, making it highly efficient for large volumes of diverse data, especially in cloud environments. By enabling data teams to load raw data quickly and transform it flexibly as needed, ELT tools accelerate data availability for analysis, support agile data modeling, and empower organizations to derive faster insights from their vast data assets.

Without ELT Tools, organizations often rely on traditional ETL processes that require data to be transformed before loading into a data warehouse, leading to slower ingestion of raw data, rigid schema definitions, and a bottleneck in adapting to new analytical requirements. This means valuable raw data is often discarded or aggregated prematurely, limiting future analytical possibilities and slowing down the data team's agility. The absence of ELT results in less flexible data architectures, increased complexity in managing transformation logic outside the target analytical platform, and slower time-to-insight, ultimately hindering an organization's ability to efficiently leverage its raw data for diverse and evolving business intelligence and machine learning needs.

Implementation	Challenges
* Identify data sources & target data warehouse/lake * Select appropriate ELT platform * Define data extraction methods & schedule * Design in-warehouse transformation logic (SQL) * Implement data quality & governance rules * Integrate with downstream BI & analytics tools	* Managing data quality and governance in raw layer * Skillset required for in-warehouse transformations * Cost optimization for cloud compute during transformation * Debugging complex SQL-based transformations * Ensuring data security for raw data in target * Handling schema evolution in source systems

Products	Platform Cost	Project Cost	Duration
Fivetran	$5-30K	$20-100K	2-6 m
Stitch Data (Talend)	$5-25K	$20-90K	2-5 m
Airbyte (Open Source)	N/A	$10-50K	1-3 m
Matillion	$10-50K	$60-250K	3-9 m
dbt (Data Build Tool)	N/A	$20-100K	2-6 m
Talend (Cloud Data Integration)	$15-60K	$70-300K	3-9 m

Reverse ETL Platforms are data integration tools designed to move cleaned, transformed, and enriched data from data warehouses or data lakes back into operational business applications, such as CRM, marketing automation, or customer support platforms. Unlike traditional ETL that moves data into the warehouse, Reverse ETL focuses on operationalizing data insights by ensuring that customer 360 views, product recommendations, or sales leads derived from analytics are immediately actionable within the tools used by sales, marketing, and support teams. By synchronizing analytical insights with operational systems, Reverse ETL empowers frontline employees with up-to-date, relevant customer intelligence, leading to more personalized customer experiences, improved operational efficiency, and better business outcomes.

Without Reverse ETL Platforms, organizations often face a significant gap between their analytical insights and operational execution, with valuable customer data and intelligence remaining trapped in data warehouses, inaccessible to frontline business users in their daily tools. This results in reactive customer engagements, missed opportunities for personalized marketing or sales, and a reliance on manual data exports or fragmented workarounds, which are prone to errors and delays. The absence of automated Reverse ETL hinders timely operationalization of data insights, reduces business agility, and ultimately prevents organizations from closing the loop between data analysis and direct customer interaction, impacting customer satisfaction and revenue growth.

Implementation	Challenges
* Identify target operational applications * Define data synchronization needs & frequency * Select a Reverse ETL platform * Map data fields between warehouse & apps * Implement robust error handling & monitoring * Plan for data volume & transformation logic	* Managing data privacy & compliance in operational apps * Ensuring data consistency across systems * Handling schema changes in source or destination * Performance and scalability for large data volumes * Overcoming API limits of destination applications * Measuring direct impact on business operations

Products	Platform Cost	Project Cost	Duration
Census	$5-30K	$20-100K	2-6 m
Hightouch	$5-25K	$20-90K	2-5 m
Grouparoo (Open Source)	N/A	$10-50K	1-3 m
RudderStack (Warehouse-Native CDP)	$10-40K	$50-200K	3-8 m
Iterable (Customer Engagement Platform)	$10-60K	$60-300K	3-9 m
Segment (CDP with Reverse ETL features)	$15-70K	$80-350K	3-10 m

API Data Gateways are central management layers that sit in front of backend data services and applications, acting as a single entry point for all API requests. They perform crucial functions such as routing requests, enforcing security policies (authentication, authorization), rate limiting, caching, and transforming data formats. By abstracting the complexity of backend data sources and providing a secure, governed interface, API Data Gateways enable consistent and controlled access to organizational data for internal and external applications, partners, and developers. This streamlines data sharing, improves data security, simplifies integration, and enhances the agility with which new data services can be exposed and managed, crucial for modern data ecosystems.

Without API Data Gateways, organizations typically expose their backend data services directly, leading to fragmented access control, inconsistent security policies, and significant challenges in managing and monitoring data consumption by various applications and users. This results in increased security vulnerabilities, slower development cycles for new data integrations, and a lack of centralized visibility into data access patterns and performance. The absence of an API gateway creates operational inefficiencies, complicates data governance, and limits the organization's ability to safely and scalably share its data assets, hindering digital transformation and partner collaborations.

Implementation	Challenges
* Inventory backend data services & APIs * Select an API gateway solution * Design API structure & authentication mechanisms * Implement security policies (rate limiting, WAF) * Integrate with identity providers & monitoring tools * Plan phased rollout & developer onboarding	* Managing complex API dependencies & versions * Ensuring high availability & scalability * Securing API endpoints from various threats * Handling data transformation & protocol translation * Performance optimization for high traffic * Skillset required for API management & security

Products	Platform Cost	Project Cost	Duration
Amazon API Gateway	$5-30K	$20-100K	2-6 m
Azure API Management	$5-25K	$20-90K	2-5 m
Google Cloud Apigee	$10-60K	$50-300K	3-9 m
Kong API Gateway (Open Source/Enterprise)	N/A / $10-50K	$30-200K	2-8 m
MuleSoft Anypoint Platform	$20-90K	$120-500K	4-12 m
Tyk API Gateway	$5-20K	$20-80K	2-5 m

Change Data Capture (CDC) Tools are technologies that identify and capture changes (inserts, updates, deletes) made to data in source databases in real-time or near real-time, and then replicate these changes to target systems like data warehouses, data lakes, or other applications. Unlike traditional batch processing, CDC ensures that only the changed data is moved, minimizing resource consumption and providing up-to-date information for analytical and operational purposes. By enabling continuous data synchronization with minimal latency, CDC tools support real-time analytics, data warehousing, database migrations, and operational data stores, ensuring data freshness and consistency across distributed systems, which is critical for timely decision-making and operational efficiency.

Without Change Data Capture (CDC) Tools, organizations typically rely on batch processing for data synchronization, which involves periodic full or incremental loads of data from source systems to analytical or operational targets. This leads to significant data latency, high resource consumption due to transferring large datasets, and a lack of real-time visibility into critical business events. The absence of CDC results in stale data for analytics, slower responses to operational changes, and complex, inefficient data pipelines that cannot support timely decision-making or real-time application needs, ultimately hindering an organization's agility and its ability to act on fresh information.

Implementation	Challenges
* Identify source databases & target systems * Select appropriate CDC tool/methodology * Configure database logging & change tracking * Design robust, fault-tolerant replication pipelines * Implement data transformation (if needed) on changes * Monitor data latency & consistency	* Managing high volume of real-time changes * Ensuring data consistency & exactly-once delivery * Handling schema evolution in source databases * Impact on source database performance * Complexity of setting up & maintaining pipelines * Disaster recovery & fault tolerance for CDC flows

Products	Platform Cost	Project Cost	Duration
Debezium (Open Source for Kafka)	N/A	$30-150K	2-7 m
Apache Flink (with CDC connectors)	N/A	$40-200K	2-8 m
Informatica PowerExchange CDC	$20-90K	$100-450K	4-12 m
Qlik Replicate (Attunity)	$15-70K	$80-350K	3-10 m
AWS Database Migration Service (DMS)	$5-30K	$20-100K	2-6 m
Oracle GoldenGate	$25-120K	$150-700K	5-16 m

5. UNIFIED METADATA REPOSITORIES

Unified Metadata Repositories are centralized systems that collect, store, and manage metadata—data about data—from across an organization's entire data landscape. This includes technical metadata (schemas, data types, lineage), business metadata (definitions, ownership, usage), and operational metadata (performance, quality metrics). By providing a single source of truth for all data assets, these repositories improve data discoverability, enhance data governance, facilitate data quality initiatives, and streamline collaboration among data professionals. They are crucial for building a comprehensive understanding of an organization's data, ensuring compliance, enabling self-service analytics, and maximizing the value derived from diverse data sources.

Without Unified Metadata Repositories, organizations struggle with "data swamps" and a lack of understanding about their data assets, where data location, lineage, definitions, and ownership are fragmented across various systems or reside only in individual tribal knowledge. This leads to inefficient data discovery, inconsistent data interpretations, duplicated efforts, and significant challenges in ensuring data quality and compliance. The absence of a centralized metadata repository results in untrustworthy analytics, increased operational risks due to poor data governance, and a slower pace of innovation, ultimately hindering an organization's ability to truly leverage its data as a strategic asset and democratize data access effectively.

Implementation	Challenges
* Inventory all data sources & types of metadata * Select a metadata management platform * Design metadata ingestion pipelines * Implement data cataloging & lineage tracking * Establish data governance rules & ownership * Integrate with data platforms & BI tools	* Extracting & integrating metadata from diverse systems * Ensuring metadata accuracy & freshness * Driving user adoption for data discovery & collaboration * Managing complex data lineage across transformations * Overcoming organizational silos & data ownership issues * Measuring clear ROI on improved data discoverability

Products	Platform Cost	Project Cost	Duration
Collibra Data Governance	$20-90K	$100-450K	4-12 m
Alation Data Catalog	$15-70K	$80-350K	3-10 m
Informatica Enterprise Data Catalog	$20-100K	$120-500K	4-13 m
Apache Atlas (Open Source)	N/A	$50-250K	3-9 m
Azure Purview	$10-50K	$60-250K	3-9 m
Dataiku (Data Catalog & Governance)	$15-70K	$80-350K	3-10 m

Message Queue Systems (or Message Brokers) are middleware technologies that enable asynchronous communication between different software applications and services. They facilitate reliable, scalable, and loosely coupled data exchange by temporarily storing messages until the receiving application is ready to process them, acting as buffers between producers and consumers of data. By decoupling applications, Message Queues prevent system overload, handle peak traffic spikes gracefully, and ensure message delivery even if consuming applications are temporarily unavailable. This technology is fundamental for building resilient, distributed, and scalable data architectures, supporting real-time data pipelines, microservices communication, and event-driven architectures crucial for modern data transformation.

Without Message Queue Systems, organizations often face challenges in building robust and scalable integrations between their distributed applications and data services, leading to tightly coupled systems that are prone to failures and performance bottlenecks. This means that if one application goes down, it can directly impact others, and peak loads can quickly overwhelm systems, resulting in data loss or service unavailability. The absence of asynchronous messaging and buffering provided by message queues limits system resilience, complicates scaling, and hinders the development of agile, event-driven architectures, ultimately impacting an organization's ability to process real-time data efficiently and build fault-tolerant data ecosystems.

Implementation	Challenges
* Identify asynchronous communication needs * Select appropriate message queue system * Design message formats & topics/queues * Implement message producers & consumers * Establish monitoring & alerting for queues * Plan for high availability & disaster recovery	* Ensuring message delivery & order guarantees * Managing high message volumes and throughput * Monitoring queue depth & consumer lag * Handling dead letter queues & message retries * Securing message transmission & access * Skillset required for distributed messaging systems

Products	Platform Cost	Project Cost	Duration
Apache Kafka (Open Source)	N/A	$30-150K	2-7 m
RabbitMQ (Open Source)	N/A	$20-100K	2-6 m
Apache ActiveMQ (Open Source)	N/A	$20-90K	2-5 m
Amazon SQS / SNS	$5-20K	$10-70K	1-4 m
Azure Service Bus / Event Hubs	$5-25K	$20-90K	2-5 m
Google Cloud Pub/Sub	$5-20K	$20-80K	2-5 m

Ai6x6™ Transformation Blueprint

Data powered 'Innovation'

"The biggest breakthroughs come when data is used to simulate what doesn't yet exist." — Fei-Fei Li , Stanford professor and AI visionary; co-director of Stanford's Human-Centered AI Institute.

Synthetic data, digital twins, and sandbox environments allow innovation to occur before reality permits it. This chapter isn't about exploration—it's about rehearsal. "Data powered Innovation" means making experimentation scalable, safe, and systematized. Mature transformation is built on testbeds where insight becomes simulated and scenarios become rehearsed. Innovation becomes low-risk, high-fidelity. Enterprises move from insight discovery to insight design. When data is used to prototype futures rather than diagnose pasts, maturity becomes predictive, not reactive.

The "Data powered innovation" chapter is crucial for data transformation's digital maturity as it emphasizes the proactive use of data to create novel solutions and explore future possibilities. A digitally mature organization leverages advanced techniques like augmented analytics, synthetic data generation, and digital twin data models to simulate scenarios, test hypotheses, and develop groundbreaking products and services. This innovative approach to data utilization, often within data sandbox platforms and experimentation labs, enables the organization to not only react to market changes but actively shape new realities and maintain a significant competitive edge.

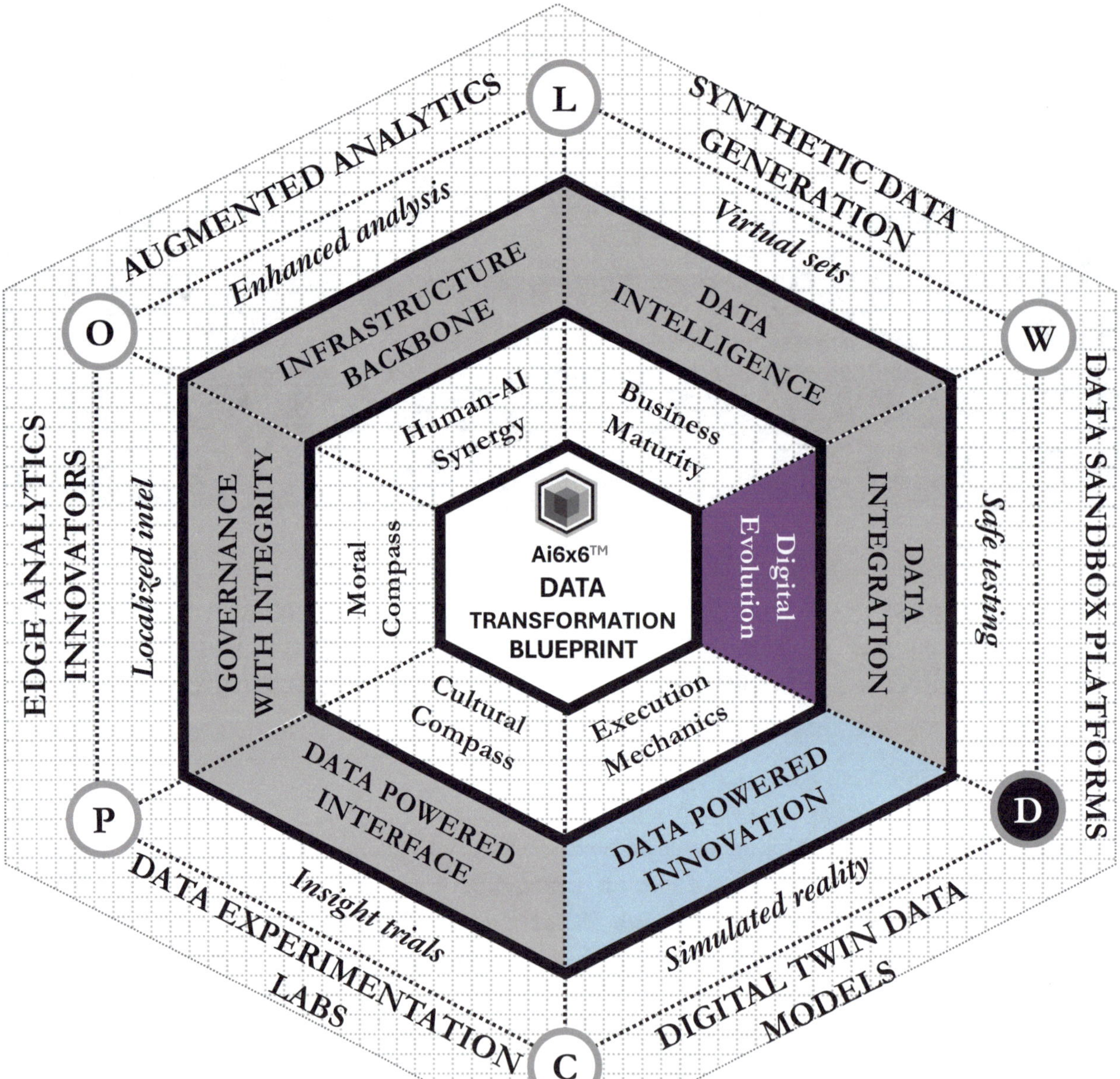

Ai6x6 DIGITAL EVOLUTION FOR **DATA LIFECYCLE | INNOVATION**

1. AUGMENTED ANALYTICS

Augmented Analytics leverages artificial intelligence (AI) and machine learning (ML) to automate data preparation, insight discovery, and insight explanation within business intelligence (BI) and analytics platforms. Instead of manual data exploration, augmented analytics proactively identifies patterns, trends, and anomalies in data, suggesting relevant correlations and even generating narratives in natural language. By democratizing advanced analytical capabilities, it empowers business users of all skill levels to derive deeper, more actionable insights faster, reducing the need for extensive data science expertise. This accelerates decision-making, improves data literacy across the organization, and maximizes the value extracted from complex datasets, fostering a more data-driven culture.

Without Augmented Analytics, organizations often face significant bottlenecks in deriving insights from their growing volumes of data, relying heavily on manual data preparation and expert data analysts to uncover valuable patterns. This leads to slower time-to-insight, missed opportunities hidden within vast datasets, and a limited ability for business users to independently explore and understand complex data relationships. The absence of AI-powered automation in analytics results in inefficient data consumption, lower data literacy across the enterprise, and a reactive rather than proactive approach to identifying critical business trends, ultimately hindering an organization's agility and its capacity to consistently make data-informed decisions.

Implementation	Challenges
* Define specific business questions for augmentation * Select a BI platform with strong augmented features * Ensure clean, well-structured data for AI * Train users on AI-driven insights & capabilities * Integrate with existing data sources & dashboards * Continuously monitor model performance & relevance	* Ensuring data quality for accurate AI insights * Trust and adoption from business users * Explaining complex AI-generated insights * Managing algorithmic bias in recommendations * Integration with diverse existing data ecosystems * Cost optimization for AI compute resources

Products	Platform Cost	Project Cost	Duration
Tableau (Ask Data, Explain Data)	$10-50K	$50-250K	3-9 m
Microsoft Power BI (Q&A, Smart Narratives)	$5-30K	$20-100K	2-6 m
Qlik Sense (Cognitive Engine)	$10-40K	$50-200K	3-8 m
ThoughtSpot (Search & AI-driven Analytics)	$15-70K	$80-350K	3-10 m
Sisense (Fusion Analytics)	$15-60K	$70-300K	3-9 m
Yellowfin BI (Assisted Insights)	$5-25K	$20-90K	2-5 m

Synthetic Data Generation is the process of algorithmically creating artificial datasets that mimic the statistical properties and patterns of real-world data without containing any original, sensitive information. This technology leverages machine learning models (e.g., Generative Adversarial Networks - GANs) to produce data that maintains the analytical utility of genuine data but guarantees privacy and confidentiality. By providing realistic, privacy-preserving datasets, Synthetic Data Generation enables safe development and testing of new AI models, software applications, and analytics without exposing sensitive customer or proprietary information. This accelerates innovation, facilitates data sharing for collaboration, and ensures compliance with strict data privacy regulations, particularly crucial in industries like healthcare and finance.

Without Synthetic Data Generation, organizations often face significant hurdles in developing and testing new data-driven applications, AI models, or analytics, due to strict privacy regulations, data security concerns, and the difficulty of accessing or sharing sensitive real-world data. This results in slower development cycles, limited experimentation, and the inability to collaborate effectively on data initiatives across teams or with external partners without risking privacy breaches. The absence of synthetic data forces organizations to either compromise on innovation, incur high costs for data anonymization, or operate in a state of constant privacy risk, ultimately hindering agile data development and compliant data utilization.

Implementation	Challenges
* Identify sensitive data for synthesis * Select a synthetic data platform/framework * Train generative models on real data * Evaluate synthetic data quality & utility * Implement privacy assurances & compliance * Integrate into dev/test pipelines	* Ensuring statistical fidelity to real data * Guaranteeing privacy & non-reversibility * Managing complexity of generative AI models * Scaling for very large datasets * Validating utility for specific use cases * Legal and ethical considerations of synthetic data

Products	Platform Cost	Project Cost	Duration
Gretel.ai	$5-30K	$20-100K	2-6 m
Mostly AI	$10-50K	$60-250K	3-9 m
Synthesized.io	$10-40K	$50-200K	3-8 m
Hazy.ai	$15-60K	$70-300K	3-9 m
Tonic.ai (Data Anonymization/Subset)	$5-25K	$20-90K	2-5 m
Python Libraries (e.g., Faker, CTGAN)	N/A	$10-70K	1-4 m

Data Sandbox Platforms are secure, isolated, and self-service environments that provide data professionals (data scientists, analysts, developers) with temporary access to datasets and computational resources for experimentation, exploration, and model development without impacting production systems. These platforms enable users to ingest, transform, analyze, and build proofs-of-concept with various tools and technologies, fostering rapid iteration and innovation. By providing a safe space for experimentation, Data Sandbox Platforms accelerate the discovery of new insights, reduce time-to-market for data products, improve data literacy, and minimize the risk of disrupting critical operations, empowering agile data innovation and a culture of continuous learning and development.

Without Data Sandbox Platforms, data professionals often face significant bottlenecks in their ability to experiment and innovate with data, as they must request access to production data, risk impacting live systems, or work with outdated data copies. This leads to slower development cycles for new models and analytics, limited opportunities for exploratory analysis, and increased risk of operational disruption. The absence of isolated, self-service data sandboxes results in a less agile data team, a slower pace of innovation, and ultimately hinders an organization's capacity to quickly test new hypotheses, develop groundbreaking data products, and stay competitive in a rapidly evolving data landscape.

Implementation	Challenges
* Define data access & security policies	* Ensuring strict isolation and data security
* Select a cloud-based or on-premise platform	* Managing compute & storage resource allocation
* Implement automated provisioning & teardown	* Integrating with diverse internal data sources
* Integrate with data catalogs & governance tools	* Providing up-to-date, relevant data copies
* Provide diverse tools & libraries for users	* Skillset required for platform administration
* Plan resource allocation & cost management	* Measuring ROI on innovation acceleration

Products	Platform Cost	Project Cost	Duration
Dataiku (Lab/Experimentation features)	$15-70K	$80-350K	3-10 m
Databricks (Workspaces)	$10-60K	$50-300K	3-9 m
Anaconda Enterprise	$10-50K	$60-250K	3-9 m
AWS SageMaker Studio / EMR Notebooks	$5-30K	$20-100K	2-6 m
Azure Machine Learning Workspaces	$5-25K	$20-90K	2-5 m
Google Cloud Vertex AI Workbench	$5-20K	$20-80K	2-5 m

Digital Twin Data Models are virtual representations of physical objects, processes, or systems, continuously updated with real-time data from their physical counterparts via sensors. These models act as living, dynamic simulations that provide comprehensive insights into performance, behavior, and potential issues, enabling proactive monitoring, predictive maintenance, and optimized operations. By creating a bridge between the physical and digital worlds, Digital Twin Data Models allow for complex simulations, scenario planning, and remote control, accelerating innovation in product design, manufacturing, smart cities, and healthcare. This technology enables organizations to make data-driven decisions that reduce downtime, improve efficiency, and develop smarter, more responsive systems in a dynamic environment.

Without Digital Twin Data Models, organizations operating complex physical assets or systems (e.g., machinery, infrastructure, production lines) lack a holistic, real-time understanding of their performance and health. This leads to reactive maintenance strategies, increased downtime, inefficient operations, and an inability to proactively identify potential failures or optimize performance through simulation. The absence of a dynamic digital replica results in higher operational costs, limited capacity for remote monitoring and control, and a slower pace of innovation in product design and process optimization, ultimately hindering an organization's ability to maximize asset utilization and maintain a competitive edge in physical-digital convergence.

Implementation	Challenges
* Identify physical assets for digital twinning * Implement IoT sensors for real-time data capture * Select a digital twin platform/framework * Develop accurate data models & simulations * Integrate with operational & analytics systems * Plan for continuous data ingestion & model updates	* High cost & complexity of sensor deployment * Ensuring data quality & accuracy from physical assets * Building accurate, real-time simulation models * Managing vast volumes of streaming IoT data * Integrating with diverse legacy operational systems * Skillset required for multidisciplinary modeling

Products	Platform Cost	Project Cost	Duration
Siemens Digital Twin (MindSphere)	$20-90K	$100-500K	4-12 m
Microsoft Azure Digital Twins	$10-60K	$50-300K	3-9 m
AWS IoT TwinMaker	$10-50K	$40-250K	3-9 m
GE Digital (Predix)	$20-80K	$100-400K	4-12 m
PTC ThingWorx	$15-70K	$80-350K	3-10 m
Dassault Systèmes (3DEXPERIENCE)	$25-100K	$150-600K	5-15 m

5. DATA EXPERIMENTATION LABS

Data Experimentation Labs are structured, agile environments within an organization dedicated to rapidly prototyping, testing, and validating new data-driven hypotheses, models, and analytical techniques. These labs provide data scientists and analysts with specialized tools, datasets, and collaborative frameworks to explore uncharted data territories, run A/B tests, and build experimental data products without the constraints of production environments. By fostering a culture of continuous learning and iterative innovation, Data Experimentation Labs accelerate the discovery of actionable insights, reduce the risk associated with deploying unproven models, and enable faster time-to-market for groundbreaking data solutions, driving competitive advantage through agile data-driven decision-making.

Without dedicated Data Experimentation Labs, organizations often struggle to rapidly test new data hypotheses, develop innovative analytical models, or validate the efficacy of data products without disrupting production systems or encountering significant bureaucratic hurdles. This leads to slower innovation cycles, increased risk in deploying unproven models directly into production, and a reactive approach to exploring new data opportunities. The absence of a structured, agile experimentation environment hinders the ability to quickly pivot on insights, limits the exploration of novel data techniques, and ultimately slows down the pace at which an organization can transform raw data into valuable, verified business intelligence and advanced AI capabilities.

Implementation	Challenges
* Define scope & governance for experiments * Provision isolated compute & storage resources * Provide access to relevant datasets & tools * Establish version control & experiment tracking * Foster collaboration & knowledge sharing * Design clear paths for successful experiments to production	* Ensuring data security & isolation for experiments * Managing compute and storage resources efficiently * Providing relevant, up-to-date datasets * Skillset required for managing diverse tools * Transitioning successful experiments to production * Measuring clear ROI on exploratory innovation

Products	Platform Cost	Project Cost	Duration
Dataiku (DSS for Experimentation)	$15-70K	$80-350K	3-10 m
Databricks (MLflow for Tracking)	$10-60K	$50-300K	3-9 m
Google Cloud Vertex AI (Experimentation)	$5-30K	$20-100K	2-6 m
Azure Machine Learning Workspaces	$5-25K	$20-90K	2-5 m
Comet ML (Experiment Tracking)	$2-15K	$10-70K	1-4 m
Domino Data Lab	$20-80K	$100-400K	4-12 m

Edge Analytics Innovators refer to the emerging technologies and specialized solutions that enable advanced analytical processing and machine learning inference to occur directly at the "edge" of the network, close to where data is generated (e.g., IoT devices, sensors, local servers), rather than sending all data to a centralized cloud. These innovators focus on optimizing models for resource-constrained environments, ensuring low-latency insights, reducing bandwidth costs, and enhancing data privacy. By empowering real-time decision-making, predictive capabilities, and autonomous operations directly at the source, Edge Analytics Innovators unlock new levels of efficiency, responsiveness, and localized intelligence for industries such as manufacturing, healthcare, smart cities, and retail, driving transformative operational changes.

Without leveraging Edge Analytics Innovators, organizations are limited to sending all raw data from their distributed devices and sensors to centralized cloud platforms for processing and analysis. This creates significant bottlenecks due to network latency, high data transfer costs, and an inability to make immediate, localized decisions critical for autonomous systems or time-sensitive operations. The absence of on-device or near-device analytical capabilities results in slower responses to critical events, missed opportunities for immediate optimization, and increased operational expenses, ultimately hindering an organization's ability to harness the full potential of its IoT data and achieve true real-time operational intelligence.

Implementation	Challenges
* Identify specific edge analytics use cases * Select appropriate edge AI hardware & software * Optimize ML models for edge deployment * Implement robust data pipelines from edge devices * Develop secure remote management & update strategies * Plan for model retraining & deployment to edge	* Limited compute/memory on edge devices * Ensuring real-time performance and accuracy * Managing distributed model deployment & updates * Network connectivity & reliability for edge devices * Data security & privacy at the edge * Skillset required for embedded AI development

Products	Platform Cost	Project Cost	Duration
AWS IoT Greengrass (Local Processing)	$5-30K	$20-100K	2-6 m
Azure IoT Edge (ML at Edge)	$5-25K	$20-90K	2-5 m
Google Cloud Vertex AI (Edge Deploy)	$5-20K	$20-80K	2-5 m
NVIDIA Jetson (Edge AI Hardware/SDK)	$1-10K	$5-50K	1-3 m
OpenVINO (Intel - Edge Inference)	N/A	$10-40K	1-3 m
FogHorn (Edge AI Software)	$10-40K	$50-200K	3-8 m

Ai6x6™ Transformation Blueprint

Data powered 'Interfaces'

"Design is not just what it looks like. It's how it works." — Steve Jobs

Apple's co-founder, universally cited for making design emotionally intuitive in technology.

Dashboards alone aren't interfaces—they're invitations. Conversational BI, immersive visualization, and data narratives make complexity touchable. "Data powered interfaces" reframes the interface as interpretive space. Transformation maturity lies in how teams feel about data—not just what they know from it. Human-centered tools don't simplify—they clarify. When exploration becomes interactive and insight becomes narrative, adoption becomes natural. The goal isn't just UX—it's executive emotional intelligence for data. Information becomes felt, not just fetched.

"Data powered interfaces" is critical for data transformation's digital maturity because it focuses on democratizing data access and making complex information intuitive and actionable for all users, not just data specialists. A digitally mature organization prioritizes user-friendly interfaces, including conversational data interfaces, live data tiles, and interactive notebooks. This human-centric approach, complemented by augmented visualization APIs and personalized data narratives, empowers employees at every level to understand and leverage data effectively. This ensures that data insights are widely adopted, leading to more informed decisions and a truly data-driven culture across the enterprise.

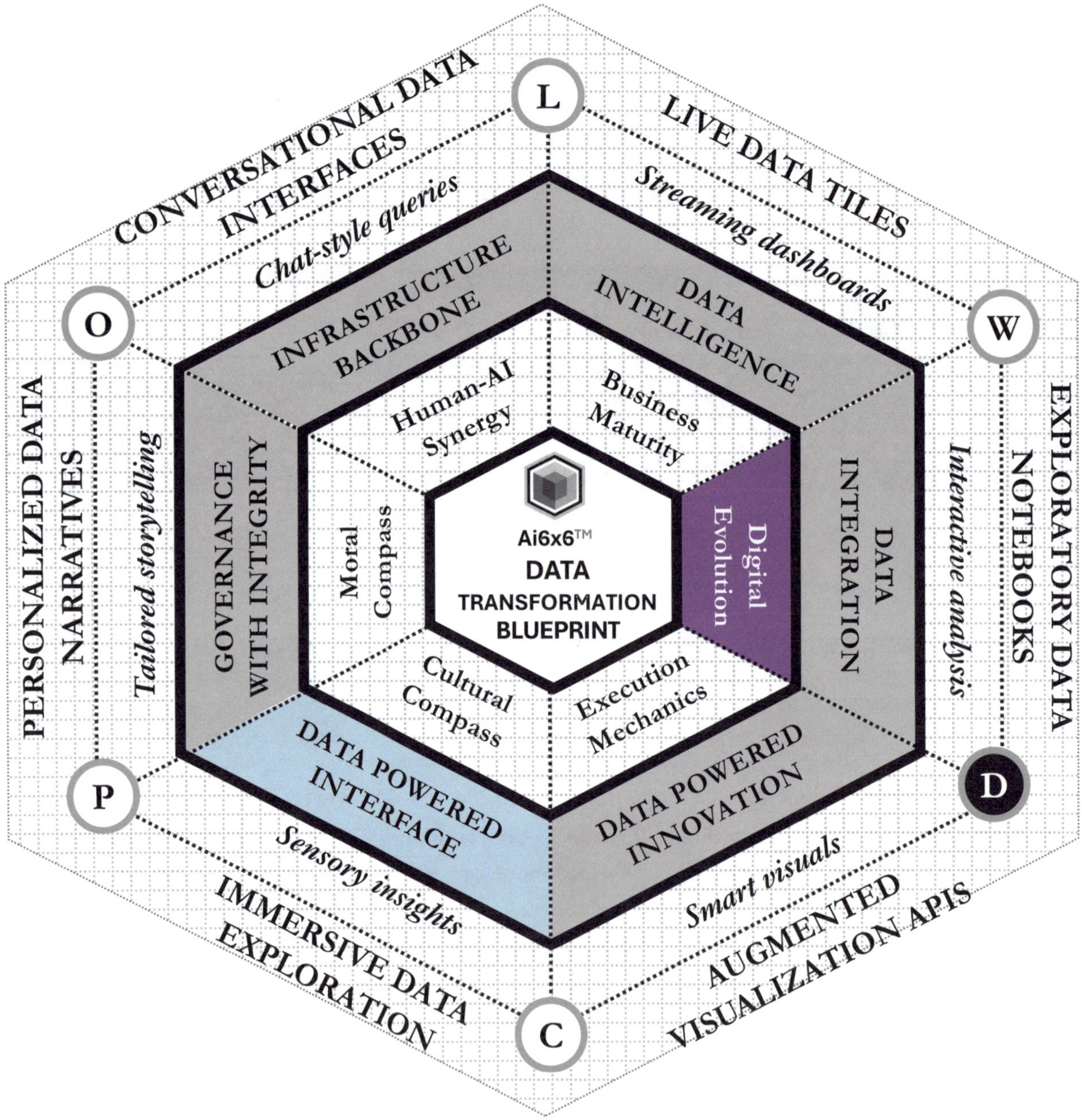

Ai6x6 DIGITAL EVOLUTION FOR **DATA LIFECYCLE | INTERFACE**

1. CONVERSATIONAL DATA INTERFACES

Conversational Data Interfaces (CDIs) allow users to interact with and query data using natural language, either through voice commands or text-based chat interfaces, eliminating the need for complex SQL queries or specialized dashboard navigation. These AI-powered interfaces leverage natural language processing (NLP) to understand user intent, translate questions into data queries, and present insights in an intuitive, human-readable format. By democratizing access to data and insights, CDIs empower a broader range of business users to ask spontaneous questions, retrieve real-time information, and make data-driven decisions faster without relying on data analysts. This enhances data literacy, accelerates self-service analytics, and fosters a more responsive, data-driven organizational culture.

Without Conversational Data Interfaces, organizations often limit data access to users proficient in complex querying languages or dependent on data analysts to generate reports, creating bottlenecks in timely decision-making and hindering data democratization. This means business users cannot spontaneously explore data or get immediate answers to their questions, leading to slower responses to market changes and missed opportunities. The absence of intuitive, natural language interaction with data results in lower data literacy across the enterprise, underutilized data assets, and a reactive rather than proactive approach to data exploration, ultimately impacting organizational agility and the ability to extract maximum value from information.

Implementation	Challenges
* Define business questions & data sources * Select a CDI platform or build custom * Train NLP models on domain-specific vocabulary * Integrate with data warehouse/BI tools * Design intuitive user experience & feedback loop * Plan for continuous model improvement & data updates	* Ensuring accuracy & context of responses * Handling ambiguity & complex queries * Integrating with diverse back-end data sources * Training AI models for specific business lexicon * User adoption & trust in AI-generated answers * Cost optimization for NLP processing

Products	Platform Cost	Project Cost	Duration
ThoughtSpot (Search & AI-driven Analytics)	$15-70K	$80-350K	3-10 m
Microsoft Power BI (Q&A feature)	$5-30K	$20-100K	2-6 m
Qlik Sense (Natural Language Interaction)	$10-40K	$50-200K	3-8 m
Yellowfin BI (Guided NLQ)	$5-25K	$20-90K	2-5 m
Akkio (AI for Business)	$1-10K	$5-30K	1-3 m
Custom NLP/LLM Integration (e.g., GPT-4)	N/A	$50-250K	3-9 m

Live Data Tiles are dynamic, interactive visual components that display real-time or near-real-time key performance indicators (KPIs), metrics, and critical information in a compact, easily digestible format, often on dashboards, portals, or mobile applications. Unlike static reports, these tiles continuously refresh with the latest data, providing an immediate snapshot of operational status, business performance, or critical events. By offering instant, always-on access to up-to-the-minute insights, Live Data Tiles empower decision-makers to react swiftly to changes, monitor operations proactively, and gain an agile understanding of dynamic business environments. This improves operational efficiency, enhances situational awareness, and enables more informed, timely responses to evolving conditions.

Without Live Data Tiles, organizations often rely on static reports or manually refreshed dashboards, leading to delayed insights and a reactive approach to business operations. This means decision-makers are working with outdated information, unable to respond swiftly to real-time changes or proactively monitor critical KPIs as they evolve. The absence of continuously updated, visually intuitive data displays results in missed opportunities for immediate intervention, reduced operational agility, and a pervasive lack of real-time situational awareness, ultimately hindering an organization's ability to optimize performance and respond effectively in fast-paced, data-rich environments.

Implementation	Challenges
* Identify critical real-time KPIs * Select a BI tool or custom development framework * Design visually intuitive and compact tiles * Implement real-time data ingestion & refresh * Ensure data accuracy & low latency * Integrate into existing dashboards/portals	* Ensuring true real-time data ingestion & processing * Maintaining low latency under high load * Visualizing complex data in a compact format * Integrating with diverse streaming data sources * Performance optimization to prevent dashboard lag * Ensuring data security for sensitive live metrics

Products	Platform Cost	Project Cost	Duration
Tableau (Live Connections)	$10-50K	$50-250K	3-9 m
Microsoft Power BI (DirectQuery/Real-time Dashboards)	$5-30K	$20-100K	2-6 m
Qlik Sense (Live Data Connections)	$10-40K	$50-200K	3-8 m
Grafana (Real-time Dashboards)	$1-10K	$5-30K	1-3 m
Looker (LookML Persistent Derived Tables)	$10-60K	$60-300K	3-9 m
Custom Web Development (JavaScript, WebSockets)	N/A	$50-250K	3-9 m

Exploratory Data Notebooks (e.g., Jupyter Notebooks, Databricks Notebooks) are interactive computing environments that combine live code (Python, R, SQL), narrative text, visualizations, and output into a single, shareable document. They serve as powerful tools for data scientists and analysts to perform iterative data cleaning, transformation, statistical analysis, model building, and visualization in a highly interactive and reproducible manner. By facilitating rapid experimentation, seamless collaboration, and transparent documentation of the entire analytical workflow, Exploratory Data Notebooks accelerate the process of data discovery, improve reproducibility of insights, and enable more agile development of data-driven solutions, fostering a dynamic and collaborative data science practice.

Without Exploratory Data Notebooks, data scientists and analysts often work in fragmented environments, using separate tools for coding, visualization, and documentation, leading to inefficient workflows, difficulty in reproducing results, and challenges in sharing analytical processes. This means slower iteration cycles for data exploration, increased risk of errors due to manual steps, and a lack of transparency in how insights are derived. The absence of a unified, interactive notebook environment hinders collaboration, slows down the development of new data models, and ultimately impacts an organization's ability to rapidly explore, validate, and operationalize valuable insights from its data assets.

Implementation	Challenges
* Select a notebook environment (cloud/local) * Integrate with data sources & compute resources * Establish version control for notebooks & code * Provide necessary libraries & toolsets * Plan for collaboration & knowledge sharing * Implement security for data access within notebooks	* Managing complex dependencies & environments * Version control and collaboration for notebooks * Securing data access within interactive sessions * Scaling compute resources for large datasets * Operationalizing notebook code into production * Ensuring reproducibility across different environments

Products	Platform Cost	Project Cost	Duration
Databricks Notebooks	$10-60K	$50-300K	3-9 m
Google Cloud Vertex AI Workbench (JupyterLab)	$5-30K	$20-100K	2-6 m
Azure Machine Learning Notebooks	$5-25K	$20-90K	2-5 m
AWS SageMaker Studio (JupyterLab)	$5-20K	$20-80K	2-5 m
JupyterHub (Open Source)	N/A	$10-70K	1-4 m
Deepnote	$1-10K	$5-30K	1-3 m

Augmented Visualization APIs are programmatic interfaces that embed AI and machine learning capabilities directly into data visualization tools and applications, enabling them to automatically generate relevant charts, suggest optimal visualization types, and highlight significant patterns or outliers. These APIs go beyond static charting by providing intelligent recommendations for data exploration and insight discovery, even for users without deep data visualization expertise. By democratizing advanced visual analytics, Augmented Visualization APIs accelerate the process of data interpretation, reduce the cognitive load on users, and ensure that data is presented in the most effective and insightful way, empowering more effective data communication and decision-making across the enterprise.

Without Augmented Visualization APIs, organizations often rely on manual selection of chart types and extensive human effort to derive insights from data visualizations, leading to suboptimal data representation, missed patterns, and a slower process of understanding complex datasets. This means business users may struggle to identify the most relevant visualizations or overlook critical insights hidden within their data, requiring constant intervention from expert analysts. The absence of AI-powered visualization capabilities results in less effective data communication, limited self-service analytics, and a slower pace of insight discovery, ultimately hindering an organization's ability to quickly and effectively interpret its data for strategic advantage.

Implementation	Challenges
* Identify target BI/Analytics platforms * Select/develop APIs for visualization augmentation * Ensure data quality & consistency for input * Train AI models for visualization recommendations * Integrate APIs seamlessly into user workflows * Plan for continuous model improvement & feedback	* Ensuring AI recommendations are accurate & relevant * Integrating with diverse existing visualization tools * Managing model complexity & inference costs * Overcoming user skepticism or over-reliance on AI * Handling large datasets for real-time recommendations * Defining clear metrics for improved insight discovery

Products	Platform Cost	Project Cost	Duration
Tableau (Extensions API, Ask Data)	$10-50K	$50-250K	3-9 m
Microsoft Power BI (Custom Visuals API, Q&A)	$5-30K	$20-100K	2-6 m
Qlik Sense (API for Cognitive Engine)	$10-40K	$50-200K	3-8 m
Sisense (Embedding SDK, AI-driven suggestions)	$15-60K	$70-300K	3-9 m
Yellowfin BI (APIs for Assisted Insights)	$5-25K	$20-90K	2-5 m
Google Charts API (Custom Integration)	N/A	$10-70K	1-4 m

Immersive Data Exploration utilizes Virtual Reality (VR), Augmented Reality (AR), and Mixed Reality (MR) technologies to create interactive 3D or spatial environments where users can intuitively interact with and visualize complex datasets. Instead of flat 2D dashboards, users can "walk through" data, manipulate models with gestures, and gain multi-dimensional perspectives that reveal patterns and relationships difficult to discern otherwise. By providing a truly spatial and engaging experience, Immersive Data Exploration enhances comprehension, fosters deeper insights, and facilitates collaborative analysis of complex data, enabling breakthroughs in fields like scientific research, urban planning, product design, and complex systems monitoring.

Without Immersive Data Exploration, organizations are limited to traditional 2D visualizations and dashboards, which can struggle to effectively represent and allow interaction with highly complex, multi-dimensional datasets. This means that subtle patterns, spatial relationships, and intricate connections within the data may remain undiscovered, hindering deep understanding and limiting the scope of analytical insights. The absence of immersive environments results in a less intuitive and engaging data experience, slower comprehension of complex data landscapes, and missed opportunities for collaborative discovery, ultimately impacting an organization's ability to extract maximum value and novel insights from its most intricate data assets.

Implementation	Challenges
* Identify complex, multi-dimensional datasets * Select VR/AR/MR hardware & software * Design intuitive spatial data visualizations * Develop real-time data streaming to immersive environment * Ensure comfortable & engaging user experience * Plan for collaborative multi-user experiences	* High cost of hardware & specialized software * Developing effective & intuitive 3D visualizations * Ensuring real-time data streaming & performance * Addressing user comfort & potential motion sickness * Skillset required for VR/AR/MR development * Integrating with existing data infrastructure

Products	Platform Cost	Project Cost	Duration
Unity 3D / Unreal Engine (Development)	$0-20K	$100-500K	4-12 m
Enthought (Scientific Data Viz)	$10-50K	$60-250K	3-9 m
WorldViz (VR Simulation Platform)	$10-40K	$50-200K	3-8 m
XR applied to existing BI (e.g., Power BI in VR)	N/A	$30-150K	2-7 m
Custom Immersive Application Dev	N/A	$200-800K+	6-18 m
HTC VIVE Pro / Oculus Quest (Hardware)	$1-5K	N/A	N/A

Personalized Data Narratives automatically generate human-readable stories, summaries, and explanations from data, tailored to individual users or specific business contexts. These AI-powered solutions go beyond presenting charts and tables by providing natural language insights, highlighting key findings, and explaining the "why" behind data trends. By transforming complex data into understandable, actionable narratives, Personalized Data Narratives democratize data literacy, enable quicker comprehension for non-technical users, and ensure that relevant insights are communicated effectively to decision-makers across different roles and departments. This leads to more consistent understanding of data, faster decision-making, and a more data-driven culture throughout the organization.

Without Personalized Data Narratives, organizations often present data insights through generic dashboards or complex reports that require significant interpretation and expertise, leading to varying levels of data comprehension across the workforce. This means that non-technical users may struggle to understand the implications of data, critical insights might be overlooked, and decision-making could be slower or based on incomplete understanding. The absence of automated, tailored data storytelling results in a less data-literate workforce, fragmented understanding of business performance, and a missed opportunity to effectively communicate data's true value, ultimately hindering the widespread adoption of data-driven decision-making.

Implementation	Challenges
* Identify data points for narrative generation * Select a Natural Language Generation (NLG) platform * Define narrative templates & personalization rules * Integrate with data sources & BI tools * Ensure accuracy & relevance of generated narratives * Plan for continuous feedback & refinement	* Ensuring generated narratives are accurate & trustworthy * Handling complex data relationships in narratives * Training NLG models for domain-specific language * Avoiding repetitive or generic narratives * Integrating with diverse data and BI platforms * User adoption and trust in AI-generated stories

Products	Platform Cost	Project Cost	Duration
Automated Insights (Wordsmith)	$10-50K	$60-250K	3-9 m
Narrative Science (Quill)	$15-70K	$80-350K	3-10 m
Arria NLG	$20-80K	$100-400K	4-12 m
Tableau (Smart Narratives feature)	$10-50K	$50-250K	3-9 m
Microsoft Power BI (Smart Narratives feature)	$5-30K	$20-100K	2-6 m
Custom NLG Development (e.g., GPT-3/4 based)	N/A	$100-500K	4-12 m

Ai6x6™ Transformation Blueprint

Governance with 'Integrity'

"Without standards, data can't be shared; without trust, it won't be shared."
— Tim Berners-Lee , Inventor of the World Wide Web; now focused on ethical data and web decentralization.

Governance isn't permission—it's precision. Metadata management, master records, and encryption keys are not just control points—they're trust indicators. "Governance with Integrity" implies more than compliance—it demands stewardship. Maturity is measured not by volume but by verifiability. Governance is not what restricts data—it's what enables it to scale ethically. Leaders embed integrity into pipelines and decisions, so data can move with confidence. Transformation thrives when access is intelligent, accountability is encoded, and information can move because it's trusted to land well.

The "Governance with Integrity" chapter is fundamental for data transformation's digital maturity as it addresses the paramount importance of data quality, security, and governance. A digitally mature enterprise establishes robust frameworks for metadata management, master data management, and data access governance to ensure data accuracy, consistency, and compliance with regulations. This unwavering commitment to data integrity, supported by encryption key management and audit logging solutions, builds trust, mitigates risks, and provides a reliable, trustworthy foundation for all data-driven initiatives, which is essential for sustained digital growth and regulatory adherence.

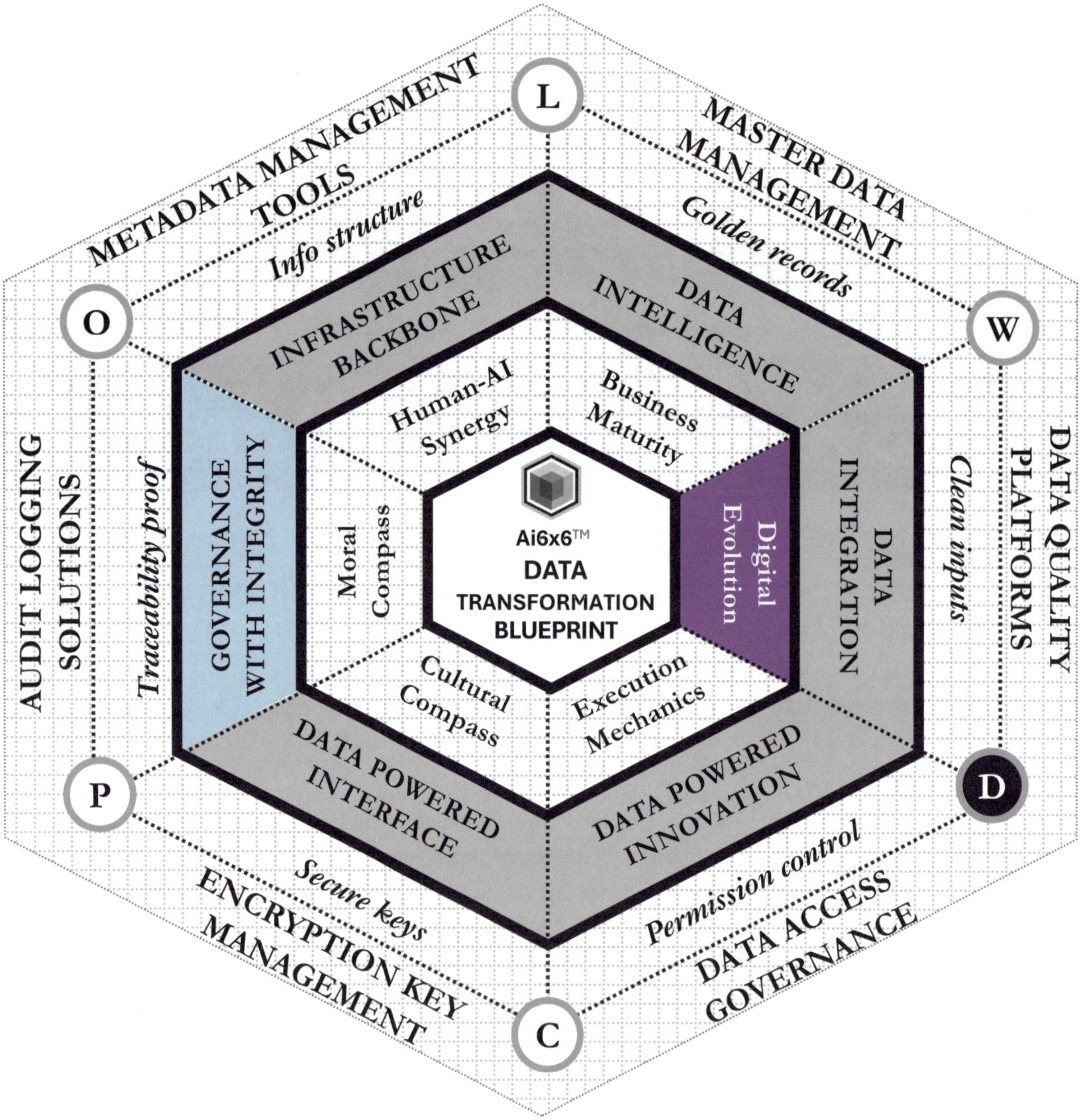

Ai6x6 DIGITAL EVOLUTION FOR **DATA LIFECYCLE | INTEGRITY**

1. METADATA MANAGEMENT TOOLS

Metadata Management Tools are specialized platforms designed to discover, collect, store, organize, and govern metadata (data about data) from across an organization's diverse data landscape. This includes technical metadata (e.g., schemas, data types, lineage), business metadata (e.g., definitions, business terms, ownership), and operational metadata (e.g., data quality metrics, usage statistics). By providing a centralized, comprehensive view of all data assets, these tools enhance data discoverability, improve data understanding, facilitate data governance and compliance, and streamline collaboration among data professionals. They are crucial for building a trusted data ecosystem, enabling self-service analytics, and maximizing the strategic value derived from an organization's information assets.

Without dedicated Metadata Management Tools, organizations often operate in a state of "data chaos," with fragmented knowledge about their data assets, unclear definitions, and undocumented data lineage. This leads to inefficient data discovery, inconsistent data interpretations, duplicated efforts across data initiatives, and significant challenges in ensuring data quality and regulatory compliance. The absence of a centralized metadata repository results in untrustworthy analytics, increased operational risks due to poor data governance, and a slower pace of innovation, ultimately hindering an organization's ability to truly leverage its data as a strategic asset and democratize data access effectively and reliably.

Implementation	Challenges
* Inventory all data sources & types of metadata * Select a metadata management platform * Design metadata ingestion pipelines * Implement data cataloging & lineage tracking * Establish data governance rules & ownership * Integrate with data platforms & BI tools	* Extracting & integrating metadata from diverse systems * Ensuring metadata accuracy & freshness over time * Driving user adoption for data discovery & collaboration * Managing complex data lineage across transformations * Overcoming organizational silos & data ownership issues * Measuring clear ROI on improved data discoverability

Products	Platform Cost	Project Cost	Duration
Collibra Data Governance	$20-90K	$100-450K	4-12 m
Alation Data Catalog	$15-70K	$80-350K	3-10 m
Informatica Enterprise Data Catalog	$20-100K	$120-500K	4-13 m
Apache Atlas (Open Source)	N/A	$50-250K	3-9 m
Azure Purview	$10-50K	$60-250K	3-9 m
Data.world	$5-30K	$20-100K	2-6 m

Master Data Management (MDM) is a technology-enabled discipline that defines and manages an organization's critical non-transactional data (master data), such as customer, product, employee, or location data, to provide a single, authoritative, and consistent view across all business systems and applications. MDM systems consolidate data from disparate sources, resolve inconsistencies, de-duplicate records, and enforce data quality rules, creating a "golden record" for each master data entity. By ensuring data consistency and accuracy across the enterprise, MDM eliminates data silos, improves operational efficiency, enhances regulatory compliance, and provides a reliable foundation for analytics, reporting, and customer experience initiatives, critical for trusted data-driven decision-making.

Without Master Data Management, organizations typically struggle with fragmented, inconsistent, and often inaccurate master data spread across various operational systems, leading to conflicting views of critical business entities (e.g., the same customer having multiple, inconsistent records). This results in operational inefficiencies, errors in reporting and analytics, poor customer experiences due to incomplete information, and challenges in complying with data privacy regulations. The absence of a single, authoritative source of master data undermines trust in business intelligence, complicates mergers and acquisitions, and ultimately hinders an organization's ability to operate efficiently, make informed decisions, and deliver consistent value across all customer touchpoints.

Implementation	Challenges
* Define critical master data domains (customer, product) * Establish data governance & ownership roles * Select a suitable MDM platform * Develop data quality rules & matching algorithms * Implement data integration & synchronization * Plan for data stewardship & ongoing maintenance	* High complexity in data matching & de-duplication * Gaining consensus on data definitions across departments * Integrating with diverse legacy source systems * Ensuring data quality & consistency continuously * Overcoming organizational resistance to data standardization * Measuring ROI on improved data consistency

Products	Platform Cost	Project Cost	Duration
Informatica MDM	$25-120K	$150-700K	5-16 m
Semarchy xDM	$15-70K	$80-350K	3-10 m
Profisee MDM	$10-60K	$50-300K	3-9 m
IBM InfoSphere Master Data Management	$20-100K	$120-600K	4-14 m
Reltio Connected Data Platform	$20-90K	$100-450K	4-12 m
Talend (Data Fabric for MDM)	$15-60K	$70-300K	3-9 m

3. DATA QUALITY PLATFORMS

Data Quality Platforms are software solutions designed to measure, monitor, cleanse, and improve the accuracy, completeness, consistency, timeliness, and validity of an organization's data. These platforms utilize various techniques such as profiling, parsing, standardization, matching, and monitoring to identify and rectify data errors at the source or during data integration. By ensuring high-quality data throughout the data lifecycle, Data Quality Platforms reduce operational errors, improve the reliability of analytics and reporting, enhance decision-making confidence, and ensure compliance with regulatory requirements. This foundation of trusted data is crucial for the success of any data transformation initiative, including AI/ML projects and digital strategies.

Without dedicated Data Quality Platforms, organizations are plagued by inaccurate, inconsistent, and incomplete data across their systems, leading to unreliable reports, flawed analytics, and misguided business decisions. This means operational processes are prone to errors, customer experiences suffer due to incorrect information, and compliance with data governance regulations becomes challenging. The absence of systematic data quality management results in significant financial losses from bad decisions, wasted resources on correcting errors, and a pervasive lack of trust in data, ultimately hindering an organization's ability to truly leverage its information assets for competitive advantage and strategic growth.

Implementation	Challenges
* Define data quality metrics & standards * Profile existing data to identify issues * Select a data quality platform * Implement data cleansing & standardization rules * Establish continuous monitoring & alerting * Integrate with data pipelines & source systems	* Gaining consensus on data quality definitions * Remediating existing poor quality data at scale * Integrating with diverse, complex data sources * Maintaining data quality continuously over time * Overcoming organizational resistance to data standards * Measuring clear ROI on data quality improvements

Products	Platform Cost	Project Cost	Duration
Informatica Data Quality (IDQ)	$20-90K	$100-450K	4-12 m
Talend Data Quality	$15-70K	$80-350K	3-10 m
IBM Watson Knowledge Catalog (Data Quality)	$15-60K	$70-300K	3-9 m
SAP Data Services (Data Quality)	$20-80K	$100-400K	4-12 m
Ataccama ONE (Data Quality & MDM)	$20-90K	$100-450K	4-12 m
Collibra (Data Quality capabilities)	$20-90K	$100-450K	4-12 m

Data Access Governance refers to the systematic approach and technologies used to define, enforce, and audit who can access what data, when, and how, across an organization's entire data estate. This involves implementing robust access control mechanisms, role-based security, data masking, anonymization, and granular permissions down to the column or row level, particularly for sensitive data. By establishing strict controls over data access, Data Access Governance ensures compliance with stringent regulations (e.g., GDPR, CCPA, HIPAA), mitigates data breach risks, protects intellectual property, and maintains data privacy. This foundational integrity component is critical for building trust in data, enabling safe data sharing, and supporting ethical data usage across diverse users and applications.

Without robust Data Access Governance, organizations face significant risks of unauthorized data access, data breaches, and non-compliance with increasingly stringent data privacy regulations, leading to hefty fines, reputational damage, and loss of customer trust. Relying on fragmented or manual access controls creates security vulnerabilities, makes auditing difficult, and limits the ability to safely share data across internal teams or with external partners. The absence of a centralized, automated data access governance framework results in a reactive security posture, hinders responsible data democratization, and ultimately jeopardizes an organization's ability to secure its most valuable asset and maintain a compliant and ethical data environment.

Implementation	Challenges
* Classify data sensitivity & regulatory requirements * Define roles, policies & access groups * Select a data access governance platform * Implement granular access controls (row/column level) * Configure data masking/anonymization for sensitive data * Establish audit logging & continuous monitoring	* Complexity of defining fine-grained access policies * Integrating with diverse data platforms & applications * Ensuring compliance with evolving privacy regulations * Managing user roles & permissions at scale * Performance impact of granular access controls * Auditing & reporting on access for compliance

Products	Platform Cost	Project Cost	Duration
Immuta	$15-70K	$80-350K	3-10 m
Okera	$10-50K	$60-250K	3-9 m
Privacera	$15-60K	$70-300K	3-9 m
BigID (Data Discovery & Access Governance)	$20-90K	$100-450K	4-12 m
OneTrust (DataGovernance & Privacy)	$20-100K	$120-500K	4-13 m
Apache Ranger (Open Source for Hadoop/Spark)	N/A	$30-150K	2-7 m

Encryption Key Management (EKM) refers to the comprehensive process of generating, storing, distributing, rotating, revoking, and securely backing up cryptographic keys used to encrypt and decrypt data. As data encryption becomes paramount for data security and privacy, EKM systems provide a centralized, highly secure platform for managing these keys across various data stores, applications, and cloud environments. By ensuring the integrity and confidentiality of encryption keys, EKM solutions protect sensitive data from unauthorized access, minimize the risk of data breaches, and facilitate compliance with stringent regulatory mandates. This foundational security component is essential for maintaining data integrity and confidentiality throughout its lifecycle, critical for any data transformation strategy.

Without robust Encryption Key Management, organizations face significant risks to data security and compliance, as the cryptographic keys used to protect sensitive data may be poorly managed, compromised, or lost. This leads to vulnerabilities where encrypted data could be exposed, or data could become irretrievable if keys are lost, resulting in severe data breaches, regulatory fines, and reputational damage. The absence of a centralized, secure EKM system creates a complex and error-prone environment for key lifecycle management, hindering an organization's ability to confidently encrypt data, meet compliance obligations, and ultimately protect its most sensitive information assets from unauthorized access.

Implementation	Challenges
* Identify data requiring encryption * Select a Key Management System (KMS) solution * Define key generation, storage & rotation policies * Integrate KMS with databases, applications & cloud services * Implement robust access controls & audit logging * Plan for disaster recovery & key backup	* Ensuring high availability & durability of keys * Managing key lifecycles (rotation, revocation) * Integrating with diverse applications & data stores * Ensuring strong access controls for keys * Performance impact of encryption/decryption * Compliance with cryptographic standards & regulations

Products	Platform Cost	Project Cost	Duration
AWS Key Management Service (KMS)	$5-30K	$20-100K	2-6 m
Azure Key Vault	$5-25K	$20-90K	2-5 m
Google Cloud Key Management Service (KMS)	$5-20K	$20-80K	2-5 m
HashiCorp Vault	$5-30K	$20-100K	2-6 m
Thales CipherTrust Manager	$15-70K	$80-350K	3-10 m
Fortanix DSM	$10-60K	$50-300K	3-9 m

Audit Logging Solutions are systems designed to automatically record and store detailed chronological records of events, activities, and changes occurring within an organization's IT systems, applications, and data environments. These logs capture who accessed what data, when, from where, and what actions were performed, providing an immutable trail of events. By centralizing and analyzing audit logs, organizations gain critical visibility into user behavior, system performance, and potential security incidents. This enables effective incident response, facilitates forensic investigations, ensures compliance with regulatory requirements, and supports internal accountability. Audit logging is fundamental for maintaining data integrity, security, and traceability throughout the entire data transformation lifecycle.

Without robust Audit Logging Solutions, organizations operate with a critical blind spot regarding activities within their IT systems and data environments, making it nearly impossible to detect unauthorized access, track data changes, or investigate security incidents effectively. This lack of visibility increases vulnerability to cyberattacks, complicates compliance with regulatory mandates requiring detailed audit trails, and hinders accountability for data access and modification. The absence of comprehensive, centralized audit logs results in delayed incident response, greater difficulty in proving compliance, and a diminished capacity to ensure data integrity and security, ultimately exposing the organization to significant operational, financial, and reputational risks.

Implementation	Challenges
* Identify critical systems for audit logging * Define events to be logged (success/failure) * Select a centralized logging platform (SIEM/ELK) * Configure log collection from diverse sources * Implement log analysis, alerting & reporting * Plan for log retention & secure storage	* Managing massive volumes of log data * Ensuring real-time log ingestion & processing * Correlating logs from disparate systems * Storing logs securely & cost-effectively * Ensuring compliance with log retention policies * Skillset required for log analysis & security operations

Products	Platform Cost	Project Cost	Duration
Splunk Enterprise Security (SIEM)	$20-100K	$120-600K	4-14 m
Elastic Stack (ELK - Elasticsearch, Logstash, Kibana)	N/A	$50-300K	3-9 m
Sumo Logic	$15-70K	$80-350K	3-10 m
Datadog (Log Management)	$10-50K	$60-250K	3-9 m
SentinelOne Singularity Data Lake	$15-60K	$70-300K	3-9 m
Exabeam Security Management Platform	$20-80K	$100-400K	4-12 m

SECTION E

BLUEPRINT SERIES

Execution Mechanics

Ai6x6™ Transformation Blueprint

Story 1 — The Manhattan Project Coordination Model (1942–1946): The Manhattan Project united scientists, engineers, and military leaders across dispersed sites under extreme secrecy and urgency. Workstreams were interdependent, requiring precise coordination without full visibility. Structured reporting, controlled interfaces, and disciplined compartmentalization enabled progress despite complexity. Human judgment guided trade-offs while machines and labs executed specialized tasks. The project demonstrated how execution succeeds when governance, clarity, and cadence align. It became a template for managing high-stakes, multi-disciplinary programs.

Story 2 — The Berlin Airlift Logistics System (1948–1949): With land routes blocked, Allied forces had to supply an entire city solely by air. Aircraft schedules, loading patterns, and turnaround times were optimized to the minute. Machines provided lift and speed, but humans orchestrated sequencing, prioritization, and contingency planning. Continuous feedback loops refined operations daily. The airlift proved that disciplined execution can overcome structural constraints. It showed how precision, rhythm, and adaptability create operational breakthroughs.

Story 3 — The Interstate Highway Construction Program (1956–1970s): The United States embarked on building a nationwide highway network requiring coordination across states, contractors, and engineering disciplines. Standardized designs ensured consistency while local teams adapted to terrain and conditions. Machines handled heavy construction, but humans managed sequencing, safety, and quality. The program relied on predictable funding, clear specifications, and phased execution. It demonstrated how large-scale delivery succeeds when structure and flexibility coexist. The highways became a symbol of disciplined national execution.

Story 4 — Toyota's Just-in-Time Logistics (1960s–1980s): As Toyota refined its production system, logistics became a critical execution layer. Parts arrived exactly when needed, minimizing inventory and waste. Machines tracked flow, but humans monitored signals, resolved disruptions, and maintained rhythm. Kanban created transparency across the value chain. The system required precision, trust, and continuous calibration. It showed how execution excellence emerges when every component of the workflow reinforces timing, quality, and flow.

Story 5 — FedEx's Overnight Delivery Network (1973–1990s): FedEx built a hub-and-spoke system capable of moving packages across the country overnight. Aircraft, sorting machines, and routing algorithms provided speed, but humans managed exceptions, weather disruptions, and customer commitments. The network operated on tight cycles that demanded flawless coordination. Data informed decisions while people ensured reliability. The system demonstrated how execution mechanics scale when tech and human oversight operate in synchronized cadence.

Story 6 — The Rise of Agile Software Delivery (2001–2015): Software teams struggled with long cycles, rigid plans, and late-stage failures. Agile introduced iterative delivery, cross-functional collaboration, and continuous feedback. Machines automated testing and deployment, while humans prioritized work, clarified requirements, and adapted to change. Execution shifted from linear phases to dynamic cycles. The approach improved predictability and responsiveness. It became a global standard for modern execution, proving that disciplined iteration outperforms rigid planning.

LESSONS FROM A DATA TRANSFORMATION PERSPECTIVE

Data becomes the backbone of execution mechanics as organizations evolve from intuition-driven decisions to structured, evidence-based operations. Early systems relied on manual reporting, while later environments integrated real-time metrics, feedback loops, and automated monitoring. Data enables predictability, exposes bottlenecks, and guides prioritization. Machines surface patterns, but humans interpret context, risk, and trade-offs. Execution maturity emerges when data flows seamlessly across functions, enabling synchronized action. The organization becomes more reliable because decisions are grounded in visibility rather than assumptions. Data transforms execution from reactive to proactive, ensuring that operations remain stable, transparent, and continuously improving.

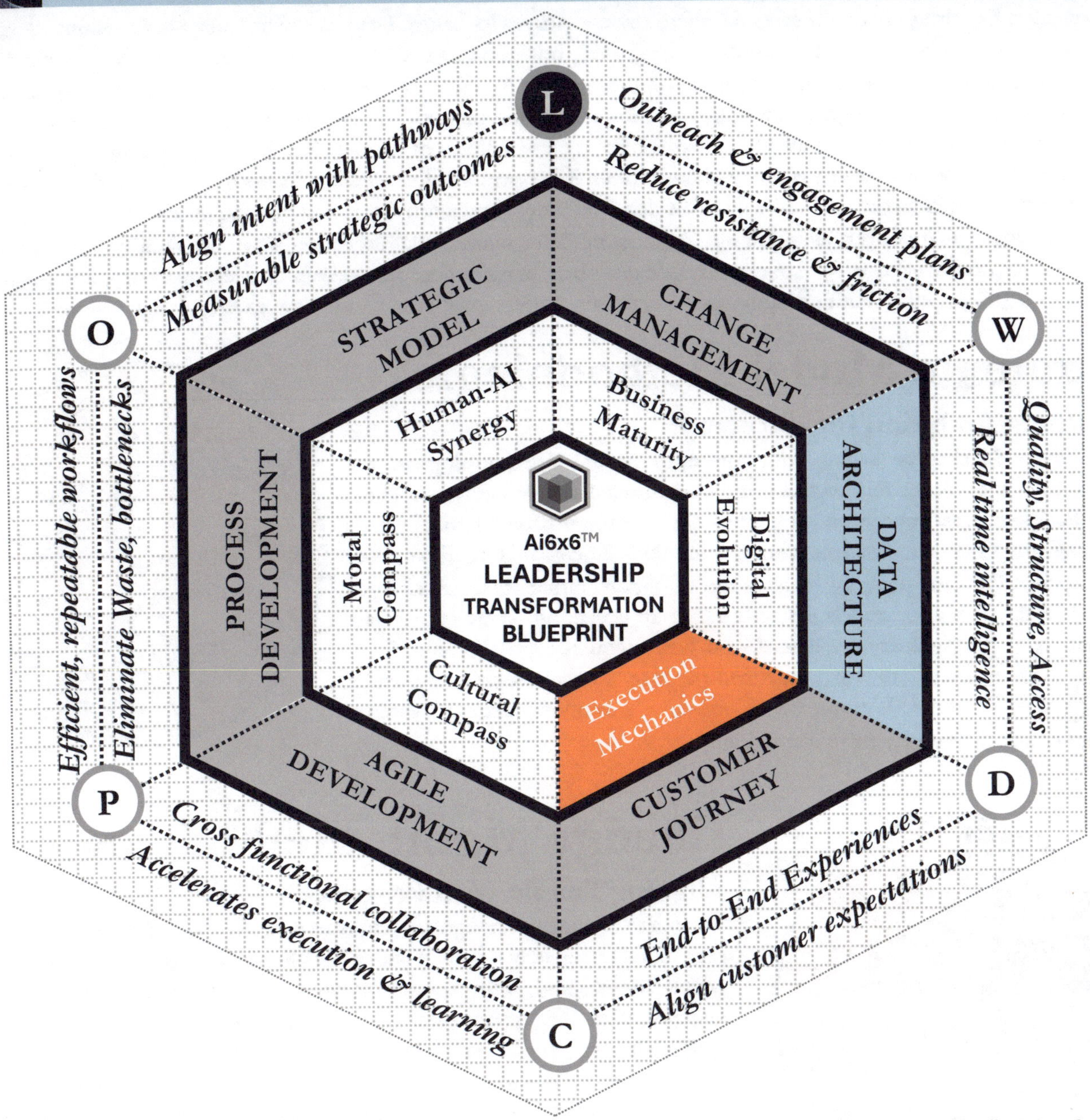

The **Ai6x6 Research Lab** identified this lens as the primary antidote to the "architectural chaos" that often cripples data-driven initiatives. We chose it because the human mind requires a disciplined, repeatable system to handle the massive processing demands of the AI era—from **Lambda/Kappa** architectures to **Data Mesh** methods. This lens isn't merely about technical frameworks; it is about the psychological certainty and mental clarity that come from a proven methodology like the **Zachman Framework** or **DAMA-DMBOK**. It is vital because it provides a mental scaffolding, turning the chaos of data processing into a series of surgical, confident strikes on "The Truth."

This lens allows the data architect's heart to remain calm while the information flows at terminal velocity through the **Data Fabric**. By relying on **Master Data Management (MDM)**, the leader replaces data ambiguity with evidence-based conviction. We believe that execution mechanics are the ultimate form of data mindfulness—the practice of being fully present in the "how" of governance so that the "what" of intelligence is achieved with excellence. This lens ensures every data stream has a defined ownership and a coherent structure, allowing the organization to operate with a "Single Source of Truth." It is the discipline that turns a fragile data pipeline into a robust engine for value creation, providing the steady footing necessary to scale intelligence across the enterprise.

Execution has always been the crucible where strategy proves its worth. Leaders can articulate bold visions, design elegant frameworks, and outline ambitious roadmaps, but none of it matters if the organization cannot execute with precision, coherence, and resilience. In the age of AI and accelerated change, **Execution Mechanics** becomes the lens that determines whether transformation becomes real or remains theoretical. It forces leaders to confront the uncomfortable truth that most failures are not failures of strategy but failures of execution. The gap between intention and impact is rarely philosophical; it is operational. Execution Mechanics exposes that gap and provides the architecture to close it. This lens reframes execution as a system—an interconnected set of methodologies that convert ambition into momentum. It recognizes that execution is not the final step of strategy but the environment in which strategy lives or dies. In a world where AI accelerates decision cycles, compresses timelines, and amplifies both strengths and weaknesses, leaders must design execution as they design vision.

Strategic Model Methodologies

Turning Ambition Into Structure

Execution Mechanics begins with **strategic model methodologies**, the structured approaches that translate ambition into actionable pathways. These methodologies are not academic exercises; they are the scaffolding that ensures strategy is grounded. Leaders who neglect this dimension find themselves improvising under pressure, relying on intuition rather than disciplined frameworks. Strategic models define the logic of execution—how choices are made, how priorities are sequenced, and how resources are allocated. They ensure that every initiative has a defined purpose, a coherent structure, and a measurable outcome. Without this foundation, execution becomes reactive, fragmented, and vulnerable to disruption.

Change Mgmt. Methodologies

Aligning People With Progress

The next dimension, **change management methodologies**, recognizes that execution is ultimately a human endeavor. No transformation succeeds unless people understand, accept, and commit to the change. Leaders who underestimate this reality discover that resistance is not a barrier but a predictable response to uncertainty. Effective change management acknowledges the emotional, cultural, and psychological dimensions of transformation. It ensures that communication is transparent, expectations are clear, and support is available. It turns change from something imposed into something shared. In this dimension, execution becomes relational—rooted in trust, empathy, and the ability to guide people toward possibility.

Data Architecture Methodologies

Building the Backbone of Execution

Data architecture provide the backbone for modern execution. In a world where decisions must be informed by evidence, organizations cannot afford fragmented data, inconsistent definitions, or unreliable pipelines. Data architecture is not a technical concern; it is a strategic one. It determines whether insights are trustworthy, whether systems can scale, and whether AI can operate responsibly. Leaders who invest in strong data architecture create the conditions for intelligence to flow freely across the enterprise. In this dimension, execution becomes intelligent—powered by clarity, consistency, and evidence.

Customer Journey Methodologies

Anchoring Execution in Value

Customer journey ensure that execution remains anchored in value creation. Too many organizations design processes around internal convenience rather than customer experience. Customer journey thinking forces leaders to see the organization from the outside in, identifying friction, gaps, and opportunities that internal perspectives overlook. It ensures that execution is not only efficient but meaningful. It aligns operations with the expectations, needs, and emotions of the people the organization serves. In this dimension, execution becomes empathetic—shaped by the reality of customers rather than the assumptions of leaders.

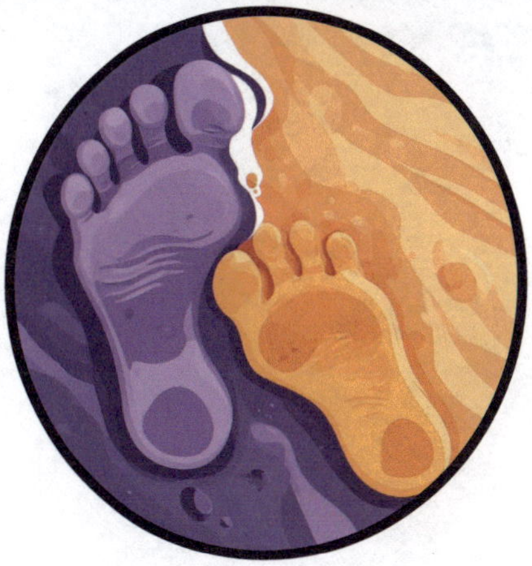

Agile Development Methodologies

Enabling Adaptation at Speed

Agile development introduce the adaptability required in a world where conditions shift rapidly. Traditional execution models assume stability, predictability, and linear progress. Agile assumes the opposite. It embraces iteration, experimentation, and continuous learning. It allows teams to adjust quickly, respond to feedback, and deliver value incrementally. Leaders who adopt agile mindsets create organizations that move with speed without sacrificing quality. In this dimension, execution becomes dynamic—capable of evolving in real time as new information emerges.

Process Improvement Methodologies

Sustaining Excellence Over Time

The final dimension, **process improvement methodologies**, completes the system by ensuring that execution evolves rather than stagnates. They provide the discipline to identify inefficiencies, eliminate waste, and refine workflows. They ensure that execution is not a one-time effort but a continuous practice. Leaders who embrace process improvement build organizations that get better over time, not simply bigger or busier. In this dimension, execution becomes self-correcting—driven by curiosity, discipline, and a commitment to excellence.

Execution Mechanics is the lens that transforms strategy into reality by ensuring that ambition is supported by structure, that change is supported by people, and that progress is supported by systems. It reveals that execution is not the final step of leadership but the continuous arena in which leadership is tested. In a world where AI accelerates both opportunity and exposure, leaders must design execution as deliberately as they design vision, building the operational muscle that allows the organization to move with coherence under pressure. This lens makes clear that execution is not a matter of effort but of engineering—of constructing the models, architectures, and behaviors that convert intention into repeatable performance. Execution becomes a living capability rather than a periodic push, a discipline that strengthens with use rather than erodes under stress.

EXECUTION MECHANICS ANALOGY – STREAMLINING AIRPORT

STRATEGIC MODEL

Flight Schedule

Establishing a universal cadence ensures every department operates with synchronized precision. The master timeline that dictates when every resource must be in place to ensure a coherent, global operation.

DATA ARCHITECTURE

Air Traffic Control

Maintaining a high-fidelity signaling system eliminates the risk of operational collisions and blind spots. The invisible infrastructure that provides the real-time telemetry needed for confident, evidence-based decisions at high velocity.

CUSTOMER JOURNEY

The Terminal Flow

Eliminating friction at every touchpoint transforms the user experience into a seamless journey of value. Redesigning the "Gate" flow to remove bottlenecks and demonstrate respect for the customer's time and needs.

AGILE DEVELOPMENT

Quick Turnaround

Compressing execution cycles minimizes the cost of delay and keeps strategic assets in constant motion. The ability to de-plane, refuel, and re-load (iterate) with extreme speed, ensuring the organization never "idles" on the tarmac.

CHANGE MANAGEMENT

Ground Crew Synergy

Cultivating organizational elasticity allows the workforce to pivot instantly without losing mechanical momentum. Leaders create psychological safety for the "crew" to adapt during disruptions, synchronized under pressure.

PROCESS IMPROVEMENT

Logistics Optimization

Embedding a habit of marginal gains compounds small refinements into a structural advantage. Identifying every millisecond lost in baggage handling or fueling to maximize the total capacity and throughput of the hub.

COMMON FRAMEWORK - TOGAF (The Open Group Architecture Framework)

Book 1: Leadership Transformation—Strategic Direction : The frameworks in this pillar ensure that all efforts are strategically aligned and that leaders are focused on the right outcomes for value creation.

1. **McKinsey 7-S Framework:** Aligns the organization's strategy with its structure, systems, staff, and skills to ensure all components support the desired transformation.
2. **Business Model Canvas:** Creates a shared understanding of how value will be created, delivered, and captured, providing a clear map for all teams.
3. **McKinsey Three Horizons of Growth:** Structures innovation efforts across short-term improvements (Horizon 1), scaling existing successes (Horizon 2), and future-proofing the business (Horizon 3), ensuring a sustainable pipeline of value.
4. **Balanced Scorecard:** Translates the strategic vision into measurable performance objectives, connecting operational activities to high-level financial and non-financial goals.
5. **SWOT Analysis:** Provides a structured way to evaluate the strategic context, enabling leaders to leverage strengths, mitigate weaknesses, seize opportunities, and counter threats.
6. **Blue Ocean Strategy:** Drives leaders to look for uncontested market space, shifting the focus from competing for value to creating new value.

Book 2: Workforce Transformation—Adoption and Engagement : This pillar addresses the human element of change, ensuring that employees are prepared, engaged, and able to sustain new ways of working.

1. **Kotter's 8-Step Change Model:** Provides a step-by-step process for leading major change, from creating a sense of urgency to solidifying new approaches in the corporate culture.
2. **ADKAR Model:** Guides change managers in ensuring that individual employees have the Awareness, Desire, Knowledge, Ability, and Reinforcement needed for successful change.
3. **Lewin's Change Management Model:** Conceptualizes change as a three-stage process: "Unfreeze" existing habits, implement the "Change," and "Refreeze" the new processes to make them stick.
4. **Bridges' Transition Model:** Focuses on the psychological journey of individuals, helping them navigate the emotional stages of an ending, a neutral zone, and a new beginning.
5. **Prosci Change Management Methodology:** Provides a structured, research-based methodology for applying change management to any project or initiative.
6. **The Satir Change Model:** Helps leaders understand and anticipate the emotional and performance impacts of change on individuals and teams.

Book 3: Data Transformation—Measurement and Insights - This pillar provides the tools and architectures to manage data as a strategic asset, ensuring that value can be reliably measured and tracked.

1. **Zachman Framework:** Offers a comprehensive, logical structure for classifying and organizing enterprise architecture, ensuring data management is part of the larger business strategy.
2. **DAMA-DMBOK:** Provides a set of best practices for managing data throughout its lifecycle, ensuring the quality and integrity of data used for value measurement.
3. **Data Mesh:** Promotes a decentralized, domain-driven approach to data ownership, empowering teams to create and manage data as a product.
4. **Lambda Architecture (or Kappa Architecture):** Establishes a technical architecture for processing large volumes of data efficiently, supporting real-time value tracking and analytics.
5. **Master Data Management (MDM) Frameworks:** Ensure that an organization's critical business data is consistent, accurate, and provides a "single source of truth" for reliable reporting.
6. **Data Fabric:** Creates a flexible, interconnected data architecture that supports seamless access and integration of data, regardless of its location.

Book 4: Customer Transformation—External Value Delivery: This pillar focuses on defining and delivering value from the customer's perspective, ensuring that transformation efforts translate to measurable customer outcomes.

1. **Customer Journey Mapping:** Visualizes the customer's end-to-end experience, identifying pain points and opportunities for value creation.
2. **Service Blueprinting:** Expands on journey mapping by detailing the front-stage and backstage processes, people, and systems that support a customer experience.
3. **Jobs to be Done (JTBD) Framework:** Helps define "value" from the customer's perspective by focusing on the functional and emotional outcomes they seek.
4. **Design Thinking Methodology:** Provides a human-centered approach to innovation that focuses on empathizing with customers to design meaningful solutions.
5. **CX (Customer Experience) Maturity Models:** Provides a benchmark for tracking and improving customer experience over time, linking strategic initiatives to better customer outcomes.
6. **Persona Development Frameworks:** Creates detailed profiles of target users, ensuring that transformation efforts are grounded in a deep understanding of customer needs.

Book 5: Product Transformation—Internal Value Delivery : This pillar establishes an agile, iterative process for continuously building and improving the products and services that deliver customer value.

1. **Scrum:** Enables cross-functional teams to deliver value incrementally in short sprints, allowing for continuous feedback and adaptation.
2. **Kanban:** Helps manage and visualize workflow, limiting work-in-progress and improving the flow of value to the customer.
3. **Lean Startup:** Applies a "build-measure-learn" feedback loop to product development, ensuring teams are constantly validating their assumptions about value creation.
4. **Scaled Agile Framework (SAFe):** Provides a system for coordinating and aligning multiple agile teams across the enterprise to deliver large-scale solutions.
5. **Feature Flags / Toggles Architecture:** Enables teams to deploy new features and capabilities quickly and safely, allowing for real-time testing of value propositions.
6. **Continuous Integration / Continuous Delivery (CI/CD) Pipeline Architecture:** Automates the software delivery process, allowing teams to deliver new value to customers efficiently and reliably.

Book 6: Operations Transformation—Efficiency and Sustainability
This pillar focuses on optimizing the internal processes and systems that support the entire value chain, ensuring efficiency and sustainability.

1. **Lean Management:** Aims to eliminate waste and optimize the flow of value through the organization's processes, generating cost savings and efficiency gains.
2. **Six Sigma (DMAIC Methodology):** Uses a data-driven approach to improve processes by defining, measuring, analyzing, improving, and controlling them.
3. **BPM (Business Process Management) Lifecycle:** Provides a structured approach to continuously improving business processes, ensuring sustained value delivery over the long term.
4. **SCOR Model (Supply Chain Operations Reference):** Offers a standard framework for managing and improving supply chain efficiency, a key driver of operational value.
5. **Digital Twin Architecture:** Creates virtual models of physical assets or processes, enabling simulation and optimization before real-world changes are made.
6. **Robotic Process Automation (RPA) Frameworks/Methodologies:** Automates repetitive, rule-based tasks, freeing up human resources for higher-value activities.

TOGAF (The Open Group Architecture Framework) is a widely recognized and comprehensive enterprise architecture methodology that provides a structured approach for developing and managing enterprise architecture. It helps organizations design, plan, implement, and govern an enterprise information architecture. At its core is the Architecture Development Method (ADM), a detailed, iterative process for developing various architectural domains: Business, Data, Application, and Technology. For leadership transformation, particularly in achieving digital maturity, TOGAF is indispensable. Digital transformation is not just about implementing new technologies; it requires aligning business strategy with IT capabilities across the entire enterprise. Without a coherent architecture, digital initiatives can become fragmented, leading to integration challenges, redundant systems, security vulnerabilities, and a lack of scalability. TOGAF provides the necessary rigor to ensure that digital investments are strategically aligned, well-integrated, and support the organization's long-term digital vision. It helps leaders understand the current state of their digital landscape, define the desired future state (e.g., cloud-native infrastructure, API-driven integration, AI-enabled applications), and create a roadmap to get there. By fostering a common language and framework for architects and stakeholders, TOGAF enables effective communication and collaboration across complex digital initiatives. It ensures that the digital transformation is not just a series of isolated projects but a cohesive, well-governed evolution of the entire enterprise, a critical component of achieving and sustaining comprehensive digital maturity. It provides the blueprint for how an organization's digital capabilities (applications, data, infrastructure) support its strategic business goals, ensuring that every technological investment contributes to a unified, adaptable, and future-ready digital enterprise. This structured approach helps avoid costly misalignments and ensures that digital initiatives are built on a solid, scalable foundation, allowing the organization to truly harness the power of technology for competitive advantage.

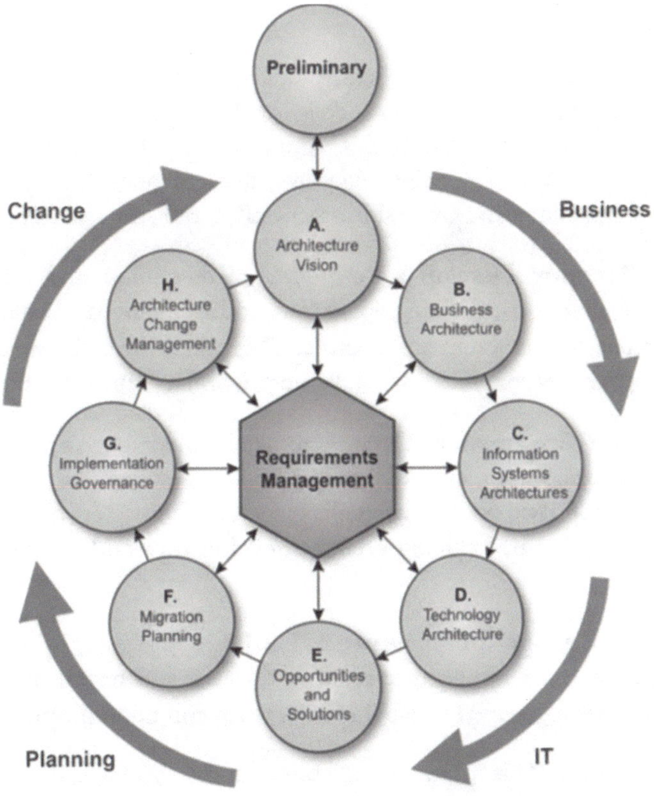

TOGAF Frameworks, Image source: medium.com

1. **Preliminary Phase: Framework and Principles:** Establish the architecture capability for the enterprise. This involves defining the organizational context for architecture work, identifying key stakeholders, establishing the architecture principles (e.g., "cloud-first," "API-driven," "data-driven decision making"), and defining the scope and governance of the architecture effort.

2. **Phase A: Architecture Vision:** Develop a high-level vision for the target architecture that aligns with business strategy and digital transformation goals. This involves identifying key stakeholders, defining the business value of the architecture work, and securing approval for the architecture development cycle.

3. **Phase B: Business Architecture:** Develop a baseline (current state) and target (future state) Business Architecture that describes the organization's processes, functions, and organizational structure. This ensures that the digital transformation efforts are aligned with and support the core business capabilities and strategic objectives, identifying where digital enablement can optimize business value, streamline workflows, and enhance customer interactions through digital means.

4. **Phase C: Information Systems Architectures (Data & Application):** Develop baseline and target Data Architecture (logical and physical data models, data governance, data flows) and Application Architecture (applications, their interactions, and how they support business functions). This phase is critical for designing integrated digital systems, ensuring data flows seamlessly across digital platforms, and supporting new digital applications that drive business value and customer engagement.

5. **Phase D: Technology Architecture:** Develop baseline and target Technology Architecture that describes the logical software and hardware capabilities required to support the Business, Data, and Application components. This includes defining infrastructure (e.g., cloud platforms, edge computing), platforms (e.g., microservices, AI/ML platforms), and security standards necessary for a robust, scalable, and resilient digital enterprise.

6. **Phase E: Opportunities & Solutions:** Identify and evaluate various implementation options and opportunities to deliver the target architecture. This involves assessing gaps between baseline and target architectures, identifying potential solutions (e.g., Commercial Off-The-Shelf (COTS), custom development, cloud migration, SaaS adoption), and defining high-level implementation projects. For digital maturity, this prioritizes digital solutions that offer the greatest business value and strategic alignment.

7. **Phase F: Migration Planning:** Develop a detailed implementation and migration plan. This involves prioritizing projects, estimating costs, defining a timeline, and outlining the sequence of activities required to transition from the baseline to the target architecture. This ensures a structured and managed approach to digital transformation implementation, minimizing disruption and maximizing efficiency.

8. **Phase G: Implementation Governance:** Oversee the implementation of the architecture. This involves establishing governance processes to ensure that projects adhere to the defined architecture, managing architectural contracts, and providing architectural guidance during development and deployment. This continuous oversight is vital for maintaining architectural integrity and consistency throughout the digital transformation.

9. **Phase H: Architecture Change Management:** Establish processes for managing changes to the architecture. This ensures that the architecture remains relevant and responsive to evolving business needs, technological advancements (e.g., new AI capabilities, quantum computing), and market shifts.

10. **Requirements Management:** This phase runs concurrently throughout all ADM phases. It involves managing the architectural requirements, ensuring they are consistent, complete, and aligned with stakeholder needs. For digital maturity, this ensures that evolving digital requirements (e.g., real-time data access, scalability for new digital services) are captured and integrated into the architectural design and implementation.

Ai6x6™ Transformation Blueprint

Data Architecture Frameworks

"If you torture the data long enough, it will confess to anything."

Ronald Coase. This highlights the need for a rigorous, well-governed architecture to ensure data integrity and prevent flawed conclusions.

Data Architecture Frameworks are the structural blueprints that guarantee data transformation efforts deliver reliable, scalable, and compliant results. Frameworks define the standardized design for all data components, from ingestion and storage (data lakes, warehouses) to processing and consumption (APIs, analytics tools). This structure ensures **data consistency, quality, and interoperability**, moving the organization beyond fragmented silos to a unified, trusted data asset. The framework governs metadata, security, and lineage, which is vital for ethical AI development and regulatory compliance (like GDPR). By enforcing a cohesive architecture, the framework maximizes the value derived from data, enabling sophisticated analytics and feeding reliable machine learning models. Without this rigorous structure, data transformation risks creating complex, untrustworthy systems, crippling the organization's ability to operate from a **single, strategic source of truth**.

The Data Architecture Framework pillar is crucial for treating data as a strategic asset, providing the necessary tools and architectures to ensure its value can be reliably measured, tracked, and maximized across the organization. Core to this is the Zachman Framework, which offers a comprehensive, logical structure for classifying enterprise architecture, integrating data management directly into the business strategy. For practical execution, the DAMA-DMBOK provides the industry standard for best practices, ensuring high quality and integrity throughout the data lifecycle. Modern architectures, like the Data Mesh, shift data ownership to individual business domains, treating data as a product, while the Data Fabric creates a flexible, interconnected layer for seamless data access regardless of its physical location. To ensure consistency for reporting, Master Data Management (MDM) Frameworks establish a "single source of truth" for critical business information. Finally, technical architectures such as the Lambda Architecture (or Kappa Architecture) are implemented to process massive data volumes efficiently, enabling real-time analytics and reliable value tracking. Collectively, these frameworks ensure the enterprise can govern, leverage, and trust its data for strategic decision-making.

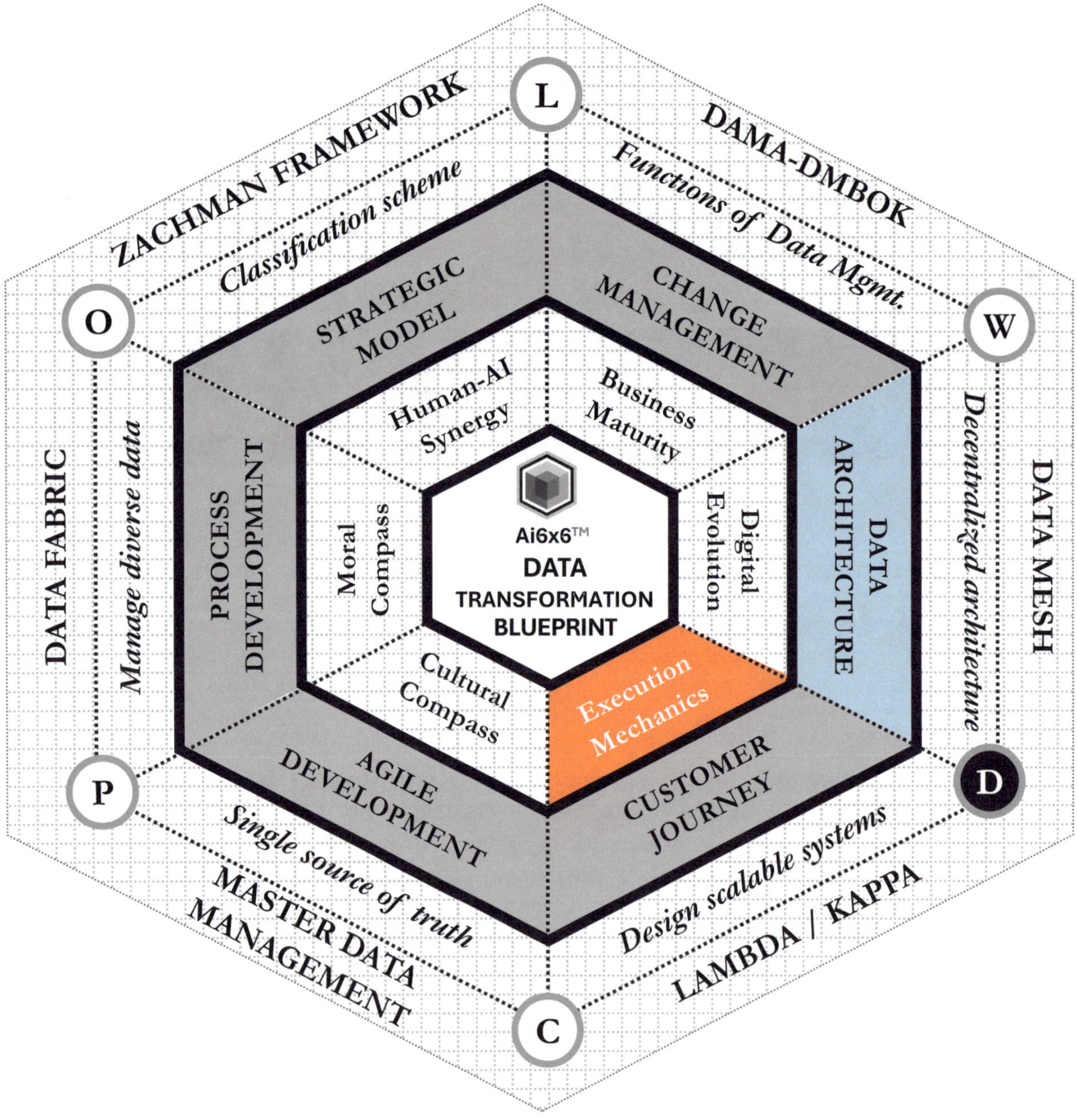

Ai6x6 DATA ARCHITECTURE FRAMEWORKS | DATA TRANSFORMATION

The Zachman Framework for Enterprise Architecture is a comprehensive ontology that provides a structured way to view and understand an enterprise. It's not a methodology but a classification scheme, a "periodic table" for architectural artifacts, organizing them into a 6x6 matrix based on six interrogatives (What, How, Where, Who, When, Why) and six perspectives (from Planner's view to Functioning Enterprise). For data transformation, this framework is invaluable because it ensures a holistic approach, preventing blind spots that often plague complex data initiatives. Digital maturity is fundamentally dependent on high-quality, accessible, and well-governed data. Without a comprehensive architectural view, data transformation efforts can become fragmented, leading to new data silos or inconsistent data definitions. The Zachman Framework compels data leaders to consider data from every angle: its strategic purpose (Why), its business meaning (What), its logical design (How), its physical storage locations (Where), its ownership and security (Who), and its lifecycle (When). By systematically addressing each cell of the matrix, organizations can build a robust data architecture that aligns with business strategy, supports diverse applications (including AI and digital products), ensures data quality and security, and provides clear data lineage. This structured approach to data architecture is critical for moving beyond basic data management to truly leveraging data as a strategic asset for digital innovation and competitive advantage, a hallmark of deep digital maturity..

Zachman	Why	How	What	Who	Where	When
Contextual	Goal List	Process List	Material List	Organisational Unit & Role List	Geographical Locations List	Event List
Conceptual	Goal Relationship	Process Model	Entity Relationship Model	Organisational Unit & Role Relationship Model	Locations Model	Event Model
Logical	Rules Diagram	Process Diagram	Data Model Diagram	Role Relationship Diagram	Locations Diagram	Event Diagram
Physical	Rules Specification	Process Function Specification	Data Entity Specification	Role Specification	Location Specification	Event Specification
Detailed	Rules Details	Process Details	Data Details	Role Details	Location Details	Event Details

ZACHMAN framework , Image source: Wikipedia.org

1. **Scope (Planner's View - Row 1):** At the highest level, define the overall strategic goals ("Why") for data transformation from an executive perspective. This includes understanding the business drivers for better data (e.g., enabling AI, personalized customer experiences, operational automation). Identify the high-level data entities ("What") that are critical to the business. This row provides a contextual, executive summary of the data landscape and its strategic importance, setting the broad direction for the transformation and aligning data initiatives with the overarching digital strategy.

2. **Business Model (Owner's View - Row 2):** Describe data in terms of business processes and concepts. This involves understanding how the business currently uses data and how it *wants* to use data in a digitally mature state. What are the key business entities (e.g., Customer, Product, Order) from a conceptual perspective, and their relationships? It focuses on the business semantics and rules governing data, independent of specific technology, ensuring data models reflect actual business operations and enable new digital capabilities.

3. **System Model (Designer's View - Row 3):** Create logical data models (e.g., Entity-Relationship Diagrams - ERDs, conceptual data flows) that define the structure and relationships of data independent of specific technology or database vendors. This level details how applications logically interact with data, including data flows and transformations, providing a technology-agnostic blueprint for data organization. For digital maturity, this ensures data is structured for seamless integration across digital systems and applications.

4. **Technology Model (Builder's View - Row 4):** Define the physical data structures (e.g., database schemas, table definitions, file formats, APIs) and the specific technologies used for data storage, processing, and access (e.g., SQL databases, NoSQL databases, data warehouses, cloud data lakes, streaming platforms). This level translates the logical design into concrete technical specifications, considering performance, scalability, security, and the specific digital platforms being used.

5. **Components (Sub-Contractor's View - Row 5):** Detail specific data components, their implementation, and configuration. This includes defining specific database tables, API definitions for data access, data integration scripts, data quality rules, and security configurations at a granular level. It's the level of detail required for developers and data engineers to build the actual data solutions that power digital products and services.

6. **Functioning Enterprise (Worker's View - Row 6):** Represent the operational data systems and how data is actually used in the day-to-day business. This includes real-time data flows, data lineage (tracking data from source to consumption), data security enforcement, and operational monitoring of data pipelines. It's the "live" view of data in action, ensuring the architecture supports actual business operations and provides the reliable, real-time data necessary for a digitally mature enterprise.

> "A fool with a tool is still a fool."
> — Grady Booch, Chief Scientist, IBM Research (Software Architecture Pioneer)

2. DAMA-DMBOK (DATA MANAGEMENT BODY OF KNOWLEDGE)

The Data Management Body of Knowledge (DAMA-DMBOK) is a comprehensive framework that defines and describes the functions of data management. It provides a common vocabulary and set of best practices for managing data as a strategic asset. For data transformation, DAMA-DMBOK is fundamental because it moves beyond just technology solutions to encompass the entire lifecycle and governance of data. Digital maturity is predicated on trustworthy, accessible, and well-governed data. Without a robust data management foundation, digital initiatives like AI, advanced analytics, or personalized customer experiences will struggle due to poor data quality, fragmented data silos, or lack of clear data ownership. DAMA-DMBOK addresses crucial areas such as Data Governance (establishing accountability), Data Quality (ensuring accuracy), Master Data Management (creating a single source of truth), Metadata Management (understanding data context), and Data Security (protecting sensitive information). By implementing the principles and practices outlined in DAMA-DMBOK, organizations can build a resilient and reliable data ecosystem. This ensures that data is not just collected, but actively managed, secured, and made available in a usable format for all digital operations and decision-making. It transforms data from a mere byproduct of operations into a strategic enabler for digital innovation and competitive advantage, a true indicator of digital maturity.

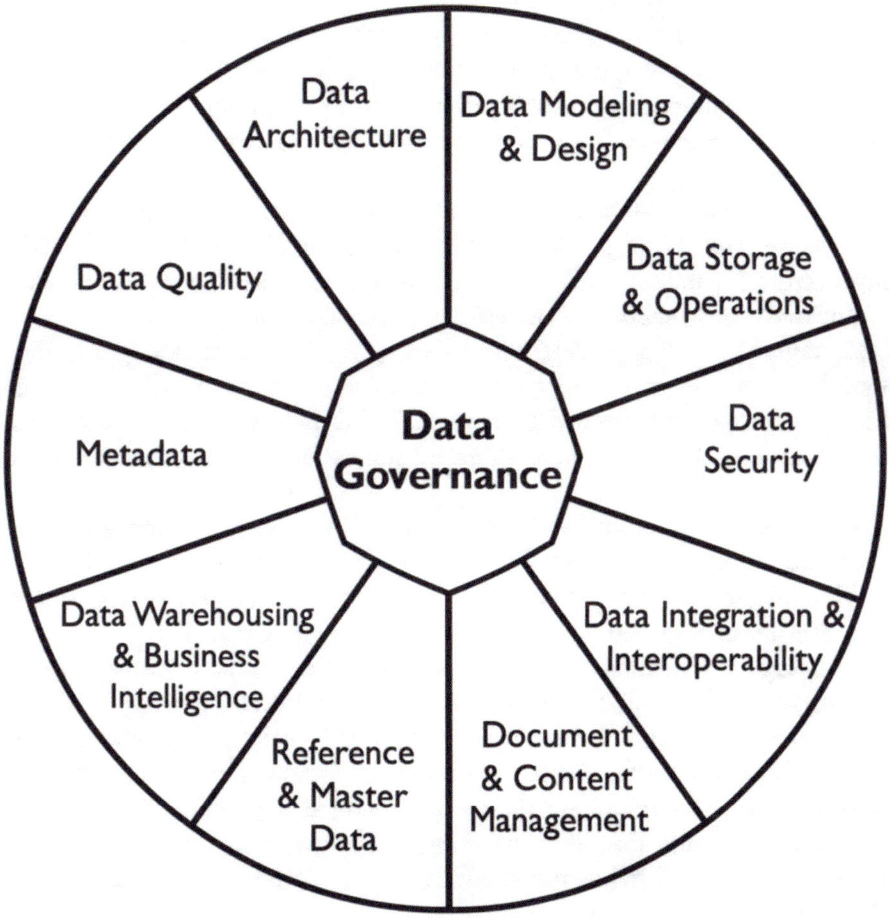

DAMA Wheel | Image source: https://datacrossroads.nl/

1. **Data Governance:** Establish policies, roles, and responsibilities for managing data across the organization. This is the foundational pillar, ensuring accountability, decision-making authority, and oversight for all data-related activities, setting the strategic direction for data transformation. For digital maturity, robust data governance ensures data is a trusted asset, enabling ethical AI development and compliant digital services.
2. **Data Architecture:** Design the overall data landscape, including conceptual, logical, and physical data models, data flows, and data integration patterns. This function defines the blueprint for how data is structured, stored, processed, and accessed across the enterprise to support business needs. A well-defined data architecture is critical for integrating new digital data sources and supporting advanced analytics platforms.
3. **Data Modeling & Design:** Create detailed conceptual, logical, and physical data models to define data structures and relationships. This involves translating business requirements into precise data definitions and database designs, ensuring data consistency and integrity. For digital maturity, accurate data models are essential for building reliable digital products and services that depend on structured information.
4. **Data Storage & Operations:** Manage the physical storage, access, and lifecycle of data assets. This includes database administration, data archiving, backup and recovery, and ensuring optimal performance and availability of data stores. In a digitally mature environment, this often involves managing data across cloud platforms, data lakes, and streaming systems with high availability and scalability.
5. **Data Security:** Implement robust measures to protect data from unauthorized access, use, disclosure, disruption, modification, or destruction. This involves defining security policies, access controls, encryption, and monitoring to ensure data privacy and integrity. For digital maturity, strong data security is paramount to maintain customer trust in digital services and comply with evolving data protection regulations.
6. **Data Quality:** Define, measure, monitor, and improve the accuracy, completeness, consistency, timeliness, and validity of data. This function ensures that data is fit for purpose and can be trusted for decision-making and operational use. High data quality is non-negotiable for reliable digital operations, accurate AI models, and effective data-driven insights.
7. **Master Data Management (MDM):** Create and maintain a consistent, accurate "single source of truth" for critical business entities (e.g., Customer, Product, Employee, Supplier). MDM eliminates data inconsistencies across disparate systems, providing a unified and reliable view of core business data, which is vital for personalized digital experiences and efficient digital operations.
8. **Metadata Management:** Manage data about data (e.g., definitions, lineage, quality rules, ownership, technical specifications) to improve understanding, trust, and usability of data assets across the organization. This enables data discovery and effective governance. In a digitally mature organization, automated metadata management allows data scientists and analysts to quickly find and understand relevant data for digital initiatives.
9. **Data Integration & Interoperability:** Combine data from disparate sources and ensure seamless data exchange between systems and applications. This involves designing and implementing ETL/ELT processes, APIs, and data virtualization techniques to create a unified data view. Seamless data integration is a cornerstone of digital maturity, enabling real-time data flows for dynamic digital services.
10. **Data Warehousing & Business Intelligence (BI):** Design and manage data repositories (data warehouses, data marts) and tools for reporting, analytics, and decision support. This function transforms raw data into actionable insights for business users. While traditional, modern BI in a digitally mature context leverages cloud data warehouses and advanced visualization for faster, more accessible insights.
11. **Document & Content Management:** Manage unstructured data assets like documents, images, videos, and web content. This ensures that all forms of organizational information are properly organized, stored, secured, and accessible. For digital maturity, this extends to managing digital content for websites, mobile apps, and AI-driven content generation.

Data Mesh, pioneered by Zhamak Dehghani, is a revolutionary, decentralized data architecture paradigm that challenges traditional centralized data lake or data warehouse approaches. It advocates for treating data as a product owned by cross-functional business domains, rather than by a central data team. This paradigm shift is incredibly relevant for data transformation, especially in the pursuit of digital maturity. In a digitally mature organization, data is not just a byproduct; it's a core asset consumed by numerous digital products, services, and AI models. Traditional centralized data teams often become bottlenecks, unable to keep pace with the diverse and rapidly evolving data needs of agile digital initiatives. Data Mesh addresses this by empowering domain teams—who understand their data best—to be responsible for its entire lifecycle, from ingestion to serving. This fosters greater accountability, improves data quality at the source, and accelerates data delivery. The self-serve data platform principle provides the necessary infrastructure and tools, abstracting away complex engineering details, allowing domain teams to easily create and expose their data products. Federated computational governance ensures consistency and compliance across the distributed landscape without imposing rigid central control. By enabling decentralized, product-oriented data ownership, Data Mesh breaks down data silos, promotes data discoverability, and drastically reduces the time it takes to get valuable data into the hands of digital product developers, data scientists, and business analysts. This agility and scalability in data delivery are essential for continuous digital innovation and a true indicator of advanced digital maturity.

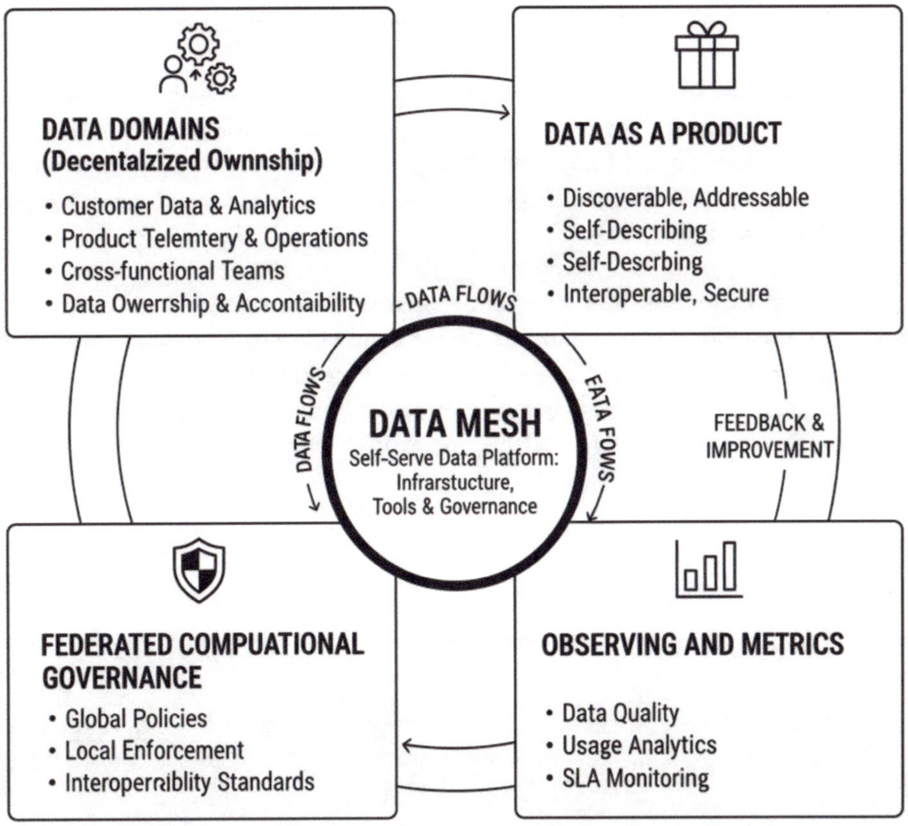

Data Mesh

3. DATA MESH (ZHAMAK DEHGHANI)

1. **Domain-Oriented Ownership:** Reorganize data ownership from centralized data teams (e.g., data platform, data warehouse) to cross-functional business domains (e.g., Sales Data Domain, Product Data Domain, Customer Support Domain). Each domain team becomes fully responsible for the entire lifecycle of its operational data, including its quality, security, and usability. This fosters accountability and expertise closer to the business context, enabling faster response to digital business needs.

2. **Data as a Product:** Each domain team treats its data as a product, with clear APIs, comprehensive documentation, defined quality standards, and service-level agreements (SLAs) for data consumers. This means data is discoverable, addressable, trustworthy, self-describing, interoperable, and secure, just like any other well-engineered product. This product-centric view of data is crucial for a digitally mature organization, where data is consumed by diverse digital applications and AI models.

3. **Self-Serve Data Platform:** Build a foundational data platform that provides reusable capabilities (e.g., data ingestion tools, storage, compute, security, monitoring, governance frameworks) for domain teams. This platform empowers domain teams to easily create, publish, and consume data products without needing deep data engineering expertise, reducing friction and accelerating the delivery of new data-driven digital capabilities. It democratizes data access and usage.

4. **Federated Computational Governance:** Establish a set of global, interoperable data governance policies and rules that are enforced automatically by the self-serve platform. This allows for decentralized execution of data management within centralized guidelines, balancing autonomy for domain teams with necessary enterprise-wide consistency, compliance (e.g., data privacy regulations), and security for digital data assets. This ensures data integrity and trust across the mesh.

5. **Enable Data Product Creation:** Focus on empowering domain teams with the tools, knowledge, and autonomy to design, build, and expose their high-quality data products. This involves providing templates, automation, and support for data modeling, quality checks, and API development, making it easier for business experts to become data product owners. This accelerates the development of new data-driven digital features and services.

6. **Foster Data Product Consumption:** Encourage internal consumers (analysts, data scientists, digital product developers, operational applications) to discover and utilize data products through the self-serve platform. This involves clear data catalogs, robust search functionalities, and easy-to-use interfaces for accessing and integrating data, promoting a data-driven culture across the organization and enabling rapid digital innovation based on readily available, trustworthy data.

> "A data mesh is a decentralized approach to data architecture that takes a domain-oriented, self-serve design."
> — Zhamak Dehghani, Creator of Data Mesh

The Lambda Architecture (and its simpler alternative, Kappa Architecture) are crucial architectural patterns for designing scalable, fault-tolerant big data systems. They address the challenge of processing massive volumes of data generated by digital interactions, providing both real-time insights and comprehensive historical analysis. For data transformation, these architectures are fundamental because they enable organizations to move beyond batch-only processing to leverage the full potential of real-time data for digital maturity. Digital maturity demands immediate responsiveness: personalized recommendations, fraud detection, dynamic pricing, and real-time operational dashboards all rely on processing data as it arrives. The Lambda Architecture achieves this by maintaining two distinct processing paths: a batch layer for accurate, historical data and a speed layer for real-time, approximate data. The Kappa Architecture simplifies this by using a single stream processing layer for both real-time and historical needs. Both architectures are designed for fault tolerance and scalability, ensuring that data pipelines remain robust and available even under extreme load or system failures, which is critical for continuous digital operations. By implementing these patterns, organizations can build data platforms that deliver timely, reliable insights to power digital products, enhance customer experiences, and optimize operational efficiency, moving beyond static reporting to dynamic, data-driven decision-making and automated responses—a clear indicator of advanced digital maturity.

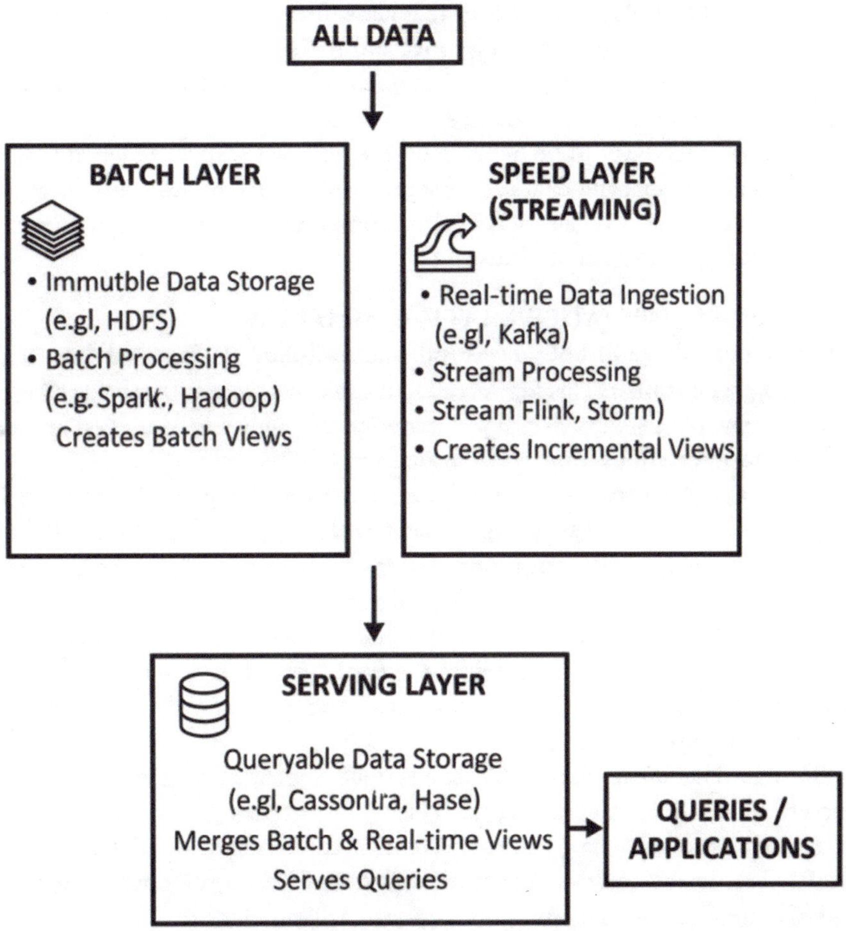

Lambda Architecture , Image source: databricks.com

1. **Batch Layer:** This layer is responsible for ingesting and storing all raw, immutable data from various digital sources (e.g., website clicks, IoT sensor data, transaction logs). It processes large volumes of historical data (e.g., nightly or hourly) to produce highly accurate, comprehensive insights and master datasets. This layer is typically robust and fault-tolerant, handling re-computation for data consistency and historical accuracy, providing the foundational analytical view for digital business.
2. **Speed Layer (Streaming):** This layer processes real-time data streams for immediate, low-latency insights. It provides approximate, real-time views of data, often used for dynamic digital dashboards, immediate alerts (e.g., fraud detection), or real-time personalization on digital platforms. It sacrifices some accuracy for speed, as it processes data as it arrives without waiting for batch computations, crucial for responsive digital experiences.
3. **Serving Layer:** This layer merges the results from both the batch layer (historical, accurate views) and the speed layer (real-time, approximate views) into a unified, queryable view. It's optimized for fast reads and serves data to applications and users, providing a comprehensive and up-to-date perspective on the data. For digital maturity, this layer ensures that digital applications and AI models have access to the most relevant and timely data.
4. **Data Ingestion:** Implement robust mechanisms to continuously feed raw data from all digital touchpoints and internal systems into both the batch and speed layers. This ensures that all data, whether historical or streaming, is captured and made available for processing in the respective layers, forming the backbone of a data-driven digital enterprise.
5. **Query & Analytics:** Provide tools and interfaces that allow users and applications to query the serving layer for insights. This includes dashboards, reporting tools, and APIs that can access the combined historical and real-time data views. For digital maturity, this means enabling self-service analytics and integrating data directly into digital products for intelligent features.
6. **Fault Tolerance:** Design each layer to be highly fault-tolerant and scalable independently. If one layer fails, the others can continue operating, and data can be reprocessed or recovered. This ensures the overall system remains robust and available even under high load or component failures, which is critical for maintaining continuous digital operations and customer trust.

KAPPA ARCHITECTURE STEPS (ALTERNATIVE, SIMPLER):

1. **Stream-First Ingestion:** All data, both historical and real-time, is ingested into a single, immutable, distributed log (e.g., Apache Kafka). This log serves as the single source of truth for all data, simplifying data management and ensuring consistency. For digital maturity, this allows for a unified approach to handling all data generated by digital interactions, from website clicks to IoT events.
2. **Stream Processing:** Data is processed in real-time directly from this unified log. Instead of separate batch and speed layers, a single stream processing engine handles all computations. If historical data needs to be re-processed or new analytics applied, the stream processor simply re-reads from the beginning of the log. This provides flexibility and agility for evolving digital analytics needs.
3. Se**rving Layer:** Processed data from the stream is stored in a serving layer (e.g., a real-time database, a search index) optimized for fast reads. This layer provides the final, queryable view of the data for applications and users. For digitally mature organizations, this enables real-time recommendations, personalized content, and immediate operational insights.
4. **Unified Data Source:** The single, immutable log simplifies data management, ensures consistency, and provides a complete audit trail for all data events. This unified approach is a key enabler for digital maturity, reducing data silos and improving data governance across complex digital ecosystems.
5. **Real-time Focus:** The architecture is inherently optimized for continuous, real-time data processing, making it ideal for applications requiring immediate insights derived from digital interactions. This responsiveness is critical for dynamic digital services and competitive advantage.
6. **Simplified Architecture:** Reduces complexity compared to maintaining separate batch and speed layers, leading to potentially faster development and easier maintenance of data pipelines. This architectural simplicity supports the rapid iteration and deployment cycles characteristic of digitally mature organizations.

Master Data Management (MDM) frameworks are essential for creating and maintaining a consistent, accurate "single source of truth" for critical business entities such as customers, products, suppliers, and locations. For data transformation, MDM is a foundational pillar because fragmented, inconsistent, or inaccurate master data undermines virtually all digital initiatives. Imagine trying to personalize a customer experience (digital maturity goal) when you have five different records for the same customer across various systems, each with conflicting information. Or attempting to optimize a supply chain with inconsistent product identifiers. MDM addresses these challenges by consolidating, cleansing, and synchronizing master data across the enterprise. It establishes clear data governance, defines authoritative data models, and implements processes to ensure data quality and integrity. By providing a unified and reliable view of core business data, MDM enables accurate analytics, trustworthy AI models, seamless omnichannel customer experiences, and efficient digital operations. It moves an organization beyond data chaos to data clarity, reducing operational inefficiencies, improving decision-making, and building customer trust in digital interactions. Achieving robust MDM is a clear indicator of an organization's commitment to data quality and its progression towards true digital maturity, where data is a trusted and actionable asset.

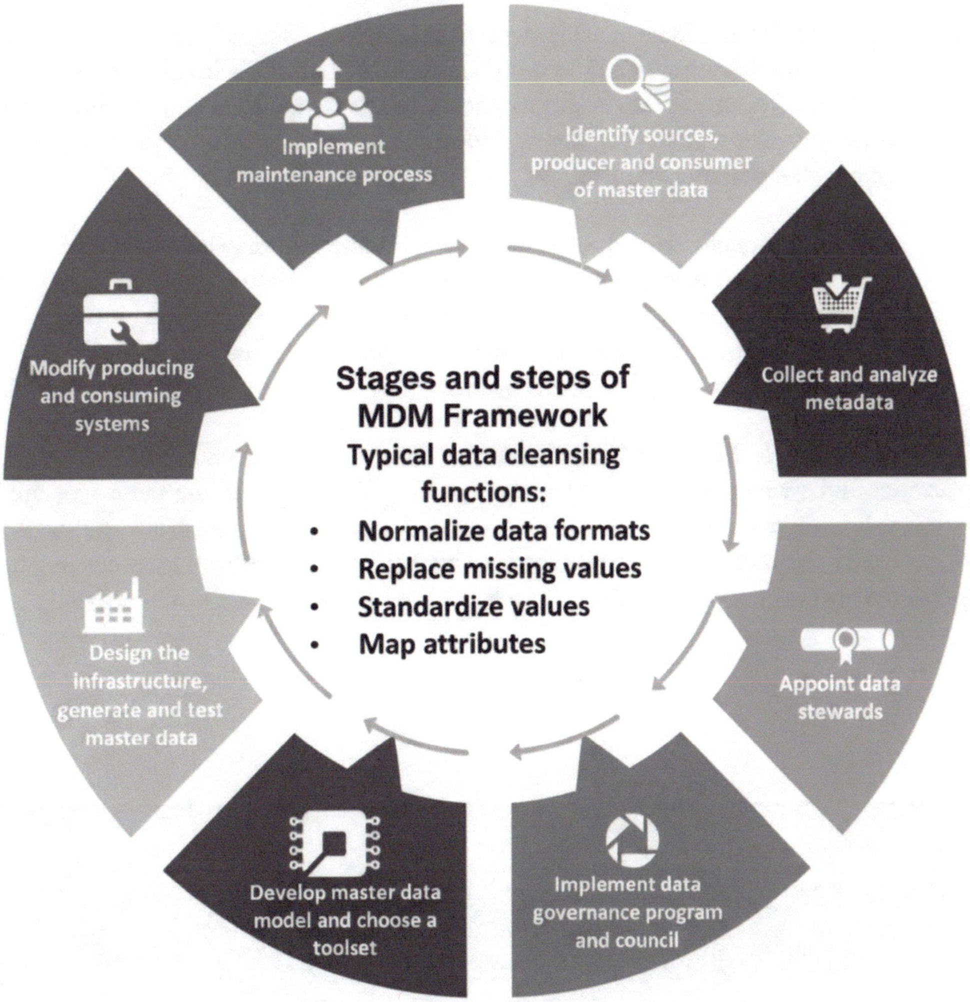

MDM Framework , Image source: Wikipedia.org

1. **Define Master Data:** Identify the critical data entities that need to be managed as master data (e.g., Customer, Product, Employee, Supplier, Location). These are the core, non-transactional entities that are shared and reused across multiple business processes and systems. For digital maturity, having a single, authoritative definition of a "customer" or "product" is essential for delivering consistent and personalized experiences across all digital channels.

2. **Data Modeling:** Develop a unified, enterprise-wide master data model that defines the structure, attributes, relationships, and business rules for these entities. This model serves as the authoritative blueprint for how master data should be represented across the organization. A robust master data model is the backbone for integrating data from disparate digital systems and ensuring data consistency for analytics and AI.

3. **Data Governance:** Establish clear ownership, policies, standards, and processes for managing master data quality, security, and access. Define roles such as "data stewards" who are responsible for the accuracy and integrity of specific master data domains, ensuring accountability. Strong MDM governance is fundamental for digital maturity, ensuring data used in digital processes and AI is trustworthy and compliant.

4. **Data Integration:** Consolidate master data from disparate source systems (e.g., CRM, ERP, e-commerce platforms, legacy systems) into a central MDM hub. This involves various techniques like data consolidation, coexistence, or registry patterns to create a unified and reconciled view of master data. Seamless integration of master data is critical for providing a 360-degree view of customers and products across all digital touchpoints.

5. **Data Quality:** Implement robust data cleansing, standardization, matching, and enrichment processes to ensure the accuracy, completeness, and consistency of master data. This step is crucial for eliminating duplicates, correcting errors, and ensuring the MDM hub provides truly reliable information. High-quality master data is non-negotiable for effective digital marketing, personalized recommendations, and accurate AI-driven insights.

6. **Data Distribution & Monitor:** Publish the trusted master data from the MDM hub to consuming applications and systems across the enterprise, ensuring everyone uses the same accurate information. Continuously monitor master data quality, usage, and compliance, evolving the MDM solution as business needs and data requirements change. This continuous feedback loop ensures that the digital organization always operates with the most reliable and up-to-date core business data.

> "Data warehousing is a process of transforming data into information and making it available to users in a timely and understandable manner."
> — Bill Inmon, "Father of Data Warehousing"

Data Fabric is an architectural concept that provides a unified, intelligent way to connect, discover, and manage diverse data sources across hybrid and multi-cloud environments. Unlike traditional data integration approaches that often involve moving all data to a central repository, a data fabric creates a logical layer that abstracts away data silos, making disparate data appear as one unified source.

For data transformation, this is a game-changer because it addresses the complexity of modern data landscapes, which are increasingly distributed across various cloud providers, on-premise systems, and SaaS applications. Achieving digital maturity requires rapid access to trustworthy data, regardless of its location or format, to power real-time analytics, AI models, and dynamic digital products. A data fabric accelerates data access by automating data discovery, integration, and transformation, often leveraging AI/ML to enhance these processes. It improves data governance across distributed landscapes by applying consistent policies and security controls. By providing a self-service experience for data consumers, it democratizes data access, empowering data scientists, analysts, and digital product developers to quickly find, understand, and use the data they need without manual intervention from central data teams. This agility in data delivery, coupled with robust governance, is a defining characteristic of a digitally mature organization that can truly leverage its data for competitive advantage.

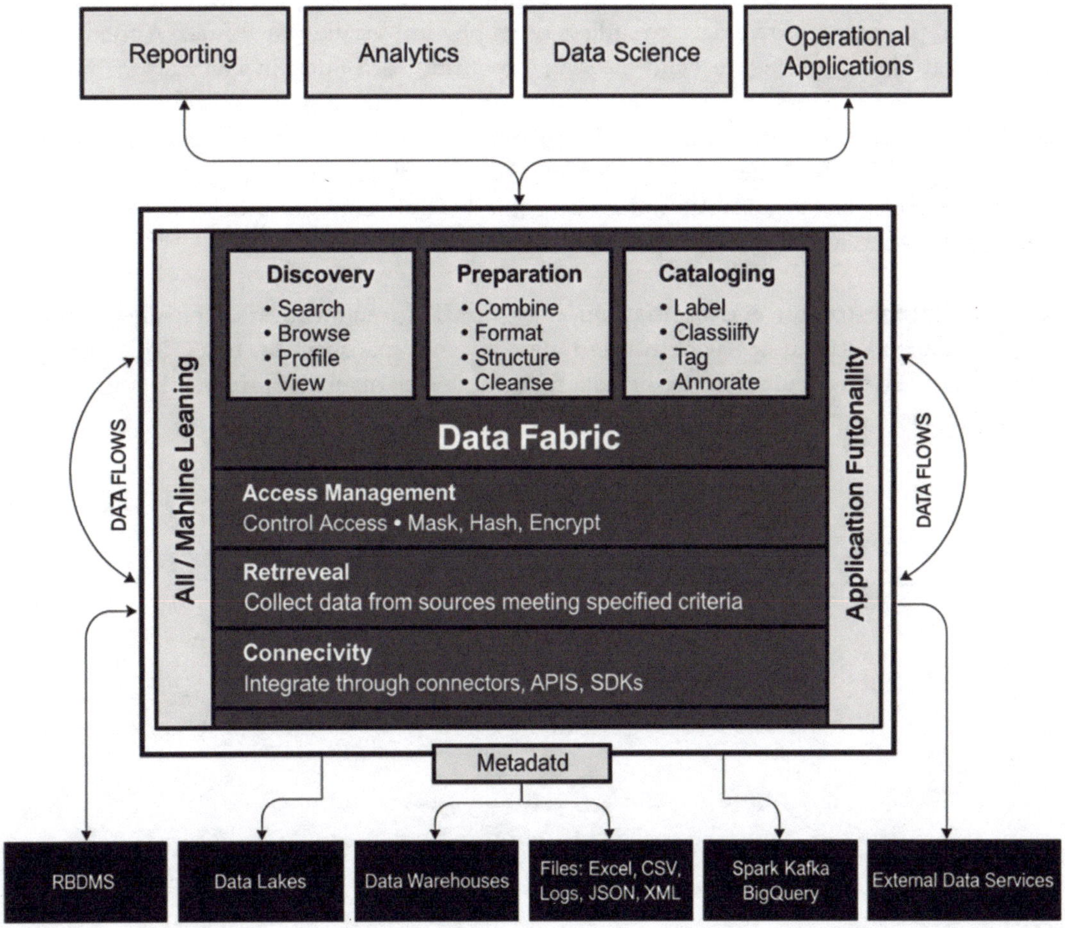

Data Fabric , Image source: Nexla.com

1. **Data Discovery & Cataloging:** Automatically discover, classify, and catalog all data assets across the enterprise, regardless of where they reside (on-premise databases, various cloud platforms, data lakes, streaming sources, applications). This creates a comprehensive, searchable inventory of available data, making it easier for data scientists and digital product teams to find and understand relevant data for their initiatives.

2. **Data Ingestion & Integration:** Connect to disparate data sources and integrate them using various techniques such as data replication, data virtualization, data streaming, or API-based integration. The goal is to create a unified, logical view of data without necessarily moving all data into one physical location. This flexibility is key for digitally mature organizations dealing with diverse and distributed data landscapes.

3. **Data Transformation & Harmonization:** Apply intelligent transformation, cleansing, and harmonization rules to prepare data for consumption. This often leverages AI/ML capabilities to automate data preparation, ensure data quality, and make disparate datasets interoperable for analytics and applications. This automation reduces manual effort and accelerates the readiness of data for digital services and AI models.

4. **Data Governance & Security:** Implement consistent, automated data governance and security policies across the entire fabric. This ensures compliance with regulations (e.g., GDPR, CCPA), enforces access controls, and protects sensitive data, regardless of its physical location or format. A robust governance layer is critical for maintaining trust and ensuring responsible use of data in a digitally mature enterprise.

5. **Data Delivery & Consumption:** Provide self-service access to data consumers through various interfaces (e.g., APIs, SQL query tools, dashboards, data science notebooks). The fabric makes data easily discoverable, understandable, and consumable by different user groups for various use cases, fostering a data-driven culture and empowering digital innovation.

6. **Intelligent Orchestration & Automation:** Use AI/ML to automate data pipelines, optimize data flows, recommend data usage patterns, and simplify data management tasks. This intelligent layer enhances the efficiency and agility of the data fabric, reducing manual effort, accelerating data delivery, and enabling more dynamic and responsive digital operations.

"The amount of data in the world is growing at an incredible pace."
— Doug Cutting, Creator of Hadoop

"In God we trust. All others must bring data."
— W. Edwards Deming, Statistician

Ai6x6™ Transformation Blueprint

The Truth-to-Trust Gap (TTG)

"If I had an hour to solve a problem and my life depended on the solution, I would spend the first 55 minutes determining the proper question to ask." --- **Albert Einstein,** the physicist whose legacy is built on challenging the fundamental questions of space and time provides the ultimate justification for the IAR. This metric measures the shift from "Digital Extraction" (assertion) to **Symbiotic Data** (inquiry). Einstein's approach validates the need for a culture that uses AI to ask deeper, more ethical questions rather than just using it to confirm pre-existing biases.

In the early days of the digital revolution, the primary challenge was "Data Scarcity." We struggled to find the signals in the noise. Today, the problem has inverted. AI provides us with high-fidelity, predictive "Truth" at a scale that far exceeds human processing power. We have the models, the accuracy, and the insights; what we lack is the Trust to act upon them. This is the Identity Chasm of the data-driven enterprise: the psychological distance between what the machine knows and what the human is willing to believe. When an AI identifies a 95% confidence signal that a market is shifting or a supply chain is breaking, and the leadership team chooses to "wait and see," the organization is paying a Validation Tax. The primary metric for this evolution is the Truth-to-Trust Gap (TTG).

The Data chasm is the gap between **Institutional Intuition** and **Algorithmic Truth**. It is the friction experienced when a data model suggests a course of action that contradicts "how we've always done it." Without identifying this chasm, organizations treat AI insights as mere suggestions rather than objective reality. Closing this gap is the only way to eliminate the "Validation Tax"—the wasted time spent double-checking a machine that is already statistically superior to human memory.

The greatest barrier to data transformation is the Expert Trap. Senior professionals often feel that their value is tied to their "Intuition"—a nebulous mix of experience and subconscious pattern matching. When a data model contradicts that intuition, the human brain perceives it as a threat to their authority. We see this manifest as "Manual Auditing," where teams waste weeks double-checking a machine's output that has already been statistically validated.

This friction creates a "Trust Lag." If the "Truth" is available in seconds, but "Trust" takes months to build, the data is essentially useless for real-time competition. To bridge this chasm, the organization must undergo a cultural shift: moving from a world where we use data to Verify our Opinions to a world where we use data to Challenge our Beliefs. This is the move from "Assertion-Led" leadership to "Inquiry-Led" orchestration.

THE RAISON D'ÊTRE:

"TO KNOW"

By closing the TTG, the organization fulfills its true purpose: To Know. Knowledge in the post-AI era is not about having all the answers; it is about having the courage to follow the evidence wherever it leads. When the "Validation Tax" is removed, the human is elevated to the role of the Moral Authority over the data. We are not blindly following the machine; we are using the machine to remove the blinders of our own biases. This allows the human to focus on the "Why" and the "So What," while the machine handles the "What" and the "When."

DEFINING THE TTG: THE LOGIC OF ADOPTION

The Truth-to-Trust Gap is a measurement of Organizational Hesitation. It asks one simple question: What is the delta between the machine's performance and the human's willingness to execute?

In a low-trust organization, the gap is wide. You have world-class AI models sitting on top of a culture that still makes decisions based on the loudest voice in the room. In a high-velocity culture, the TTG is narrow. The organization has built "Bridges" of transparency and ritual that allow them to move as soon as the signal is clear. The goal is to minimize the "Validation Tax" so that "Truth" results in immediate "Action."

THE OBJECTIVE: TOWARD UNIVERSAL TRUTH

The goal of the TTG intervention is to reach a state of Equilibrium we call Universal Truth. In this state, there is no "Data Team" and "Business Team"—there is only the organization acting on a shared reality.

Universal Truth allows the firm to move at the speed of the market. By closing the gap, you aren't just becoming "Data-Driven"; you are becoming "Truth-Aligned." You are allowing the "Logic" of the data to settle the "What," so the "Wisdom" of the leader can decide the "How." This is the only way to maintain a competitive edge in a world where the speed of insight is the only sustainable advantage.

CASE STUDY	THE CHASM REALITY	ACTION & RESULT
1: A GLOBAL RETAILER faced a crisis when seasoned buyers ignored AI-driven inventory alerts during a shift in trends. Buyers relied on "market feel," stocking heavy winter apparel during a warm quarter, leading to massive surplus.	The firm suffered from **Institutional Intuition**, where senior experts used "gut feel" as a shield against machine signals, creating an expensive "Validation Tax" on every order.	They implemented **Trigger-Action Mapping**, where specific signals automatically initiated markdowns or shipping holds. This reduced excess inventory by 25% and increased full-price sell-through by 18%.
2: A MAJOR AIRLINE struggled with fuel hedging as volatility increased. Executives were "Selective Believers," trusting data only when it confirmed growth targets and dismissing "Red Flag" signals as glitches.	Leadership was trapped in **Confirmation Bias**, cherry-picking data to justify existing plans while ignoring high-fidelity signals that suggested a need for a radical pivot in fuel strategy.	The CEO mandated **Blind Trust Experiments**, forcing three major tactical shifts based solely on machine signals. These shifts saved $40M in fuel costs, proving algorithmic truth was more reliable than consensus.
3: A PUBLIC UTILITY was paralyzed by "Dashboard Fatigue." Despite collecting millions of data points, maintenance crews were still dispatched based on manual phone complaints rather than predictive alerts.	This was a **Black Box Fear** scenario. Field managers didn't understand how AI predicted failures, so they defaulted to the "Safe" manual process they had used for 30 years, ignoring high-fidelity truth.	The utility built **Explainability Bridges**, using simple visuals to show crews exactly *why* the AI flagged a specific transformer. With the "Black Box" opened, crews trusted signals, reducing outages by 35%.
4: A GLOBAL PHARMA FIRM lost ground in drug discovery because senior scientists were "Intuition Anchors." They insisted on reviewing every molecular combination, viewing AI simulations as "shortcuts" lacking rigor.	Scientists were caught in the **Expert Trap**. They felt their value was tied to the "manual hunt" for breakthroughs, leading to intentional "Feature Ghosting" of the new AI acceleration platform.	Leadership shifted to **Inquiry-First Leadership**, measuring scientists by the quality of questions they asked the AI rather than manual tests. This shortened a key oncology drug's discovery phase by 14 months.
5: A REGIONAL TELECOM saw churn skyrocket despite a state-of-the-art predictive model. Sales managers were "Anecdotal Overriders," ignoring "High Risk" alerts because "the customer sounded happy on a golf outing."	The culture was rooted in **Selective Hearing**. One positive personal interaction was allowed to override thousands of data points showing declining usage, creating a massive gap in defensive strategy.	They installed **Algorithmic Guardrails** requiring a formal "Data Justification" to dismiss any churn alert. By forcing the team to confront Algorithmic Truth, the company saved 12,000 at-risk accounts in six months.
6: A GLOBAL LOGISTICS HUB faced delays because dispatchers were "Manual Monitorers." Even though AI optimized routes in real-time, dispatchers manually "tweaked" routes every morning to suit familiar roads.	Dispatchers were clinging to **Control Illusion**. They believed their manual adjustments made the system "better," when in fact every human touch introduced new delays and broke the optimized flow.	The hub ran **Shadow Mode Testing**, showing a side-by-side comparison of "Human" vs. "Machine" routes. Once the machine proved 22% faster, they removed manual editing, achieving a 99% on-time rate.

ASSIGNMENT : THE DATA CULTURE DIAGNOSTIC AUDIT

Closing the Truth-to-Trust Gap requires moving beyond technical accuracy and addressing the psychological hurdles of data adoption. We often build better models when we should be building better rituals. To bridge the Identity Chasm, you must identify where the "Expert Trap" is causing teams to ignore the machine's truth in favor of historical comfort. This audit is designed to expose the "Validation Tax" in your organization. On the left are ten common Symptoms of a high Truth-to-Trust Gap. On the right are ten Strategic Interventions designed to build the "Bridges" necessary for Universal Truth.

The Assignment: Match the symptom of data-hesitation to the specific intervention required to reach Equilibrium.

THE SYMPTOMS (THE DIAGNOSTIC)	MATCH	THE INTERVENTIONS (THE REMEDY)
1. The Validation Tax: High-fidelity AI insights are ignored until they can be "manually verified" by a human team.		**A. The Evidence Ritual:** Mandating that every strategic meeting begins with an AI-surfaced question that challenges the status quo.
2. Selective Hearing: Leaders only use data that supports their pre-existing strategy, ignoring contradictory signals.		**B. Shadow Mode Testing:** Running the AI in the background and comparing its "Truth" against human "Gut" over 30 days.
3. The Expert Trap: Senior specialists reject AI models because the machine's logic doesn't match their "30 years of experience."		**C. Trust-Reporting:** Adding "Adoption Velocity" to the data team's KPIs, not just "Model Accuracy."
4. Anecdotal Override: A single "bad experience" with the data is used to justify dismissing the entire system.		**D. Inquiry-First Leadership:** Training executives to ask "What does the data say?" before offering an assertion.
5. The Black Box Fear: Staff refuse to act on insights because they don't understand *how* the AI reached the conclusion.		**E. Explainability Bridges:** Using "glass-box" models that visualize the "Why" behind the "What" for human reviewers.
6. Dashboard Fatigue: Teams have 100+ charts but no clear protocol on which signal triggers an immediate resource shift.		**F. Trigger-Action Mapping:** Pre-defining exactly which data thresholds will result in an automatic, human-approved pivot.
7. The Goalpost Shift: When the AI is right, it's "obvious"; when it's wrong, it's "dangerous"—a double standard for trust.		**G. Cognitive Bias Training:** Explicitly teaching teams to recognize "Confirmation Bias" and the "Sunk Cost Trap" in data review.
8. Data Siloing: Different departments "hide" their data to maintain a monopoly on truth and organizational power.		**H. The Transparency Compact:** Guaranteeing that the machine's "Truth" will be accessible to all levels, not just the C-Suite.
9. The Consensus Stall: Everyone agrees the data is right, but no one wants to be the "first" to authorize the change.		**I. Algorithmic Accountability:** Holding humans responsible for *ignoring* a valid signal, just as they are held for making a bad call.
10. Latent Literacy: Staff have the tools but lack the basic skills to interpret a "Probability Score" vs. a "Fact."		**J. Augmented Literacy Bootcamps:** Immersive training focused on "Statistical Intuition" rather than just "Tool Training."

ANSWER MATCH: 1-B, 2-G, 3-D, 4-C, 5-E, 6-F, 7-I, 8-H, 9-A, 10-J

BLUEPRINT SERIES

SECTION F

Cultural Compass

Ai6x6™ Transformation Blueprint

Story 1 — The Post-War Japanese Industrial Ethos (1945–1960): Japan emerged from the war with limited resources but a collective commitment to national rebuilding. Companies embraced a culture of discipline, apprenticeship, and shared responsibility. Workers identified deeply with their organizations, reinforcing loyalty and craftsmanship. Leaders cultivated humility and long-term thinking. This cultural foundation enabled rapid industrial recovery. It demonstrated how shared values can accelerate capability even before systems mature.

Story 2 — NASA's Mission Culture During Apollo (1961–1972): NASA built a culture defined by precision, accountability, and collective purpose. Engineers and astronauts operated with a shared belief that failure was unacceptable, yet learning was essential. Communication was open, direct, and cross-disciplinary. Rituals, reviews, and norms reinforced psychological safety and technical rigor. This culture enabled teams to navigate uncertainty and complexity. It became a global example of mission-driven cohesion.

Story 3 — The Civil Rights Movement's Organizing Ethos (1954–1968): The movement relied on a culture of discipline, nonviolence, and moral clarity. Communities trained in de-escalation, collective action, and strategic communication. Leaders modeled courage and restraint, shaping norms that guided behavior under extreme pressure. Shared values created unity across diverse groups. This cultural coherence amplified impact far beyond organizational structure. It showed how culture can mobilize change at societal scale.

Story 4 — The Nordic Social Trust Model (1970s–1990s): Nordic countries cultivated cultures rooted in transparency, equality, and civic trust. Institutions operated with high accountability and low hierarchy. Citizens expected fairness and contributed to collective welfare. This cultural fabric enabled strong public services, stable governance, and economic resilience. Trust reduced friction in decision-making and collaboration. It demonstrated how cultural norms can become strategic national assets.

Story 5 — The Rise of Silicon Valley's Innovation Culture (1980s–2000s): Silicon Valley developed a culture that celebrated experimentation, risk-taking, and rapid iteration. Informal networks, open information flow, and meritocratic ideals shaped behavior. Failure was reframed as learning, attracting talent willing to push boundaries. Venture capital reinforced speed and ambition. This culture accelerated technological breakthroughs and entrepreneurial growth. It showed how cultural norms can amplify innovation beyond structural advantages.

Story 6 — The Globalization of Corporate Diversity & Inclusion (2000s–2020s): Organizations worldwide recognized that diverse teams outperform homogeneous ones when supported by inclusive cultures. Companies invested in psychological safety, equitable processes, and representation. Employees gained platforms to voice perspectives and challenge norms. Inclusion became a driver of creativity, retention, and ethical integrity. This cultural shift reshaped leadership expectations and organizational identity. It demonstrated how culture evolves to meet societal and competitive demands.

LESSONS FROM A DATA TRANSFORMATION PERSPECTIVE

Cultural maturity shapes how organizations interpret, govern, and act on data. Over time, institutions learned that transparency, trust, and shared norms determine whether data becomes a tool for alignment or a source of fear. Mature cultures encourage open information flow, constructive challenge, and evidence-based decision-making. They balance accountability with psychological safety, enabling teams to surface issues without blame. Data becomes meaningful when culture supports honesty, clarity, and collective responsibility. Maturity is achieved when data is not manipulated, hidden, or politicized but used to strengthen learning, integrity, and performance. Culture becomes the lens through which data gains credibility.

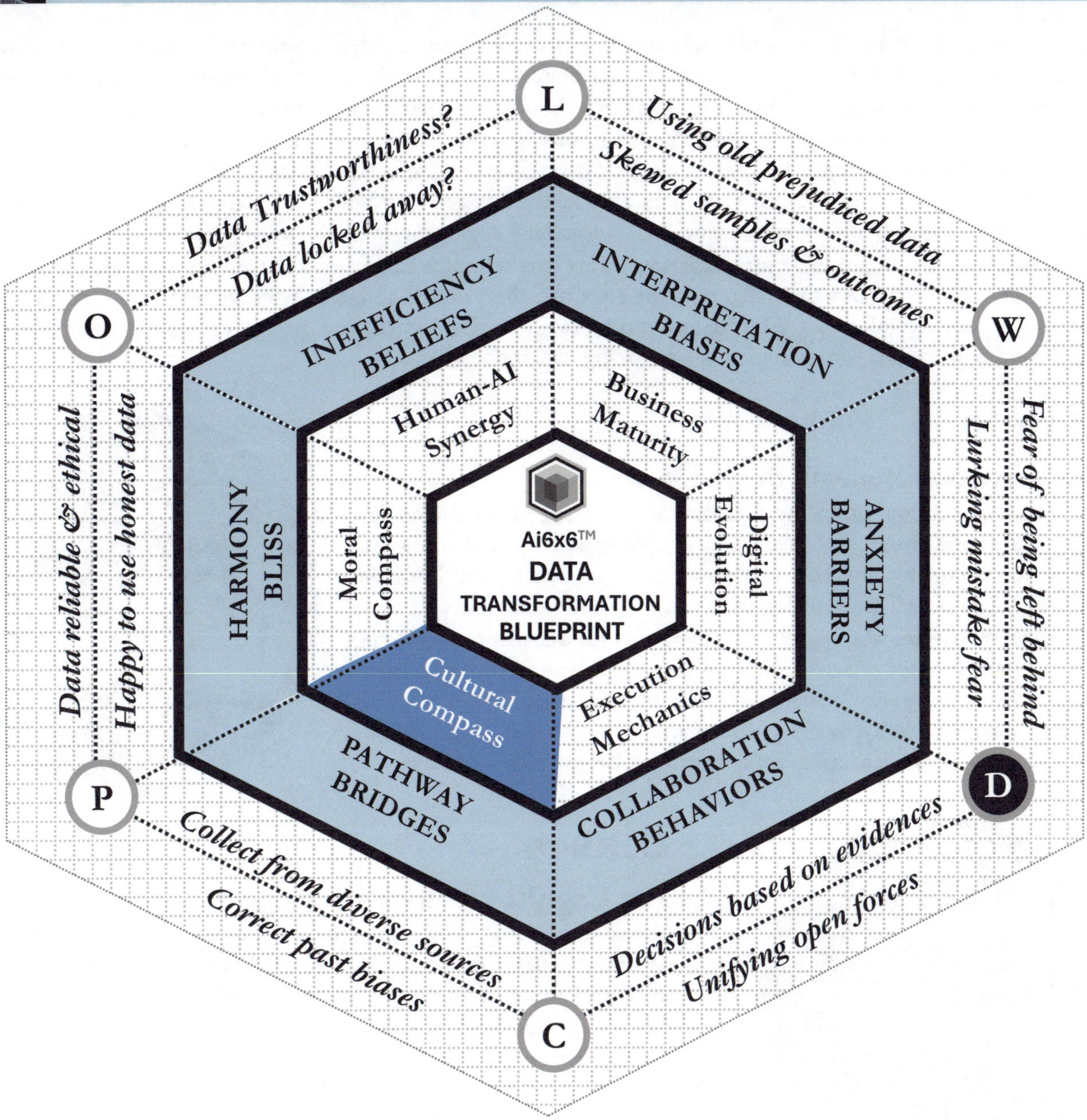

The **Ai6x6 Research Lab** chose this lens because the ultimate utility of data is found in the collective belief and trust of the people who use it. We reframed this dimension as the stewardship of "Data Integrity Bliss," recognizing that culture is the invisible force that either validates or sabotages the data strategy. It is vital because it addresses the invisible shadows—the **Data Anxiety**, **Inefficiency**, and **Interpretation Biases**—that live in the minds of the people. This lens is the bridge between the high-tech architecture of the data mesh and the human reality of daily collaboration, moving the workforce from skepticism to a state of data fluency.

We selected it to ensure leaders build a data-literate and harmonious community rather than just a complex repository. By dismantling structural barriers and building "Data Integrity Bridges," the leader creates a "greenhouse" where information is treated as a shared language of truth. The Lab believes that in an age of AI-generated content, human integrity is the only self-sustaining source of institutional trust. This lens forces a look at how biases impact the collective understanding of information. When data culture is treated as a discipline, the organization achieves a state of "bliss"—a high-fidelity hum where every decision-maker feels empowered, valued, and fundamentally connected to the organization's data reality.

Culture has always been the invisible architecture of an organization, shaping how people think, behave, and make decisions long before any formal strategy takes hold. In the age of AI, this architecture becomes even more consequential. Technology may accelerate capability, but culture determines whether that capability becomes meaningful progress or wasted potential. The **Cultural Compass** lens forces leaders to confront the truth that transformation is not sustained by systems alone; it is sustained by the beliefs people hold, the biases they carry, the barriers they navigate, the behaviors they exhibit, the bridges they build, and the moments of profound satisfaction that anchor their commitment. Without cultural alignment, even the most sophisticated AI strategy collapses under the weight of human resistance. This lens reframes culture as a dynamic system rather than a static identity. It recognizes that culture is not a set of values on a wall but a living pattern of interpretation, interaction, and intention. As AI reshapes workflows, decision cycles, and power dynamics, leaders must understand how cultural forces amplify or undermine transformation.

Beliefs

The Narratives Beneath the Surface

Beliefs and assumptions form the first dimension of the Cultural Compass. They are the quiet forces that shape how people interpret change, how they respond to uncertainty, and how they judge the intentions behind leadership decisions. When beliefs are outdated or misaligned, they distort perception and create friction. Leaders who ignore this dimension discover that people do not resist technology—they resist the meaning they attach to it. Addressing beliefs requires leaders to surface the narratives that operate beneath the surface and reshape them with clarity, honesty, and purpose. In this dimension, cultural truth-telling becomes a leadership act rather than a communication tactic.

Biases

The Cognitive Patterns That Shape Judgment

Biases and mental shortcuts represent the second dimension. These cognitive patterns influence how people evaluate information, assign trust, and make decisions. In an AI-enabled environment, biases can distort both human judgment and the interpretation of machine-generated insights. Addressing them requires intentional design, transparent dialogue, and environments where people feel safe challenging their own assumptions. Leaders must create conditions where cognitive unlearning becomes possible. In this dimension, cultural evolution is driven by awareness, humility, and the willingness to question long-held mental models.

Barriers

The Structural Forces That Restrict Change

Barriers and systemic impediments form the third dimension. These are the structural forces—policies, hierarchies, incentives, and legacy processes—that constrain behavior even when people are willing to change. Leaders often misinterpret resistance as emotional when it is structural. Removing barriers is not about demanding more effort; it is about redesigning the environment so that the desired behaviors become the natural ones. In this dimension, organizational de-friction becomes a strategic priority rather than an operational afterthought, enabling people to move with clarity rather than struggle against constraints.

Behaviors

The Visible Expression of Culture

Behaviors and observable actions represent the fourth dimension. Culture is not what people say; it is what they do consistently. In the AI era, behaviors must evolve to reflect curiosity, adaptability, collaboration, and responsible experimentation. Leaders must model these behaviors visibly and repeatedly. They must demonstrate that learning is valued, that questioning is encouraged, and that accountability is shared. In this dimension, behavioral signaling becomes a powerful tool for shaping norms, reinforcing the idea that culture is enacted through daily choices rather than declared through slogans.

Bridges

The Pathways That Connect People and Purpose

Bridges and cultural pathways form the fifth dimension. These are the connections that allow ideas, trust, and collaboration to flow across the organization. Bridges are built through shared language, shared rituals, and shared purpose. They transform isolated teams into aligned communities. Leaders who invest in cultural pathways create environments where people feel connected to something larger than their role. In this dimension, collective identity-building becomes a catalyst for cohesion, enabling the organization to move as a unified whole rather than a collection of disconnected parts.

Bliss

The Emotional Anchor of Commitment

Bliss and profound satisfaction complete the compass. Bliss is not about comfort; it is about meaning. It is the sense of fulfillment that emerges when people feel their work matters, their contributions are recognized, and their growth is supported. Bliss is the emotional anchor that sustains commitment through uncertainty. Leaders who cultivate it create cultures where people do not merely comply—they care. In this dimension, purposeful fulfillment becomes a strategic asset, strengthening resilience and deepening engagement.

The Cultural Compass lens reminds leaders that transformation is not sustained by technology or process alone. It is sustained by the human experience of change. Culture determines whether people interpret transformation as threat or opportunity, whether they respond with resistance or curiosity, and whether they move in alignment or fragmentation. In a world where AI accelerates both progress and pressure, leaders must treat culture as the primary operating system of the enterprise. When leaders align beliefs, challenge biases, remove barriers, shape behaviors, build bridges, and cultivate meaning, they create cultures capable of evolving continuously. They build environments where people feel psychologically safe, intellectually stimulated, and emotionally invested. Culture becomes the force that amplifies strategy rather than undermining it. It becomes the stabilizing anchor in moments of disruption and the accelerating engine in moments of opportunity.

The False Climate — BELIEFS

Dismantling "Weather Myths" ensures that unrealistic, universal expectations don't stifle individual growth. Exposing the false belief that all "plants" require the same sun, water, or timing—allowing for specialized care.

The Invasive Weeds — BIASES

Identifying and removing hidden mental shortcuts prevents old habits from choking out innovation. Surgically exposing the leadership judgment and systemic biases that create "shadow" cultures and block progress.

The Hard Clay — BARRIERS

Breaking through structural resistance allows the organizational roots to expand and stabilize. Removing the deeply embedded obstacles like data, automation and AI anxiety that act as physical walls against digital growth.

Calibrated Care — BEHAVIORS

Executing specialized nurturing routines ensures each distinct "species" receives what it needs to thrive. Moving away from "one-size-fits-all" mode to model specific watering, feeding, and light-balancing behaviors required for a diverse ecosystem.

The Trellis System — BRIDGES

Constructing pathways for upward growth provides the support needed for teams to reach their full potential. Building the specific leadership and development paths that allow the workforce to climb through the maturity stages.

The Full Harvest — BLISS

Securing a state of permanent organizational harmony results in a self-sustaining ecosystem of excellence. Reaching the ultimate destination where the "Cultural Soul" operates in a high-fidelity hum, delivering constant value.

Ai6x6™ Transformation Blueprint

Data Inefficiency Beliefs

"Without big data analytics, companies are blind and deaf, wandering out onto the web like deer on a freeway." - Geoffrey Moore

Geoffrey Moore (b. 1946) is an American organizational theorist and management consultant. He is a leading expert on high-tech marketing and the author of the influential book Crossing the Chasm.

Data inefficiency beliefs are negative assumptions about data that prevent an organization from leveraging it as a strategic asset. These beliefs can be deeply ingrained, often from past experiences with inaccurate or difficult-to-access data. A common belief is that "data is unreliable," which leads to a culture where decisions are based on instinct rather than evidence. Another limiting belief is that "only analysts can understand data," which creates silos and prevents a data-literate culture from forming. The journey to data transformation begins by challenging these beliefs and cultivating a new mindset that views data not as a threat but as a powerful strategic tool. This requires a commitment to data accuracy, transparency, and education, ensuring every employee feels empowered to use data to make informed decisions.

Serving as the intelligence layer of the Execution Core, this chapter explores the Data Inefficiency Beliefs that underpin successful data transformation and deliver true Foresight via Human-AI Synergy. It establishes that effective data utilization is a cultural commitment, not merely a technical deployment. The core conviction is Data-Driven Beliefs, mandating that evidence and empirical facts must guide critical decisions over intuition or hierarchy. The book dedicates significant attention to governing machine intelligence, establishing principles for Algorithmic Trust Beliefs and Data Ethics Beliefs to ensure models are reliable, unbiased, and used responsibly within legal and moral bounds. Furthermore, it formalizes the strategic potential of Predictive Beliefs, leveraging data science and advanced modeling to anticipate market shifts and future opportunities, thereby moving the organization from reactive to proactive. Finally, it addresses the necessary safeguards through Data Security Beliefs and culminates in cultivating Data-as-an-Asset Beliefs, treating data not as a system byproduct, but as a critical, monetizable resource that powers the future enterprise strategy.

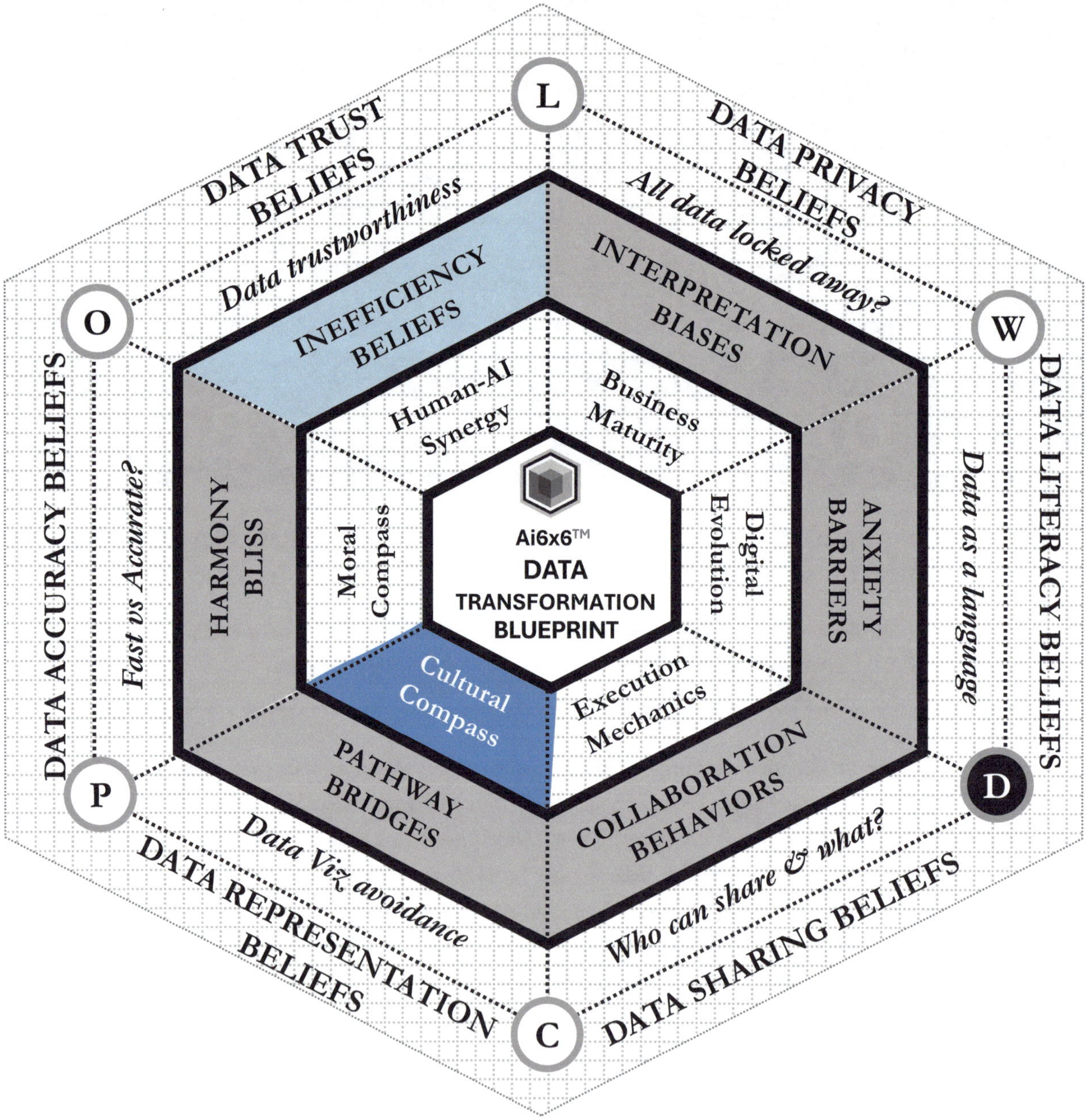

CULTURAL COMPASS FRAMEWORK | DATA INEFFICIENCY | BELIEFS

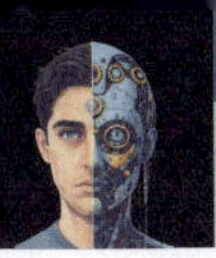

1. DATA TRUST BELIEFS | Trustworthiness of Data

If you fundamentally believe that "gut feel" always trumps the dashboard, you will never achieve a true AI transformation. This belief often stems from past experiences with "dirty data," leading to a daily skepticism that halts automated decision-making at its tracks. For an executive, this means you are paying for expensive AI infrastructure while still making decisions based on the loudest voice in the room or personal intuition. To evolve, you must shift your belief from seeing data as a "suggestion" to seeing it as the primary source of truth. Without foundational trust, your AI models are merely expensive window dressing that your team will ignore the moment a difficult or counter-intuitive insight is generated by the complex system.

DO'S FOR THE LEADER	DON'TS FOR THE LEADER
Use data from trusted, clean sources.	Assume all data is untrustworthy.
Audit data for accuracy regularly.	Blindly trust data without checking.
Be transparent about data limits.	Hide data problems from the team.
Use clear data governance rules.	Create confusing data policies.
Connect data to business results.	Let data exist without a clear purpose.
Communicate data integrity efforts.	Ignore data quality issues entirely.

SITUATION	SUGGESTED ACTION
Employees hesitate to use datasets because they believe underlying information is incomplete or unreliable	Establish transparent data-quality scoring, publish validation checks, and communicate lineage clearly to strengthen confidence in using organizational datasets for decisions
Teams rely on personal judgment instead of data because they distrust system-generated outputs	Explain model logic, share accuracy benchmarks, and provide examples where data-driven decisions improved outcomes to rebuild trust in analytical outputs
Managers question dashboard insights because they assume metrics are outdated or inconsistently refreshed	Implement automated refresh schedules, display update timestamps, and communicate data-pipeline reliability to reinforce trust in operational dashboards
Employees avoid cross-functional data because they assume other teams maintain lower quality standards	Create unified quality protocols, publish shared definitions, and conduct cross-team audits to ensure consistent trustworthiness across all organizational datasets
Leaders doubt predictive analytics because they believe historical data cannot represent current realities	Show model retraining cycles, highlight recent performance improvements, and demonstrate how updated inputs strengthen predictive accuracy over time
Teams resist automation because they assume algorithmic decisions lack transparency and fairness	Provide explainability tools, document decision rules, and review automated outputs regularly to reinforce trust in algorithmic processes and governance

Viewing data privacy as a "legal hurdle" rather than a competitive advantage is a belief that exposes you to massive strategic risk. In the AI era, if you believe privacy is just about checking compliance boxes, you will fail to build the "data dignity" required to gain consumer trust. This belief leads to a daily culture of cutting corners, which eventually results in data leaks or ethical scandals that can destroy your brand overnight. Transforming this belief means seeing privacy as a core value that protects the integrity of your training sets. When you believe privacy is paramount, you create a secure "data moat" that allows for more bold experimentation because the ethical foundation is rock-solid and respected by both users and employees.

DO'S FOR THE LEADER	DON'TS FOR THE LEADER
Prioritize data privacy protections.	Treat data privacy as an afterthought.
Follow all data privacy laws.	Ignore privacy regulations for speed.
Be transparent about data use.	Use personal data without consent.
Anonymize data where possible.	Collect and store too much personal data.
Empower customers with data control.	Make it hard for people to control data.
Train staff on privacy best practices.	Assume everyone knows privacy rules.

SITUATION	SUGGESTED ACTION
Employees hesitate to use shared datasets because they fear sensitive information may be exposed unintentionally	Establish clear privacy classifications, provide anonymization tools, and reinforce responsible data-handling practices to strengthen confidence in safe organizational data usage
Teams restrict access excessively because they assume broader visibility automatically increases privacy risks	Define role-based access standards, communicate safeguards clearly, and demonstrate how controlled transparency supports collaboration without compromising privacy
Managers avoid collecting useful data because they believe privacy concerns outweigh potential benefits	Provide privacy-by-design guidelines, highlight compliant use cases, and show how responsible collection strengthens insight generation while maintaining trust
Employees resist new analytics tools because they assume personal information will be monitored unnecessarily	Communicate monitoring boundaries, publish data-use policies, and demonstrate how tools focus on operational insights rather than individual surveillance
Leaders delay data initiatives because they fear regulatory violations from improper handling practices	Offer compliance training, implement automated safeguards, and provide legal guidance to ensure initiatives progress confidently within regulatory expectations
Teams avoid cross-functional data sharing because they believe privacy rules prohibit most collaboration	Clarify permissible sharing scenarios, provide secure exchange mechanisms, and reinforce that privacy and collaboration can coexist through structured governance

The assumption that data is "only for the math people" creates a massive intelligence bottleneck across the entire organization. If you believe your broader workforce is incapable of understanding data, you will centralize power in a small data science team that becomes overwhelmed and disconnected from business reality. This belief prevents the democratization of insights. In your daily life, it means you aren't getting data-driven ideas from the people closest to the customer. Transformation requires believing that every employee can and must be data-literate. By shifting this belief, you turn your entire workforce into a sensor network, where every person can interpret signals and making informed, data-backed adjustments to their daily operations, regardless of their technical background.

DO'S FOR THE LEADER	DON'TS FOR THE LEADER
Invest in data literacy training.	Assume all employees are data literate.
Provide easy-to-use data tools.	Rely on complex, inaccessible tools.
Foster a culture of data curiosity.	Punish people for asking basic questions.
Use simple data visualizations.	Use confusing charts and graphs.
Assign data mentors to help others.	Let people struggle with data alone.
Reward data-driven decisions.	Base decisions on gut feelings alone.

SITUATION	SUGGESTED ACTION
Employees avoid analytical tools because they believe data skills require advanced technical backgrounds	Provide foundational training, offer guided practice sessions, and highlight examples of non-technical employees successfully building confidence with data-driven tasks
Teams rely on intuition because they assume data interpretation is too complex for everyday decision-making	Introduce simple frameworks, provide contextual explanations, and demonstrate how basic analysis strengthens routine operational choices across functions
Managers hesitate to assign data-related tasks because they believe only specialists can handle them effectively	Create tiered literacy programs, clarify expected competencies, and show how distributed data capability accelerates organizational performance
Employees misinterpret metrics because they assume definitions are obvious and universally understood	Publish standardized definitions, provide usage examples, and reinforce shared understanding through regular data-literacy refreshers
Teams struggle to evaluate dashboards because they assume visualizations automatically reflect correct insights	Teach interpretation techniques, explain common pitfalls, and encourage validation steps before acting on displayed metrics
Leaders underestimate employees' ability to learn data skills, limiting development opportunities	Highlight learning progress, provide structured pathways, and reinforce that data capability grows through practice, not innate

4. DATA SHARING BELIEFS | The Walled Garden

The belief that "data is power" often leads to data hoarding, where departments protect their databases like private fiefdoms. If you believe that keeping data private within your silo increases your department's value, you are actively sabotaging the organization's AI potential. AI thrives on cross-functional datasets; hoarding creates a fragmented view that leads to hallucinating models and missed correlations. In your daily executive role, this manifests as "turf wars" over API access. Transforming this belief involves recognizing that data's value is only unlocked when it is connected. You must move from a "need to know" mindset to a "duty to share" mindset to fuel the large-scale models that drive modern competitive advantage and allow for a truly unified enterprise intelligence.

DO'S FOR THE LEADER	DON'TS FOR THE LEADER
Share data openly across the company.	Hoard data in department silos.
Create a central data repository.	Let teams keep their own separate data.
Establish clear data access rules.	Make it hard to access needed data.
Encourage cross-functional teams.	Create data competition between groups.
Promote data as a shared asset.	Treat data as a private resource.
Be transparent about data origins.	Hide where your data comes from.

SITUATION	SUGGESTED ACTION
Teams restrict access because they believe sharing data reduces control and increases operational risk	Define sharing protocols, demonstrate secure exchange methods, and highlight benefits of transparency for collaboration and decision quality
Employees hesitate to share insights because they assume other teams may misuse or misinterpret information	Provide context templates, encourage clarification dialogue, and reinforce shared accountability for responsible interpretation
Managers keep datasets siloed because they believe only their function understands the underlying context	Establish cross-functional data reviews, document assumptions, and show how broader visibility strengthens organizational alignment
Teams avoid collaborative analysis because they assume shared data will expose performance weaknesses	Normalize learning-focused reviews, emphasize improvement over blame, and highlight examples where transparency accelerated progress
Employees resist shared dashboards because they believe visibility invites unnecessary scrutiny	Communicate purpose clearly, define usage boundaries, and reinforce that transparency supports alignment rather than micromanagement
Leaders limit data access because they assume openness increases compliance risk	Provide governance guidelines, implement automated safeguards, and demonstrate how structured sharing supports both compliance and collaboration

5. DATA REPRESENTATION BELIEFS

Assuming that data visualization is just "making things look pretty" is a belief that leads to poor strategic communication. If you believe the representation of data is secondary to the calculation, you will find your board meetings bogged down in confusion rather than focused on action. Data representation is about cognitive efficiency. In your daily life, poor visualization leads to misinterpretation and "decision fatigue" as you struggle to find the signal in the noise of dense spreadsheets. Transformation requires believing that how data is presented is as important as the data itself. By valuing high-fidelity, intuitive visualization, you ensure that complex AI outputs are translated into clear, actionable narratives that can be understood and acted upon instantly by the entire leadership team.

DO'S FOR THE LEADER	DON'TS FOR THE LEADER
Use clear and simple visualizations.	Create misleading or confusing charts.
Ensure visuals are accessible to all.	Use visuals that are hard for some to see.
Standardize your company's reports.	Let every department make their own charts.
Use dashboards that tell a clear story.	Show data without context or a narrative.
Get feedback on your data visuals.	Assume your charts are easy to understand.
Link data to a business objective.	Show data just for the sake of it.

SITUATION	SUGGESTED ACTION
Employees distrust visualizations because they believe charts oversimplify complex realities and hide nuance	Provide layered views, explain design choices, and reinforce how visual summaries support clarity without eliminating access to detailed data
Teams prefer dense tables because they assume visuals lack precision and analytical rigor	Demonstrate accuracy controls, show comparative insights, and highlight scenarios where visuals accelerate understanding and decision-making
Managers request overly complex dashboards because they believe sophistication signals analytical maturity	Promote simplicity standards, explain cognitive load principles, and show how streamlined visuals improve comprehension and actionability
Employees misinterpret charts because they assume colors and scales carry universal meaning	Provide visualization guidelines, standardize palettes, and teach interpretation techniques to reduce ambiguity
Teams resist new visualization formats because they believe traditional styles are inherently more trustworthy	Share evidence of improved clarity, run side-by-side comparisons, and encourage experimentation with modern formats
Leaders question visual insights because they assume aesthetics overshadow analytical accuracy	Document data sources, show calculation logic, and reinforce that strong visuals enhance—not replace—rigorous analysis

If you believe that "speed is more important than accuracy," you are building your AI house on sand. While agility is vital, a belief that sacrifices data integrity for rapid reporting leads to a "garbage in, garbage out" cycle that can derail a company in the AI era. Daily, this manifests as making pivots based on "directional" data that is fundamentally flawed. To transform, you must believe that accuracy is the non-negotiable prerequisite for speed. When your data is accurate, you can move faster because you don't have to waste time questioning the results. Shifting this belief ensures that your AI transformations are built on a bedrock of reliable evidence, preventing the costly errors that come from high-speed, low-accuracy decision-making.

DO'S FOR THE LEADER	DON'TS FOR THE LEADER
Prioritize data quality and accuracy.	Prioritize speed over accurate data.
Implement data validation processes.	Let bad data enter your systems.
Use single sources of truth.	Allow multiple versions of the same data.
Correct data issues as they arise.	Let data errors compound over time.
Communicate data quality issues.	Hide or ignore bad data.
Invest in data cleansing software.	Rely on manual data correction.

SITUATION	SUGGESTED ACTION
Teams delay decisions because they believe data must be perfectly accurate before any action	Define acceptable accuracy thresholds, communicate decision timelines, and reinforce that timely insight often outweighs marginal precision
Employees rush analysis because they assume speed is always more valuable than accuracy	Provide quality checkpoints, highlight risks of errors, and reinforce balanced expectations for timely and reliable outputs
Managers reject preliminary insights because they believe only fully validated data is trustworthy	Introduce staged validation, explain confidence levels, and show how early signals support proactive decision-making
Teams over-engineer data pipelines because they assume perfection is required for credibility	Establish pragmatic quality standards, prioritize critical fields, and reinforce iterative improvement over exhaustive upfront design
Employees hesitate to share early findings because they fear inaccuracies will damage credibility	Normalize provisional insights, label confidence levels clearly, and encourage transparent communication about assumptions
Leaders push for rapid reporting because they assume speed demonstrates operational excellence	Align expectations with analytical realities, communicate trade-offs, and reinforce that sustainable performance requires balanced accuracy and timeliness

Ai6x6™ Transformation Blueprint

Data Interpretation Biases

"It is a capital mistake to theorize before one has data. Insensibly one begins to twist facts to suit theories, instead of theories to suit facts." - Arthur Conan Doyle (Sherlock Holmes)

Sir Arthur Conan Doyle (1859–1930) was a British writer and physician. He is globally celebrated for creating the character of the masterful detective Sherlock Holmes, known for his powers of observation and logical deduction.

Data interpretation biases are the cognitive shortcuts that can corrupt the analysis and understanding of data. These biases can lead to flawed conclusions and poor decision-making. Selection Bias, for example, occurs when data is not representative of the entire population, leading to conclusions applicable only to a narrow group. Another critical bias is Algorithmic Bias, where biases are embedded in the very algorithms used to automate decisions. This can lead to discriminatory outcomes in hiring or loan applications. Overcoming these biases requires a commitment to a rigorous process of auditing and validating data and analytical tools. As Sherlock Holmes noted, it is a "capital mistake to theorize before one has data."

Serving as the intelligence layer of the Execution domain, this chapter explores the Data Interpretation Biases—systematic errors embedded in data pipelines and machine learning models—that compromise data integrity, hinder Foresight, and prevent true Human-AI Synergy. It establishes that "objective" data often reflects subjective human errors and historical prejudices. The chapter dives into sampling errors like Selection Bias (unrepresentative data) and the critical problem of Historical Bias, where past societal prejudices are amplified and automated by models, leading to discriminatory outcomes. It addresses data quality issues via Measurement Bias (inaccurate collection) and the flaws in the architecture itself through Algorithmic Bias (errors in model design and training logic). Understanding the data source is key, leading to analysis of Training Bias (imbalanced training datasets) and the overarching problem of Systemic Data Bias, which requires holistic, enterprise-wide mitigation strategies to ensure data is a reliable, ethical, and predictive asset.

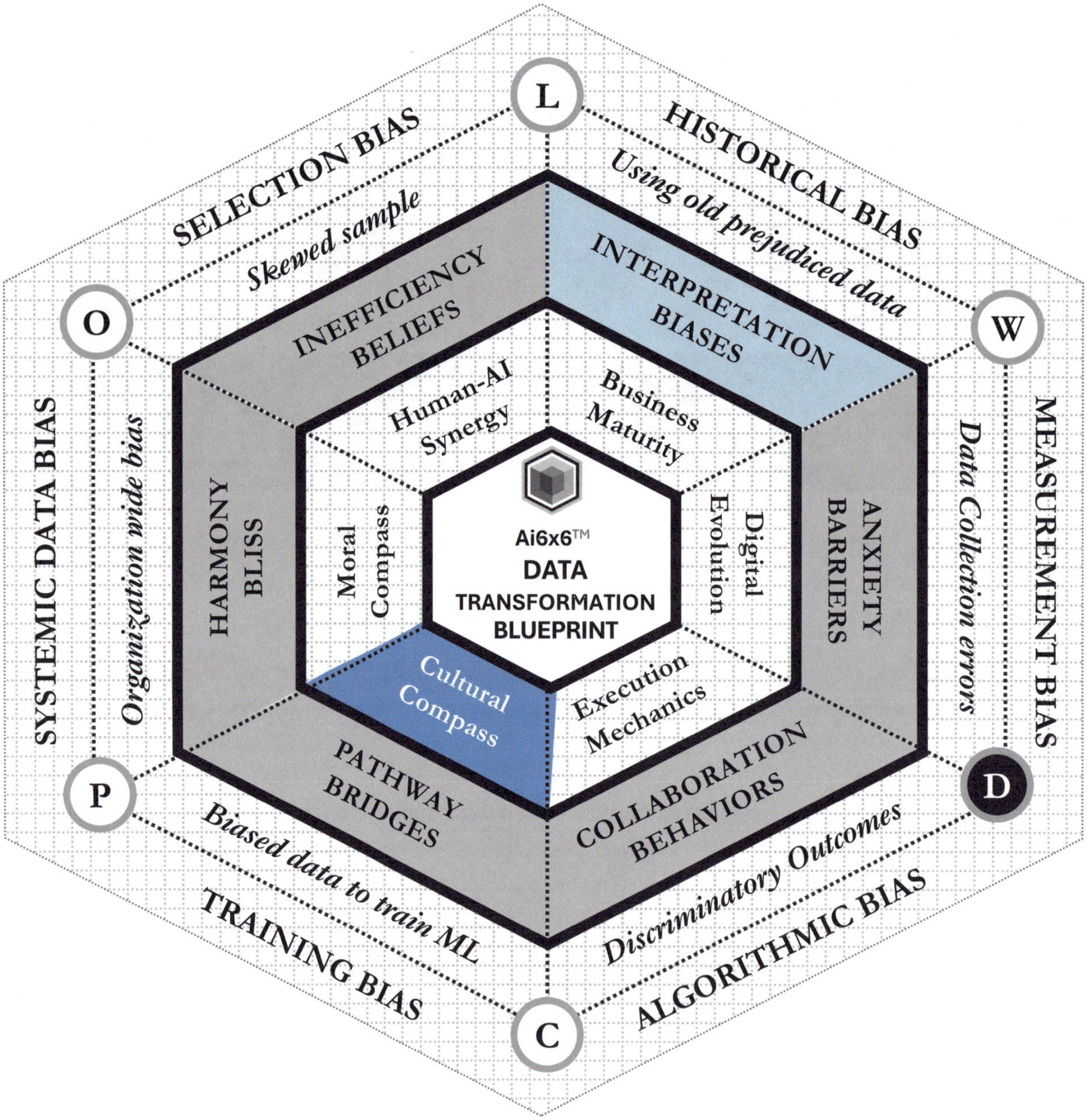

CULTURAL COMPASS FRAMEWORK FOR DATA LIFECYCLE | BIASES

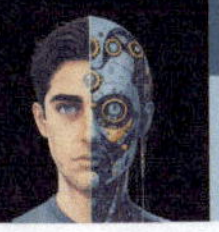

1. SELECTION BIAS | The Skewed Sample

This bias occurs when the data you use to train your models isn't representative of the real-world population you serve. In an executive context, this means your "customer insights" might only be reflecting your most vocal or tech-savvy users, leaving a massive silent majority ignored. Daily, this leads to product roadmaps that solve problems for the wrong people. Transformation requires a rigorous audit of your data sources to ensure you aren't optimizing for a narrow slice of reality. By recognizing selection bias, you move from "anecdotal AI" to "representative AI," ensuring that your strategic decisions are based on a comprehensive, unbiased view of the entire market landscape rather than just a convenient subset of easily accessible but misleading data points.

DO'S FOR THE LEADER	DON'TS FOR THE LEADER
Ensure data is representative.	Use data that isn't representative.
Collect data from diverse sources.	Rely on a single source of data.
Be transparent about data limitations.	Claim your data is all-encompassing.
Use stratified sampling.	Cherry-pick data that fits your needs.
Ask, "who is missing from our data?"	Ignore a large part of the population.
Test your data against other sources.	Assume your data is a perfect sample.

SITUATION	SUGGESTED ACTION
Employees rely on limited samples because they assume small datasets reflect broader population patterns	Expand sampling methods, validate representativeness, and reinforce that decisions require inclusive data capturing full population diversity
Teams exclude certain groups unintentionally because they believe available data already covers all perspectives	Conduct coverage audits, identify missing segments, and adjust collection strategies to ensure complete population representation
Managers trust convenience samples because they assume easily accessible data is sufficient for decisions	Require sampling justification, compare against population benchmarks, and reinforce disciplined data-selection practices
Employees overlook underrepresented segments because they assume low volume equals low importance	Highlight impact of missing data, quantify risks, and adjust weighting to ensure all segments influence decisions
Teams generalize findings from narrow pilots because they assume early results apply universally	Validate across diverse groups, run replication tests, and reinforce that pilot insights require broader confirmation
Leaders prioritize high-performing segments because they assume they represent organizational norms	Analyze full distributions, review outliers, and ensure decisions reflect entire population patterns rather than selective subsets

Using old data to train modern AI models often bakes in the prejudices and limitations of the past.6 If you're hiring or lending algorithms are trained on ten-year-old data, they will likely replicate the social biases of that era. As a leader, this bias creates a "reputation trap," where your AI makes discriminatory decisions that reflect poorly on your current values. Daily, this manifests as "automated unfairness" that is hard to trace. Overcoming historical bias requires a "temporal audit" of your data. You must realize that data is not neutral; it reflects the time it was collected. Transformation means intentionally correcting these historical echoes to ensure your AI reflects the inclusive, forward-thinking organization you are building today for a better future.

DO'S FOR THE LEADER	DON'TS FOR THE LEADER
Audit old data for past prejudices.	Use old data that reflects past biases.
Train models with new, fair data.	Train models on historically biased data.
Adjust algorithms for past unfairness.	Let historical unfairness persist.
Acknowledge historical context.	Use data without understanding its past.
Correct for systemic biases.	Ignore systemic issues in your data.
Use diverse data to balance past views.	Rely on a single, old dataset.

SITUATION	SUGGESTED ACTION
Employees rely on legacy datasets containing outdated assumptions that no longer reflect current realities	Audit historical sources, identify embedded prejudices, and update datasets to align with present-day conditions
Teams use old performance patterns because they assume past behavior predicts future outcomes reliably	Introduce recency weighting, incorporate new variables, and reinforce continuous dataset refresh cycles
Managers trust long-standing metrics without questioning whether they encode historical inequities	Review metric origins, evaluate fairness, and redesign indicators to reflect equitable modern standards
Employees reuse inherited models because they assume previous generations validated them thoroughly	Reassess model logic, test against current data, and document adjustments to remove outdated biases
Teams benchmark against historical norms that reflect past exclusionary practices	Establish new baselines, incorporate diverse perspectives, and ensure comparisons reflect inclusive modern expectations
Leaders make strategic decisions using historical trends shaped by outdated societal structures	Integrate contemporary datasets, validate assumptions, and reinforce forward-looking analysis grounded in current realities

3. MEASUREMENT BIAS | The Unseen Pitfall

Measurement bias happens when the way data is collected systematically distorts the results.7 For example, if your "productivity data" only measures keystrokes, you are ignoring the value of strategic thinking and creative problem-solving. In your daily life as an executive, this leads to rewarding the wrong behaviors and missing the true drivers of success. This bias turns your AI into a tool for "optimizing the irrelevant." To transform, you must ensure that your data collection methods are aligned with your true strategic goals. This requires a deep dive into the "how" of data gathering, ensuring that your sensors and metrics are measuring what matters, rather than just what is easy to track or quantify in a basic digital format.

DO'S FOR THE LEADER	DON'TS FOR THE LEADER
Use a clear, standardized metric.	Use an unclear or subjective metric.
Ensure all data collectors are trained.	Let people collect data on their own.
Test tools for accuracy.	Use uncalibrated or flawed tools.
Gather feedback on your methods.	Assume your measurement is perfect.
Use multiple methods for comparison.	Rely on only one method of measurement.
Document your collection process.	Have an undocumented, messy process.

SITUATION	SUGGESTED ACTION
Employees trust flawed metrics because they assume collection tools capture information consistently across contexts	Validate instruments regularly, document limitations, and adjust methods to ensure accurate measurement
Teams rely on inconsistent survey responses because they assume participants interpret questions uniformly	Standardize wording, pilot test instruments, and refine phrasing to reduce ambiguity and improve reliability
Managers overlook device variability because they assume all sensors perform identically across environments	Calibrate equipment, monitor drift, and implement quality checks to ensure consistent measurement accuracy
Employees misinterpret results because they assume data collection processes are error-free	Provide transparency about collection methods, highlight known issues, and encourage cautious interpretation
Teams aggregate inconsistent inputs because they assume volume compensates for measurement flaws	Clean datasets rigorously, apply correction factors, and reinforce disciplined preprocessing standards
Leaders act on biased indicators because they assume measurement errors are negligible	Quantify potential error ranges, communicate uncertainty, and adjust decisions to reflect realistic confidence levels

This is the "ghost in the machine"—biases embedded in the logic of the code itself. Often, these are unintended consequences of how an algorithm prioritizes certain variables. For an executive, algorithmic bias is a massive liability; it can lead to systematic exclusion of certain demographics from your services.8 Daily, this manifests as "black box" decisions that no one can explain but that feel inherently "off" or unfair. Transforming this requires a commitment to "algorithmic transparency." You must demand that your tech teams stress-test models for fairness. By proactively hunting for algorithmic bias, you protect your company from the legal and ethical fallout of automated discrimination, ensuring your AI acts as a fair and objective partner in long-term growth.

DO'S FOR THE LEADER	DON'TS FOR THE LEADER
Test algorithms for discrimination.	Let algorithms produce unfair results.
Use diverse training data.	Use biased data to train algorithms.
Be transparent about how algorithms work.	Use a black box algorithm.
Use human oversight on AI decisions.	Let AI make all decisions alone.
Regularly audit algorithms.	Assume your algorithm is fair.
Adjust algorithms that show bias.	Ignore signs of algorithmic bias.

SITUATION	SUGGESTED ACTION
Employees trust algorithmic outputs blindly because they assume automated decisions are inherently objective	Provide explainability tools, review decision pathways, and reinforce that algorithms require oversight to ensure fairness
Teams overlook disparate impacts because they assume performance metrics reflect equal treatment across groups	Conduct fairness audits, compare subgroup outcomes, and adjust models to reduce discriminatory patterns
Managers deploy models quickly because they assume accuracy guarantees ethical behavior	Integrate ethical reviews, test for unintended consequences, and reinforce responsible deployment standards
Employees misunderstand algorithm limitations because they assume models capture all relevant context	Document constraints clearly, provide training, and encourage contextual judgment alongside automated insights
Teams reuse existing models because they assume prior validation ensures ongoing fairness	Revalidate regularly, monitor drift, and update models to reflect evolving data patterns
Leaders trust vendor algorithms without questioning embedded assumptions or training data quality	Require transparency documentation, evaluate datasets, and test outputs independently to ensure fairness and reliability

5. DATA TRAINING BIAS | The Glass Half Full

Training bias occurs when the "examples" you give an AI to learn from are skewed. If you only show a sales AI examples of successful deals from one specific region, it will fail to understand the nuances of a global market. Daily, this leads to AI tools that are "locally brilliant but globally incompetent." For an executive, this bias results in rigid strategies that cannot scale effectively. To transform, you must ensure your training sets are as diverse as the world you operate in. This requires a "balanced diet" of data. By feeding your models a wide variety of scenarios—including failures and edge cases—you create a more robust, intelligent AI that can navigate the complex, unpredictable realities of a global business environment.

DO'S FOR THE LEADER	DON'TS FOR THE LEADER
Use balanced datasets for AI.	Use skewed, biased data to train AI.
Clean data for unfairness before training.	Assume all data is clean.
Monitor for bias during training.	Don't check for bias until it's a problem.
Use diverse teams to build AI.	Let a homogeneous team build the AI.
Document your training data.	Use undocumented, untracked data.
Test the AI for fair outcomes.	Assume the AI is fair by default.

SITUATION	SUGGESTED ACTION
Employees trust model outputs without questioning whether training data reflects diverse real-world conditions	Audit datasets for representation gaps, rebalance samples, and reinforce disciplined validation before deploying models broadly
Teams reuse historical datasets because they assume past information is sufficient for training modern systems	Introduce updated sources, incorporate contemporary patterns, and reinforce continuous dataset refresh cycles
Managers overlook skewed training inputs because they assume algorithms automatically correct for imbalance	Provide fairness testing tools, document assumptions, and require explicit bias-mitigation steps during model development
Employees rely on vendor-provided datasets because they assume external sources guarantee neutrality	Evaluate dataset composition, test subgroup performance, and require transparency documentation from vendors
Teams train models on convenience data because they assume availability equals suitability	Define quality standards, enforce sampling requirements, and validate representativeness before training
Leaders underestimate training bias risks because they assume accuracy metrics reflect fairness	Compare performance across demographic groups, highlight disparities, and adjust models to ensure equitable outcomes

This is the most dangerous bias because it is woven into the very fabric of your organization's data infrastructure. It's the cumulative effect of years of biased collection, storage, and interpretation practices. Daily, it means that even your "raw" data is already tainted by old-world thinking. As a leader, systemic bias makes true transformation feel impossible because the foundation is flawed. Overcoming this requires a "data reconstruction" project. You must be willing to tear down and rebuild biased data architectures. Transformation means acknowledging that your systems aren't neutral pipes; they are active shapers of reality. By cleaning the systemic bias, you unlock the true power of AI to provide objective, groundbreaking insights that were previously hidden by legacy departmental thinking.

DO'S FOR THE LEADER	DON'TS FOR THE LEADER
Have a broad audit of all systems.	Let biases exist throughout your data.
Create a company-wide data ethics team.	Have no one to watch for data ethics.
Ensure fair data collection processes.	Collect data in a biased way.
Promote data ethics at all levels.	Treat data ethics as a niche issue.
Review all processes for bias.	Only check one or two data systems.
Be proactive about finding and fixing bias.	Be reactive after a problem occurs.

SITUATION	SUGGESTED ACTION
Employees rely on legacy systems containing structural biases embedded across multiple datasets and processes	Conduct system-wide audits, identify recurring patterns, and implement coordinated remediation across data pipelines
Teams inherit biased definitions because they assume long-standing terminology accurately reflects organizational reality	Redefine metrics collaboratively, document updated standards, and reinforce shared understanding across functions
Managers trust enterprise dashboards without recognizing that upstream systems encode historical inequities	Trace data lineage, highlight bias sources, and adjust pipelines to ensure fair representation
Employees overlook systemic issues because they assume individual datasets operate independently	Map data flows, identify cross-system dependencies, and address bias at structural integration points
Teams normalize skewed patterns because they assume long-term consistency indicates correctness	Compare against external benchmarks, validate assumptions, and recalibrate indicators to reflect equitable expectations
Leaders underestimate systemic bias impact because they assume isolated fixes resolve broader issues	Implement organization-wide governance, monitor trends, and reinforce continuous improvement across all data systems

Data Anxiety Barriers

"Measurement is the first step that leads to control and eventually to improvement. If you can't measure something, you can't understand it. If you can't understand it, you can't control it. If you can't control it, you can't improve it." - H. James Harrington

H. James Harrington (1929–2024) was a business expert and consultant. He is a leading figure in quality management and is known for his work in business process improvement, earning numerous awards for his contributions.

Data anxiety barriers are the cultural and psychological impediments that prevent employees from engaging with data. These barriers are often rooted in a fear of what data might reveal. An Exposure Barrier is a reluctance to share data due to the fear that mistakes will be revealed and scrutinized. A Power Loss Barrier is the belief that data is a tool for a select few at the top, leading to resistance from managers who fear that democratizing data will dilute their authority. The Unknown Barrier is a fundamental resistance to new insights that challenge the status quo. Overcoming these barriers requires a shift in culture, from one of judgment and blame to one of transparency and learning. As H. James Harrington's quote suggests, if you can't measure something, you can't improve it.

This component explores Data Anxiety Barriers—the systemic issues in technology and data management—that compromise data integrity, hinder Foresight, and prevent true Human-AI Synergy. The core challenge lies in managing Legacy Systems, outdated technology that cannot easily integrate, scale, or process modern data formats efficiently, acting as a brake on transformation. This results in Data Inaccessibility, where valuable information is locked in disparate, proprietary, and isolated sources across the enterprise. The subsequent threat is Data Quality Issues, where unreliable, incomplete, or inconsistent data leads to faulty analysis and dangerously skewed decisions. This section addresses the structural void of Lack of Data Governance, meaning no clear rules, standards, or ownership exist for managing data integrity and use. Furthermore, it details the constant challenge of Security and Privacy Concerns, ensuring that the enterprise's data transformation efforts comply with regulations and protect customer trust. Finally, the component addresses Inadequate Tooling, the absence of the right software, platforms, or cloud services necessary for modern data ingestion, warehousing, and machine learning operations.

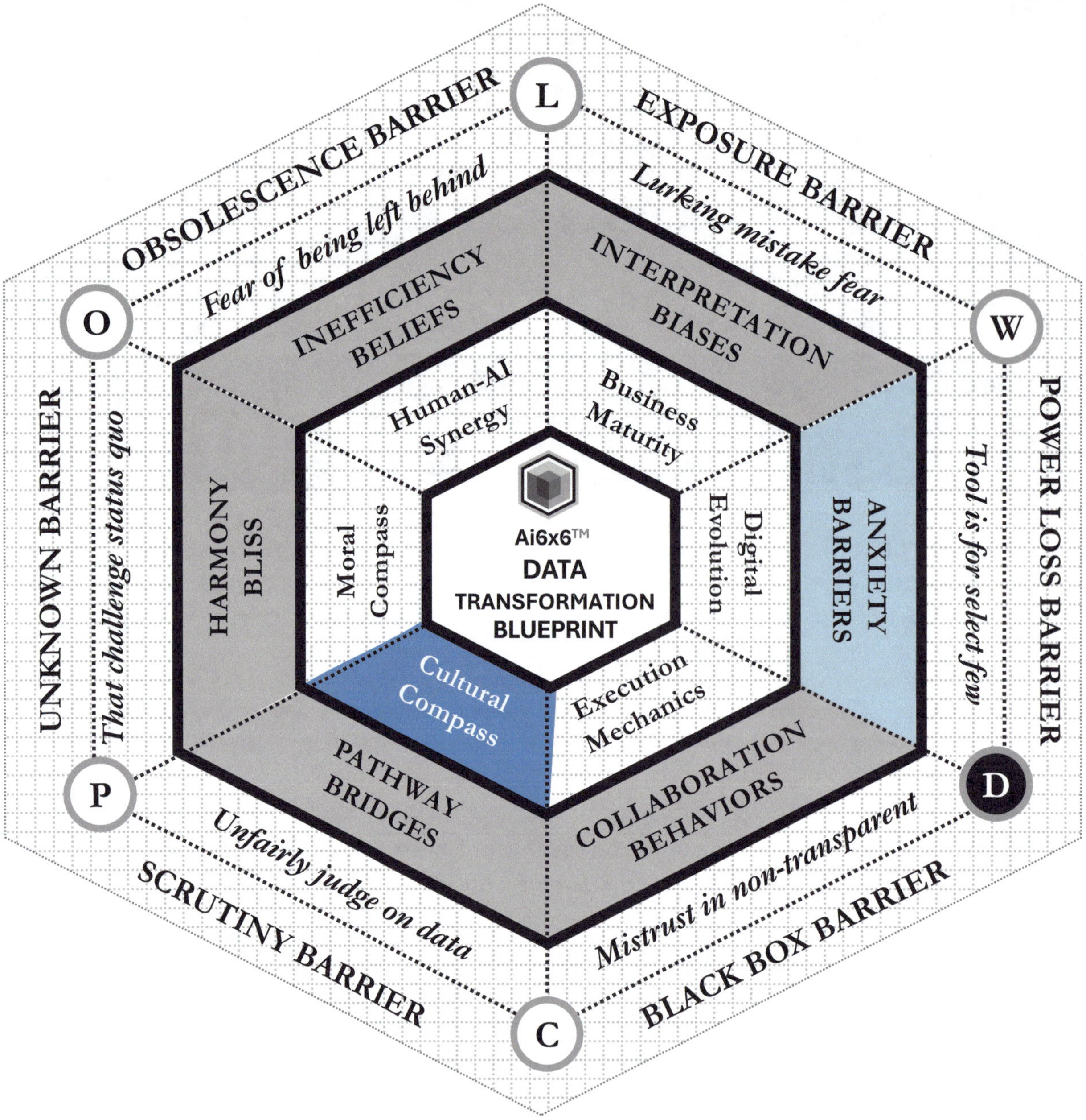

CULTURAL COMPASS FRAMEWORK FOR DATA LIFECYCLE | BARRIERS

1. OBSOLESCENCE BARRIER | The Blight of Distrust

The fear that "AI will replace me" is a massive barrier to workforce transformation. When employees—including middle managers—fear that data technology will make their skills irrelevant, they will subtly sabotage the rollout of new tools. In your daily operations, this manifests as low adoption rates and "shadow systems" where people stick to old manual ways. To break this barrier, you must reframe AI as "augmented intelligence" rather than "replacement intelligence." As an executive, your job is to show how data tools free people from drudgery to focus on high-value human creativity. By addressing this fear head-on, you move your workforce from resisting technology to embracing it as a vital partner in their own career evolution and personal productivity.

DO'S FOR THE LEADER	DON'TS FOR THE LEADER
Stay updated with data technology.	Fear being left behind by data tech.
Invest in new data tools.	Stick to old, inefficient tools.
Train staff on new technologies.	Let people struggle with outdated tech.
Be a leader in data innovation.	Be reactive to new data tech.
Create a roadmap for tech upgrades.	Have no plan for technology updates.
Learn from others' tech mistakes.	Make the same mistakes as others.

SITUATION	SUGGESTED ACTION
Employees avoid new tools because they fear rapid technological change will outpace their skills	Provide structured learning pathways, highlight achievable progress, and reinforce that capability grows through consistent practice
Teams resist modernization because they assume existing systems are easier to maintain and understand	Demonstrate efficiency gains, provide transition support, and reinforce long-term benefits of updated technologies
Managers hesitate to adopt advanced analytics because they fear appearing unprepared or uninformed	Offer leadership-focused training, normalize learning curves, and reinforce that growth is expected at all levels
Employees disengage from data initiatives because they assume new tools require expert-level proficiency	Provide beginner-friendly resources, offer guided practice, and highlight success stories from non-technical colleagues
Teams delay upgrades because they assume modernization introduces unnecessary complexity	Communicate simplification benefits, provide hands-on demonstrations, and reinforce improved usability of modern systems
Leaders worry about organizational readiness because they assume employees cannot adapt quickly enough	Implement phased rollouts, provide coaching, and highlight early adopters who demonstrate successful transitions

Data is a mirror, and many people are afraid of what it will show about their work. The exposure barrier is the reluctance to share data because it might reveal past mistakes, inefficiencies, or "padding" in reports. Daily, this leads to people "massaging" the numbers before they reach your desk, giving you a distorted view of reality. This fear of exposure kills the transparency required for AI to work. To transform, you must create a "no-blame" data culture. When you treat data as a tool for improvement rather than a weapon for punishment, people will be more willing to share the "ugly" truths. This honesty is the only way to get the accurate datasets needed to train effective, high-impact AI models.

DO'S FOR THE LEADER	DON'TS FOR THE LEADER
Foster an open, blameless culture.	Be reluctant to share data.
Promote a learning from mistakes culture.	Punish people for making mistakes.
Share data for collaboration.	Hide data for fear of being wrong.
Use data to improve, not to judge.	Use data to unfairly judge performance.
Be transparent with data and insights.	Let departments hide their data.
Create a psychologically safe workplace.	Make people feel unsafe to share.

SITUATION	SUGGESTED ACTION
Employees hide datasets because they fear errors will damage credibility or professional reputation	Normalize error correction, provide supportive review processes, and reinforce learning-focused data culture
Teams avoid sharing insights because they assume imperfections will be judged harshly by others	Establish constructive feedback norms, highlight collaborative improvement, and reinforce psychological safety
Managers restrict visibility because they fear exposing inconsistencies in their function's reporting	Provide cross-functional review frameworks, emphasize transparency, and reinforce shared accountability for accuracy
Employees hesitate to publish dashboards because they worry minor inaccuracies will overshadow value	Implement quality checks, communicate confidence levels, and reinforce that early visibility supports iterative improvement
Teams avoid collaborative analysis because they fear exposing gaps in their understanding	Provide training, encourage joint exploration, and reinforce that learning together strengthens capability
Leaders discourage open sharing because they assume transparency increases scrutiny without benefit	Highlight alignment advantages, demonstrate improved decision quality, and reinforce that openness strengthens organizational trust

3. POWER LOSS BARRIER | The Turf War

Many leaders believe that "knowledge is power" and that by democratizing data, they are losing their influence. This belief creates a barrier where executives or department heads restrict data access to maintain their "expert" status. Daily, this manifests as information silos and slow decision-making as everyone waits for the "gatekeeper" to provide a report. To transform, you must shift the definition of power from "owning information" to "facilitating insights." When data is shared, the entire organization becomes more powerful. By breaking this barrier, you empower your teams to act with autonomy and speed, shifting your role from a bottlenecked expert to a strategic orchestrator who leads through a unified, data-informed vision that benefits everyone involved in the mission.

DO'S FOR THE LEADER	DON'TS FOR THE LEADER
Decentralize data access.	Believe data is a tool for a select few.
Empower all employees with data.	Keep data access limited to a few.
Use data to democratize decisions.	Let one person have all the data power.
Teach data literacy to everyone.	Keep data a secret from your staff.
Promote data as a shared company asset.	Treat data as a source of personal power.
Break down data silos.	Maintain data silos and exclusivity.

SITUATION	SUGGESTED ACTION
Employees avoid data tools because they believe only specialists should control analytical capabilities	Provide accessible training, decentralize access, and reinforce that data empowerment strengthens organizational performance
Teams rely on gatekeepers because they assume data ownership belongs exclusively to technical groups	Establish shared governance, clarify responsibilities, and encourage collaborative data stewardship across functions
Managers resist democratization because they fear losing influence over information-driven decisions	Highlight benefits of distributed insight, reinforce leadership's strategic role, and show how shared access accelerates outcomes
Employees hesitate to request data because they assume permission must come from senior experts	Publish access guidelines, simplify request processes, and reinforce transparency around data availability
Teams avoid cross-functional collaboration because they assume data authority resides within isolated silos	Create shared workspaces, align definitions, and reinforce collective ownership of organizational data assets
Leaders underestimate the value of broad data access because they assume centralization ensures quality	Demonstrate improved decision speed, highlight successful decentralized use cases, and reinforce governance that supports both access and accuracy

4. DATA BLACK BOX BARRIER | The Red Tape

Mistrust of complex, non-transparent AI systems leads to a "Black Box Barrier," where people refuse to follow a data insight because they don't understand how the machine reached it. If an algorithm says "fire this person" or "close this store" without explanation, any reasonable leader will hesitate. Daily, this leads to a "manual override" of your best AI tools. To transform, you must prioritize "Explainable AI" (XAI). This means building bridges of understanding between the math and the business mission. When people can see the "logic" behind the data, they are far more likely to trust and act on it. Breaking the black box barrier is essential for creating the psychological safety required for deep human-AI collaboration.

DO'S FOR THE LEADER	DON'TS FOR THE LEADER
Use transparent data systems.	Mistrust complex, non-transparent systems.
Document all data processes.	Use systems no one understands.
Explain how data is collected and used.	Use a black box system without docs.
Use simple, clear visualizations.	Use visualizations that are confusing.
Empower people to understand the data.	Let people feel disconnected from the data.
Make system logic easy to follow.	Use a complex, hidden system.

SITUATION	SUGGESTED ACTION
Employees distrust automated outputs because they cannot understand how systems generate recommendations	Provide explainability tools, document logic clearly, and reinforce transparency as a core system requirement
Teams avoid advanced analytics because they assume complexity hides potential errors or biases	Offer model walkthroughs, highlight validation steps, and encourage open discussion about system limitations
Managers question algorithmic decisions because they believe hidden processes reduce accountability	Establish review protocols, clarify decision pathways, and reinforce shared responsibility for oversight
Employees hesitate to adopt new tools because they assume non-transparent systems undermine autonomy	Communicate system purpose, show user control points, and reinforce that tools support—not replace—human judgment
Teams misinterpret outputs because they assume system logic is too technical to understand	Provide simplified explanations, offer training, and reinforce that comprehension is essential for responsible use
Leaders delay deployment because they assume complexity increases organizational risk	Demonstrate safeguards, highlight governance controls, and reinforce that transparency mitigates risk effectively

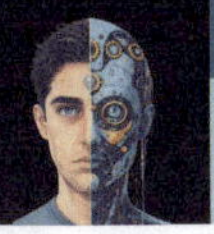

5. DATA SCRUTINY BARRIER | Fear of unfair judgement

The fear that data will be used to unfairly judge performance leads to a workforce that is defensive and risk-averse. If employees feel that every keystroke or minute of downtime is being scrutinized by a "big brother" AI, they will focus on "looking busy" rather than being productive. Daily, this leads to a toxic culture of surveillance that kills innovation. To overcome this, you must explicitly state—and show—that data is used to optimize systems, not to persecute individuals. Transformation requires using data to support and coach, rather than just to police. When employees see data as a tool that helps them succeed, the scrutiny barrier falls, and they begin to use those same insights to drive their own professional growth and development.

DO'S FOR THE LEADER	DON'TS FOR THE LEADER
Use data for improvement, not for blame.	Fear data will be used to judge.
Use objective, fair metrics.	Use unfair or biased metrics.
Focus on the process, not the person.	Publicly shame people based on data.
Create a safe place for mistakes.	Punish people for making errors.
Be transparent about how data is used.	Keep data use a secret.
Use data to help people, not to harm.	Use data as a tool for punishment.

SITUATION	SUGGESTED ACTION
Employees avoid dashboards because they fear metrics will be used punitively rather than constructively	Communicate purpose clearly, emphasize improvement, and reinforce supportive performance practices
Teams resist reporting because they assume transparency increases exposure to criticism	Normalize learning-focused reviews, highlight collaborative problem-solving, and reinforce psychological safety
Managers hesitate to share performance data because they fear negative comparisons across functions	Provide context guidelines, encourage balanced interpretation, and reinforce shared accountability for outcomes
Employees underreport issues because they assume data will highlight individual shortcomings	Establish non-punitive reporting norms, celebrate early detection, and reinforce improvement over blame
Teams manipulate metrics because they assume leadership values numbers more than underlying reality	Align incentives with accuracy, communicate expectations, and reinforce integrity as a core performance principle
Leaders avoid granular insights because they assume detailed data invites unnecessary scrutiny	Highlight benefits of precision, demonstrate improved decision quality, and reinforce that transparency strengthens organizational trust

New data insights often challenge the "status quo" and long-held industry dogmas.10 The "Unknown Barrier" is the resistance to these insights because they make people feel uncomfortable or wrong. Daily, this looks like executives dismissing an AI finding as "an anomaly" or "outlier" simply because it contradicts their years of experience. This barrier prevents the organization from discovering the "hidden opportunities" that AI is uniquely designed to find. To break this, you must foster a culture of "curious skepticism." Encourage your team to lean into the weird data. By valuing the "unknown" as a source of potential competitive advantage, you ensure that your transformation isn't just a digital version of the past, but a genuine evolution into new market territories.

DO'S FOR THE LEADER	DON'TS FOR THE LEADER
Be open to new data and insights.	Resist new data that challenges the status quo.
Encourage challenging old assumptions.	Hold on to old ways of thinking.
Test new data against old beliefs.	Dismiss new data without a second thought.
Use data to drive innovation.	Fear new data insights and ideas.
Create a culture of learning and growth.	Let people get stuck in their old habits.
Promote intellectual honesty.	Reject new information out of hand.

SITUATION	SUGGESTED ACTION
Employees reject unfamiliar insights because they believe new data threatens established practices	Provide context, explain methodology, and reinforce that evolving insights strengthen long-term performance
Teams ignore contradictory findings because they assume existing knowledge is inherently more reliable	Encourage evidence-based review, validate assumptions, and reinforce openness to updated information
Managers resist new metrics because they fear changing long-standing evaluation frameworks	Demonstrate benefits, provide transition support, and reinforce alignment with strategic goals
Employees question emerging trends because they assume unfamiliar patterns indicate analytical errors	Explain detection logic, show historical parallels, and reinforce that novelty often signals opportunity
Teams avoid experimentation because they assume new insights introduce unnecessary risk	Provide safe testing environments, highlight successful pilots, and reinforce learning-driven adaptation
Leaders dismiss disruptive findings because they assume stability is preferable to change	Present scenario analyses, highlight competitive implications, and reinforce strategic value of embracing new information

Ai6x6™ Transformation Blueprint

Data Collaboration Behaviors

"Never doubt that a small group of thoughtful, committed people can change the world. Indeed, it is the only thing that ever has." - Margaret Meade

Margaret Mead (1901–1978) was a pioneering American cultural anthropologist. She is famous for her fieldwork in the South Pacific and her books, including Coming of Age in Samoa, which explored gender and culture.

Data collaboration behaviors are positive and proactive actions that promote a healthy, transparent, and collaborative data culture. These are the practices that empower every employee to use data effectively, breaking down silos and fostering shared responsibility. Transparency Behavior is the practice of sharing data and insights openly across the organization, empowering every employee to make informed decisions. Experimentation Behavior encourages bold innovation with data, fostering a culture where teams are not afraid to test new hypotheses and learn from what doesn't work. Shared Power Behavior involves decentralizing data access and decision-making, ensuring that the right information reaches the right people at the right time. As Margaret Meade's quote suggests, "a small group of thoughtful, committed people can change the world." These behaviors collectively build a new reality where data is a shared resource, a tool for innovation, and a language that unites the organization.

DATA COLLABORATION BEHAVIORS

This framework explores Data Collaboration Behaviors—the systematic actions necessary to utilize data infrastructure, achieve Foresight, and realize true Human-AI Synergy. The fundamental behavior is Data-Driven Behavior, where decisions are rigorously based on evidence and objective analysis rather than reliance on intuition or personal preference. This requires Curious Behavior, actively asking "why," digging into data, and investigating patterns to find deeper insights and uncover innovation opportunities. A crucial check is Skeptical Behavior, constantly questioning assumptions, methodology, and the integrity of the data being used before accepting its conclusions as fact. Furthermore, all data actions must be governed by Ethical Behavior, ensuring data is used responsibly, fairly, and in strict compliance with all privacy regulations and organizational values. Organizational memory is built through Documenting Behavior, meticulously recording data processes, models, and analytical decisions for transparent future reference and audit. Finally, the team must demonstrate Learning Behavior, actively seeking to understand and master new analytical tools, platforms, and machine learning techniques to stay current.

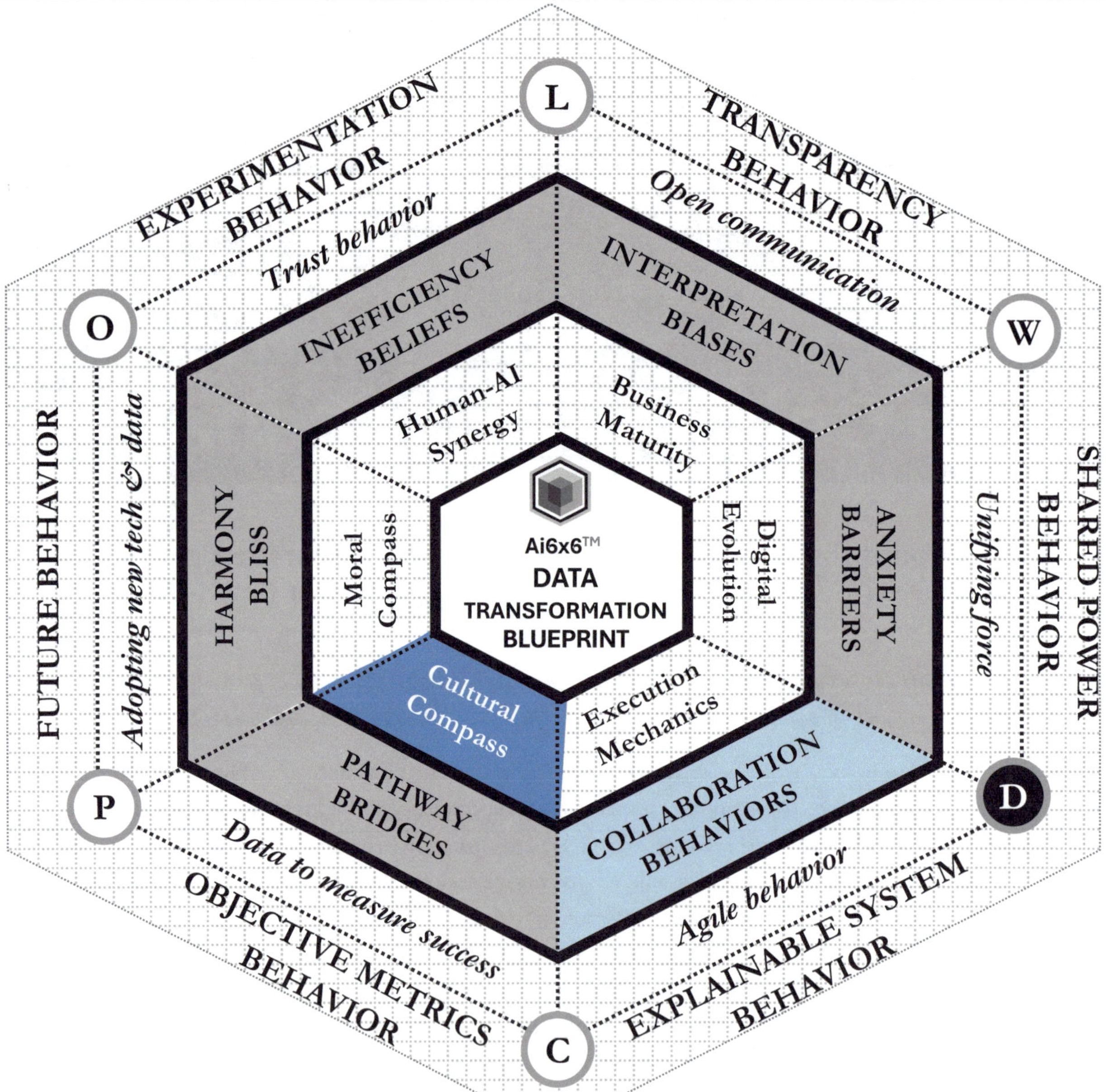

CULTURAL COMPASS FRAMEWORK FOR DATA LIFECYCLE

1. EXPERIMENTATION BEHAVIOR | The Trust Accelerator

In a data-transformed organization, experimentation isn't a special event; it's a daily behavior. Employees at all levels feel empowered to run "A/B tests" on their processes and ideas. This behavior turns your company into a "living lab" where data determines what scales and what fails. As an executive, you support this by providing "sandboxes"—safe environments where people can play with data without breaking the main system. When experimentation is a habit, the organization becomes incredibly resilient and adaptive. You stop guessing what the market wants and start "asking" the data through constant, low-cost trials. This behavior ensures that your AI models are always being fed with fresh, real-world evidence of what works in the current and ever-changing business environment.

DO'S FOR THE LEADER	DON'TS FOR THE LEADER
Encourage bold data innovation.	Punish people for making mistakes.
Create a safe place for new ideas.	Create a culture of fear and blame.
Provide resources for data experiments.	Keep a tight control on all resources.
Celebrate learning from data.	Only celebrate ideas that succeed.
Use data to test new theories.	Rely on intuition instead of data.
Promote a culture of curiosity.	Make people feel stupid for asking.

SITUATION	SUGGESTED ACTION
Teams explore new datasets confidently because experimentation is encouraged and failure is treated as learning	Promote **safe testing environments**, highlight iterative progress, and reinforce that experimentation strengthens organizational adaptability and long-term capability
Employees propose unconventional ideas because they trust leadership will evaluate them fairly and thoughtfully	Establish **idea-review forums**, communicate evaluation criteria, and reinforce that creative thinking is essential for innovation and competitive advantage
Managers support pilot projects because they believe experimentation accelerates insight generation and improvement	Provide **structured pilot frameworks**, allocate resources, and highlight successful experiments that delivered measurable value
Teams test multiple approaches before committing because they value evidence-driven exploration	Encourage **comparative analysis**, document outcomes, and reinforce that diverse trials strengthen decision quality
Employees share experimental findings openly because they trust colleagues to build constructively on early insights	Create **knowledge-exchange rituals**, recognize contributors, and integrate learnings into broader organizational practices
Leaders champion experimentation publicly because they understand innovation requires continuous exploration and curiosity	Highlight **leadership sponsorship**, celebrate bold attempts, and reinforce that experimentation is a strategic expectation

Sharing data and insights openly across the organization is a behavior that eliminates silos and fosters collective intelligence. Daily, this looks like "open dashboards" where any employee can see the company's key performance metrics in real-time. This transparency builds trust and alignment. When everyone sees the same data, the "politics" of information disappears. For you, it means your leadership is based on shared facts rather than personal persuasion. Transparency behavior ensures that the "right hand" always knows what the "left hand" is doing, allowing for a much more synchronized response to challenges. It also invites "eyes-on" peer review of data, which naturally increases the overall accuracy and integrity of the entire organizational data system by crowdsourcing the validation process.

DO'S FOR THE LEADER	DON'TS FOR THE LEADER
Share data and insights openly.	Hoard data in department silos.
Provide clear, objective feedback.	Give vague and unhelpful feedback.
Be honest about data limitations.	Overstate your data's accuracy.
Use clear data visualizations.	Use complex, confusing charts.
Make all data processes open.	Have a secret data process.
Communicate key decisions with data.	Keep data from your teams.

SITUATION	SUGGESTED ACTION
Teams share dashboards openly because they believe transparency strengthens alignment and decision quality	Establish **open-access norms**, clarify usage expectations, and reinforce that visibility accelerates collaboration
Employees communicate data assumptions clearly because they value shared understanding across functions	Use **standardized context templates**, encourage clarification dialogue, and reinforce that transparency reduces misinterpretation
Managers publish performance metrics consistently because they believe openness builds trust and accountability	Implement **regular reporting rhythms**, provide interpretation guidance, and reinforce constructive use of insights
Teams document data lineage because they understand transparency improves reliability and confidence	Maintain **lineage repositories**, highlight dependencies, and reinforce disciplined documentation
Employees share early insights because they trust colleagues to interpret provisional findings responsibly	Communicate **confidence levels**, encourage collaborative refinement, and reinforce learning-focused interpretation
Leaders model transparency by sharing strategic data that informs organizational direction	Provide **strategic context briefings**, explain rationale, and reinforce openness as a leadership expectation

3. SHARED POWER BEHAVIOR | The Unifying Force

Decentralizing data access and decision-making is a behavior that pushes authority to where the information is. In a transformed workforce, the person closest to the data has the power to act on it. This behavior eliminates the "executive bottleneck" and allows for real-time adjustments at the edges of the company. Daily, this means you spend less time on routine approvals and more time on high-level strategy. Shared power behavior builds a sense of ownership and accountability across the workforce. It ensures that your AI transformation isn't just a tool for the C-suite, but a capability that empowers every individual to contribute to the organization's success, making the whole company more responsive, faster, and intellectually diverse in its problem-solving approach.

DO'S FOR THE LEADER	DON'TS FOR THE LEADER
Decentralize data access.	Keep data access for a select few.
Empower all with data decision-making.	Have a single person make all decisions.
Promote data literacy for all.	Keep data knowledge a secret.
Break down data silos.	Let departments keep their own data.
Use data to democratize the company.	Use data as a source of power.
Give people the power to question.	Punish anyone who questions the data.

SITUATION	SUGGESTED ACTION
Employees access data independently because they trust decentralized systems to support informed decisions	Provide role-based access, maintain governance controls, and reinforce empowerment as a performance expectation
Teams make decisions collaboratively because data access is distributed across functions	Establish shared decision protocols, align definitions, and reinforce collective ownership of insights
Managers encourage autonomy because they believe distributed data strengthens agility and responsiveness	Provide decision-making frameworks, clarify boundaries, and reinforce trust in team capability
Employees contribute insights confidently because they know their analysis carries equal weight in discussions	Promote inclusive contribution norms, recognize diverse perspectives, and reinforce shared analytical responsibility
Teams avoid bottlenecks because data is accessible without relying on centralized gatekeepers	Maintain self-service platforms, provide training, and reinforce efficient workflows
Leaders support distributed authority because they understand shared power accelerates innovation and execution	Communicate empowerment expectations, highlight success stories, and reinforce decentralized decision culture

Actively choosing to use and develop transparent, easy-to-understand data systems is a behavior that builds "digital trust." This looks like rejecting "black box" vendors in favor of partners who provide "Explainable AI" solutions. Daily, it means your teams spend time documenting the "logic" behind their models so that non-technical stakeholders can understand them easily. This behavior ensures that AI remains a tool for humans, not a replacement for them. When systems are explainable, adoption rates skyrocket because people feel in control. As an executive, this behavior protects you from the risks of "automated errors" by ensuring that there is always a clear, human-understandable path back to how a specific data-driven decision was made, fostering long-term reliability.

DO'S FOR THE LEADER	DON'TS FOR THE LEADER
Use transparent data systems.	Use a black box system.
Document how systems work.	Have no documentation for systems.
Make system logic easy to follow.	Use a complex, hidden system.
Use simple, clear data visuals.	Use confusing, complicated visuals.
Explain how data is collected.	Keep your data collection a secret.
Empower people to understand the data.	Let people feel disconnected from the data.

SITUATION	SUGGESTED ACTION
Employees trust automated insights because system logic is clearly documented and easy to understand	Provide explainability dashboards, highlight decision pathways, and reinforce transparency as a system requirement
Teams adopt new tools quickly because explanations accompany every major analytical output	Offer contextual tooltips, provide examples, and reinforce comprehension as essential for responsible use
Managers rely on models confidently because they understand underlying assumptions and limitations	Publish assumption summaries, review constraints, and reinforce informed oversight
Employees validate outputs effectively because systems reveal how inputs influence final recommendations	Provide input-impact visualizations, encourage verification, and reinforce analytical rigor
Teams collaborate more effectively because shared explanations reduce confusion and misinterpretation	Maintain interpretation guides, align terminology, and reinforce consistent understanding
Leaders support system adoption because transparency reduces risk and strengthens organizational trust	Highlight governance safeguards, communicate reliability, and reinforce transparency as a strategic advantage

5. OBJECTIVE METRICS BEHAVIOR | The Catalyst of Ownership

A commitment to using unbiased, verifiable data to measure success is a behavior that replaces "subjective judgment" with "objective truth." This looks like setting KPIs that are based on hard data rather than "feel-good" metrics or personal biases. Daily, this behavior ensures that promotions, bonuses, and project funding are based on actual impact. It creates a "meritocracy of data" that is incredibly motivating for high-performers. For you, it means your organizational "map" is accurate. By consistently adhering to objective metrics, you remove the emotional drama from performance reviews and strategic pivots, ensuring that the organization remains focused on the cold, hard realities of market performance and operational efficiency, regardless of internal political dynamics or personal favoritism.

DO'S FOR THE LEADER	DON'TS FOR THE LEADER
Commit to using unbiased data.	Use biased data to measure success.
Create clear, standardized metrics.	Use vague, subjective metrics.
Use metrics to help, not to harm.	Use data as a tool for punishment.
Focus on the process, not the person.	Blame people for bad metric results.
Review all metrics for fairness.	Assume all of your metrics are fair.
Be transparent about how you measure.	Keep your metrics a secret.

SITUATION	SUGGESTED ACTION
Teams evaluate performance consistently because objective metrics guide decisions rather than subjective impressions	Establish standardized measurement frameworks, align definitions, and reinforce disciplined interpretation across all functions
Employees trust evaluations because metrics are transparent, consistent, and clearly connected to organizational goals	Publish metric explanations, provide context, and reinforce shared understanding of performance indicators
Managers prioritize evidence over assumptions because objective data strengthens decision quality and fairness	Use data-driven review processes, highlight analytical reasoning, and reinforce accountability for unbiased evaluation
Teams resolve disagreements quickly because objective metrics provide a shared foundation for alignment	Maintain common dashboards, clarify interpretation rules, and reinforce collaborative problem-solving
Employees improve performance confidently because metrics highlight clear development opportunities	Provide targeted feedback, link insights to growth plans, and reinforce continuous improvement expectations
Leaders communicate progress effectively because objective metrics create clarity and organizational trust	Share performance summaries, explain trends, and reinforce transparency in reporting practices

6. FUTURE BEHAVIOR | Eager willingness

Proactively adopting new data technologies and insights before they become industry standards is the "behavior of a leader." This looks like investing in "edge" technologies today to prepare for tomorrow. Daily, this behavior involves encouraging your team to stay "dangerously curious" about the next wave of AI evolution. It ensures that your organization never falls into the "obsolescence trap." By modeling "future behavior," you send a clear message that the transformation is never "finished"—it is a permanent state of being. This mindset keeps your company at the forefront of the AI era, ensuring you are the one disrupting the market rather than the one being disrupted by it. It requires a relentless focus on what's next, rather than just optimizing what's now.

DO'S FOR THE LEADER	DON'TS FOR THE LEADER
Adopt new data technologies.	Stick to old, inefficient data tools.
Be a leader in data innovation.	Fear new technologies and insights.
Invest in new data skills.	Let people's skills become obsolete.
Use data to forecast future trends.	Ignore new data about the future.
Create a roadmap for tech upgrades.	Have no plan for future technology.
Learn from other companies' mistakes.	Make the same mistakes as others.

SITUATION	SUGGESTED ACTION
Employees explore emerging tools because they believe innovation strengthens long-term organizational capability	Provide guided experimentation, highlight successful adoption stories, and reinforce curiosity as a strategic expectation
Teams integrate new insights quickly because they value forward-looking analysis over outdated assumptions	Establish trend-monitoring routines, share scenario models, and reinforce proactive adaptation
Managers champion modernization because they understand future-ready systems improve agility and competitiveness	Communicate technology roadmaps, provide training, and reinforce leadership responsibility for transformation
Employees embrace evolving data practices because they trust the organization to support continuous learning	Offer skill-building pathways, celebrate progress, and reinforce growth-oriented culture
Teams adopt predictive tools confidently because they understand how future insights strengthen planning	Provide model interpretation guides, explain assumptions, and reinforce responsible usage
Leaders invest in advanced capabilities because they recognize future-focused data drives strategic advantage	Highlight innovation outcomes, align investments with goals, and reinforce long-term vision

Ai6x6™ Transformation Blueprint

Data Pathway Bridges

"In God we trust. All others must bring data." - W. Edwards Deming

W. Edwards Deming (1900–1993) was an American statistician and management consultant. He is widely regarded as a leading figure in the quality management movement and a key influence on Japanese manufacturing after World War II.

Data pathway bridges are formal frameworks and processes designed to overcome negative data patterns and ensure data accuracy, fairness, and security. They are the strategic tools that turn a vision for a data-driven organization into a secure, ethical, and highly effective reality. A Representative Data Bridge is a mechanism for ensuring data is collected from diverse sources, thereby mitigating selection bias. The Historical Fairness Bridge is a process for auditing historical data to correct for past prejudices. The Fair Algorithms Bridge is a critical framework for testing algorithms for discriminatory outcomes, ensuring that automation is a force for good. As W. Edwards Deming famously said, "In God we trust. All others must bring data." These bridges are the practical tools that provide a clear road map for navigating the difficult journey of data transformation, ensuring the organization has the support and tools it needs to succeed.

Serving as the intelligence layer of the Execution domain, this framework explores **Data Pathway Bridges**—the systematic technical connections necessary to achieve Foresight and realize true Human-AI Synergy. Data Pathway Bridges are crucial, employing tools and standards (like ETL/ELT pipelines) to connect disparate, siloed data sources into a unified, accessible environment. API Bridges enable applications and services to communicate and exchange data seamlessly, avoiding manual intervention and creating real-time data flow across the enterprise. Data Governance Bridges translate policy into practice, ensuring automated enforcement of rules for data quality, lineage, and ownership. Security Bridges deploy layered protocols and technologies to safeguard data at rest and in transit, maintaining compliance and customer trust. Analytics Bridges ensure the right tools, dashboards, and visualization platforms are connected to the data layer, allowing analysts to interpret complex information quickly. Finally, AI/ML Pipeline Bridges manage the entire flow of data from ingestion and training through to model deployment and monitoring, ensuring models are always current and performing reliably.

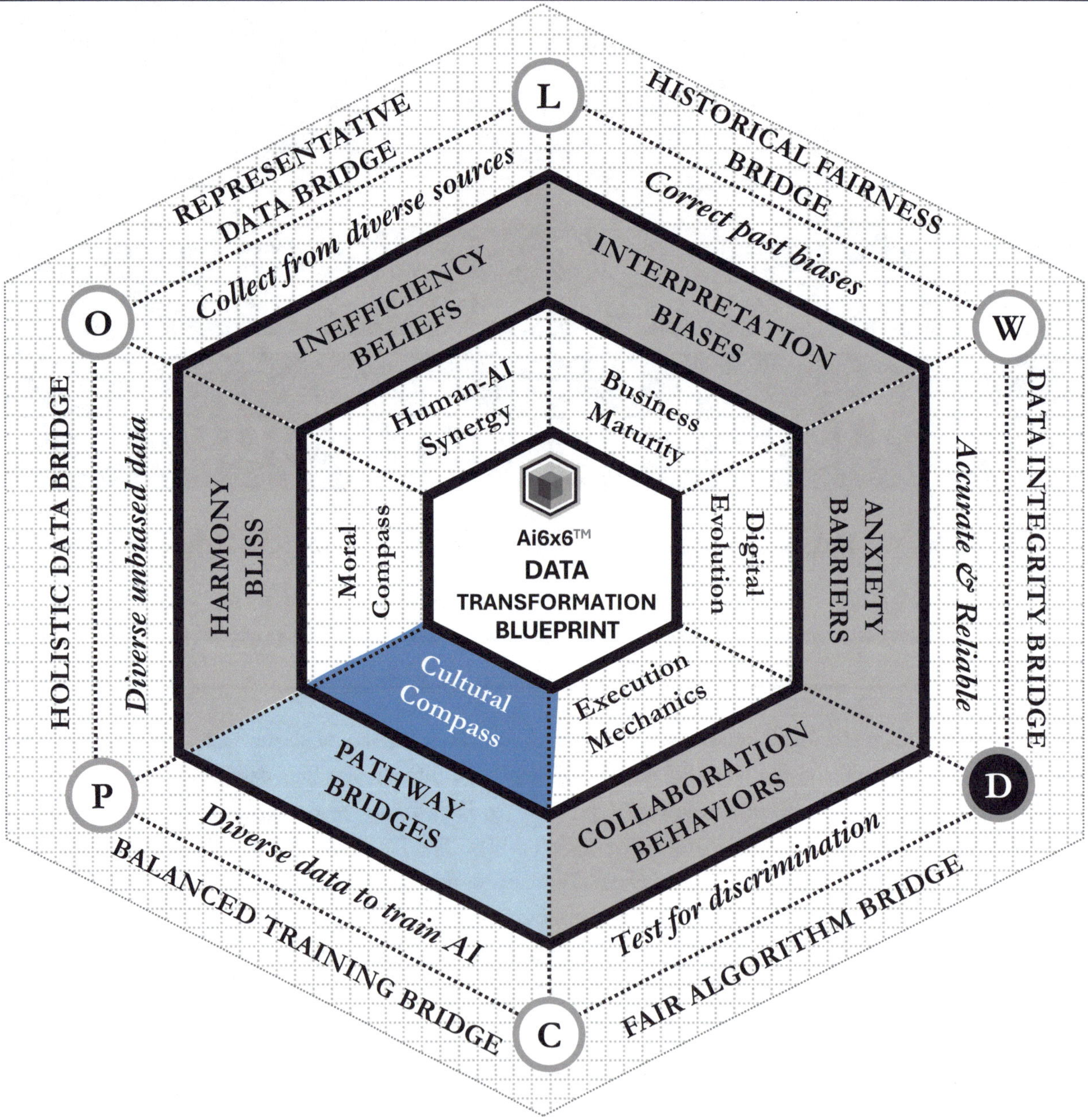

CULTURAL COMPASS FRAMEWORK FOR DATA LIFECYCLE

1. REPRESENTATIVE DATA BRIDGE | The Reality Check

This bridge ensures that your AI is built on a foundation of true diversity. It is a systematic process for sourcing data from a wide variety of demographics, geographies, and scenarios. In your daily operations, this bridge manifests as "data sourcing audits" that look for missing voices. By building this bridge, you cross from the island of "biased insights" to the mainland of "market reality." It ensures that your products and services work for everyone, not just a privileged subset. For an executive, this bridge is a major "de-risking" tool; it prevents the embarrassing and costly failures that occur when an AI model is launched into a diverse world after being trained in a narrow, homogeneous and unrepresentative vacuum.

DO'S FOR THE LEADER	DON'TS FOR THE LEADER
Ensure data is from diverse sources.	Rely on data from only one source.
Create a process for data sampling.	Use a biased sampling method.
Ask, "who is missing from this data?"	Ignore a large part of the population.
Use data that reflects all groups.	Use data that is skewed to a few.
Correct for any selection bias.	Assume your data is perfectly clean.
Test your data against other sources.	Only use your own data for insights.

SITUATION	SUGGESTED ACTION
Teams expand data collection because they understand diverse sources strengthen insight accuracy and fairness	Establish inclusive sampling standards, validate coverage, and reinforce representativeness as a core requirement
Employees identify missing segments because they value complete and equitable data representation	Conduct coverage audits, highlight gaps, and adjust collection strategies
Managers prioritize broad sourcing because they recognize narrow datasets distort organizational understanding	Provide source diversity guidelines, compare alternatives, and reinforce balanced input
Teams validate assumptions carefully because they understand representativeness affects decision quality	Use population benchmarks, test subgroup patterns, and reinforce disciplined evaluation
Employees advocate for additional sources because they see value in capturing overlooked perspectives	Encourage source proposals, evaluate feasibility, and integrate diverse inputs
Leaders support representative data efforts because they understand fairness requires comprehensive coverage	Allocate collection resources, monitor progress, and reinforce accountability

2. HISTORICAL FAIRNESS BRIDGE | The Project Pivot

This bridge is a framework for auditing and "cleaning" historical datasets to correct for past prejudices. It recognizes that "raw data" is often "biased data." Daily, this looks like using "de-biasing algorithms" and human oversight to adjust old training sets, so they reflect modern ethical standards. By building this bridge, you ensure that your AI is a tool for progress rather than a "reinforcement machine" for the past. It allows you to leverage the value of historical data without inheriting its flaws. For the executive, this bridge provides the ethical "green light" needed to scale AI systems across sensitive areas like HR, finance, and customer service with total confidence in their fairness, integrity, and social responsibility.

DO'S FOR THE LEADER	DON'TS FOR THE LEADER
Audit historical data for biases.	Use old data that reflects past biases.
Adjust data for past unfairness.	Let old biases continue to persist.
Acknowledge the historical context.	Use data without understanding its past.
Use diverse data to correct for bias.	Rely on a single, old, and biased source.
Remove data that is unfair.	Keep using data that is biased.
Be a leader in ethical data use.	Be passive about data ethics.

SITUATION	SUGGESTED ACTION
Teams review legacy datasets because they understand historical patterns may encode outdated prejudices	Conduct bias audits, document findings, and update datasets to reflect equitable standards
Employees question inherited metrics because they recognize past definitions may not align with modern fairness	Redesign evaluation criteria, validate assumptions, and reinforce inclusive measurement
Managers reassess long-standing models because they understand historical data may distort predictions	Retrain models with updated inputs, test subgroup performance, and reinforce fairness expectations
Teams adjust baselines because they realize historical norms may reflect exclusionary practices	Establish new benchmarks, incorporate diverse perspectives, and reinforce equitable comparisons
Employees challenge outdated assumptions because they understand fairness requires continuous reevaluation	Facilitate review sessions, highlight improvements, and reinforce learning culture
Leaders support historical audits because they recognize fairness strengthens organizational trust and credibility	Provide governance oversight, allocate resources, and reinforce long-term commitment

3. DATA INTEGRITY BRIDGE | The Grounding Lens

A robust framework for ensuring data is accurate, reliable, and "clean" is the bridge that allows for trusted AI. This looks like automated "data hygiene" processes and clear "stewardship" roles for every dataset. Daily, this bridge ensures that the reports on your desk are 100% accurate. It eliminates the "data doubt" that slows down strategic execution. By building this bridge, you turn data from a "messy liability" into a "precision asset." This integrity is the prerequisite for any advanced AI application; without it, your models will hallucinate, and your predictions will fail. This bridge provides the "bedrock of truth" that allows your entire organization to move with the speed and confidence required in the competitive AI era.

DO'S FOR THE LEADER	DON'TS FOR THE LEADER
Ensure data is accurate and reliable.	Use bad data to make decisions.
Implement a data quality framework.	Have no plan for data quality.
Correct bad data at the source.	Let data problems compound over time.
Use a single source of truth.	Allow multiple versions of the same data.
Be transparent about data quality.	Hide data problems from your staff.
Invest in tools to check for integrity.	Rely on manual data checks.

SITUATION	SUGGESTED ACTION
Teams validate datasets consistently because they understand integrity strengthens every downstream decision	Establish quality checkpoints, document lineage clearly, and reinforce disciplined verification practices across all data pipelines to maintain organizational accuracy
Employees correct errors quickly because they value maintaining trustworthy and dependable information across systems	Create error-reporting channels, encourage rapid fixes, and reinforce shared responsibility for accuracy throughout the organization
Managers prioritize data hygiene because they recognize clean inputs drive reliable insights and operational confidence	Implement maintenance routines, monitor quality trends, and reinforce proactive stewardship to ensure dependable information
Teams align definitions carefully because they understand consistency improves organizational clarity and decision-making	Publish standard glossaries, review terminology regularly, and reinforce shared understanding across all functions
Employees track source reliability because they know integrity depends on trustworthy origins and transparent documentation	Maintain source evaluations, highlight risks, and reinforce careful selection to protect data quality
Leaders support integrity initiatives because they understand accuracy strengthens credibility and long-term performance	Allocate governance resources, monitor compliance, and reinforce sustained commitment to organizational data excellence

This bridge is a systematic process for testing algorithms for discriminatory outcomes before they are deployed. It involves "adversarial testing"—trying to break the model to see if it behaves unfairly toward specific groups. Daily, this looks like a "Fairness Scorecard" for every new AI tool. By building this bridge, you protect the organization's reputation and legal standing. It ensures that your "automated decisions" are as ethical as your "human decisions." For an executive, this bridge is the "ethical guardrail" that allows you to embrace the efficiency of AI without sacrificing the values of the company. It turns "algorithmic fairness" from a vague goal into a measurable, manageable business process that builds long-term brand equity.

DO'S FOR THE LEADER	DON'TS FOR THE LEADER
Test algorithms for discrimination.	Let algorithms produce unfair results.
Use balanced training data.	Use biased data to train your AI.
Be transparent about how algorithms work.	Use a black box algorithm.
Use human oversight on AI decisions.	Let AI make all decisions alone.
Regularly audit algorithms for bias.	Assume your algorithm is fair.
Explain how the algorithm arrived at an answer.	Don't document the algorithm's process.

SITUATION	SUGGESTED ACTION
Teams test models regularly because they understand fairness requires continuous evaluation across different demographic groups	Conduct bias assessments, compare subgroup outcomes, and reinforce corrective action expectations to maintain equitable algorithmic performance
Employees question unexpected patterns because they recognize fairness issues may appear subtly within model outputs	Provide diagnostic tools, encourage investigation, and reinforce analytical curiosity to uncover potential disparities
Managers review algorithmic decisions because they understand oversight prevents discriminatory impacts across populations	Establish review committees, document findings thoroughly, and reinforce accountability for ethical model behavior
Teams validate training data carefully because they know fairness depends on representative and balanced inputs	Audit dataset composition, rebalance samples, and reinforce inclusive sourcing practices
Employees escalate concerns quickly because they trust fairness issues will be addressed transparently and promptly	Maintain escalation pathways, communicate resolutions clearly, and reinforce openness around model governance
Leaders champion fairness testing because they understand ethical algorithms strengthen organizational trust and credibility	Provide governance oversight, allocate resources, and reinforce ethical standards across all analytical initiatives

This bridge ensures that AI models are trained on diverse, balanced datasets that include plenty of "edge cases" and "non-obvious" examples. It's the process of "synthetic data generation" and "data augmentation" to fill in the gaps. Daily, this bridge means your AI is smarter and more robust. It doesn't get "confused" by a new market shift or a unique customer profile. By building this bridge, you create an AI that is "street smart"—capable of handling the messy, unpredictable variety of real-world business. For the executive, this bridge provides the "operational resilience" needed to trust AI in high-stakes environments, knowing that the model has been trained to handle the full spectrum of possibility, not just the standard "happy path."

DO'S FOR THE LEADER	DON'TS FOR THE LEADER
Use diverse datasets to train AI.	Use skewed, biased data to train AI.
Clean data for unfairness before training.	Assume all data is clean and fair.
Monitor for bias during training.	Don't check for bias until it is a problem.
Use diverse teams to build AI.	Let a single group build the AI.
Document your training data.	Use untracked, undocumented data.
Test the AI for fair outcomes.	Assume the AI is fair by default.

SITUATION	SUGGESTED ACTION
Teams source diverse datasets because they understand balanced training improves model fairness and predictive accuracy	Establish representation targets, validate coverage, and reinforce inclusive data practices across all training pipelines
Employees identify gaps early because they value complete and equitable training inputs for reliable models	Conduct coverage reviews, highlight missing segments, and adjust sourcing to strengthen dataset diversity
Managers prioritize dataset diversity because they recognize narrow inputs distort predictions and reduce fairness	Provide sourcing guidelines, compare alternatives, and reinforce balanced selection practices
Teams test subgroup performance because they understand fairness requires evaluating multiple perspectives consistently	Use performance dashboards, analyze disparities, and reinforce corrective adjustments to improve equity
Employees advocate for additional sources because they see value in capturing overlooked patterns and perspectives	Encourage source proposals, evaluate feasibility, and integrate diverse inputs into training datasets
Leaders support balanced training because they understand diverse data strengthens long-term AI reliability and trustworthiness	Allocate collection resources, monitor progress, and reinforce accountability for inclusive training practices

Connecting data from disparate sources—CRM, ERP, Social, Supply Chain—into a single "unbiased picture" is the bridge to true strategic insight. This is the "Data Mesh" or "Fabric" that breaks down silos. Daily, this bridge allows you to see the "hidden connections" between different parts of your business. It turns "isolated data points" into "comprehensive narratives." By building this bridge, you ensure that your AI isn't just optimizing a single department but is optimizing the entire company as a unified system. For an executive, this bridge provides the "God's-eye view" of the organization, allowing for the kind of complex, multi-variable decision-making that is the hallmark of a truly transformed, AI-ready leader operating at peak strategic efficiency.

DO'S FOR THE LEADER	DON'TS FOR THE LEADER
Connect data from all sources.	Use data that is siloed and separate.
Create a single, complete view of data.	Have a fragmented view of the data.
Use a central data warehouse.	Let each team have its own data.
Promote a culture of data sharing.	Hoard data in departments.
Use tools to connect systems.	Let systems operate in isolation.
Create a single source of truth.	Let different systems have different data.

SITUATION	SUGGESTED ACTION
Teams integrate multiple sources because they understand holistic data strengthens insight accuracy and overall completeness	Build integration pipelines, align schemas, and reinforce cross-system consistency to support unified analysis
Employees identify contradictions quickly because they value unified views across interconnected datasets	Conduct reconciliation checks, document discrepancies, and reinforce collaborative resolution practices
Managers prioritize connected systems because they recognize siloed data limits organizational understanding and decision quality	Implement integration roadmaps, provide resources, and reinforce long-term alignment across platforms
Teams validate cross-source consistency because they understand holistic accuracy requires careful comparison and verification	Use consistency dashboards, analyze mismatches, and reinforce disciplined evaluation across datasets
Employees advocate for unified platforms because they see value in seamless data access and improved usability	Support platform consolidation, gather requirements, and reinforce user-centered integration
Leaders invest in integration efforts because they understand holistic data improves strategic decision-making and organizational foresight	Allocate governance oversight, monitor progress, and reinforce enterprise-wide adoption of integrated systems

Ai6x6™ Transformation Blueprint

Data Harmony Bliss

"Data that is loved tends to survive." - Kurt Bollacker

Kurt Bollacker is a computer scientist and researcher. He is known for his work in machine learning and digital libraries, and he was the technical director of The Internet Archive, where he focused on data preservation.

Data Harmony bliss is the ultimate harmonious state where data is a source of profound satisfaction and success. It is the culmination of a successful data transformation, where the organization has transcended its negative beliefs and biases and successfully dismantled the barriers that once held it back. Blissful Clarity is the joy of having transparent, accurate, and easy-to-understand data that empowers every member of the team. Blissful Insights is the satisfaction of using data to generate new, innovative ideas that drive the business forward. As Kurt Bollacker's quote suggests, "Data that is loved tends to survive." In this state, data is no longer a source of anxiety or confusion but a powerful tool that enables the entire organization to operate with confidence, creativity, and integrity. This bliss is not about the data itself, but about the profound impact it has on the organization's ability to learn, innovate, and grow.

Serving as the intelligence layer of the Execution domain, this framework explores **Data Harmony Bliss**, detailing the strategic outcomes generated by systematic data utilization, securing Foresight and achieving Human-AI Synergy. The core outcome is Data-Driven Decisions, where all significant organizational choices are made with confidence, backed by verifiable evidence and objective analysis, minimizing reliance on guesswork. This leads to Actionable Insights, where data analysis doesn't just report history, but provides clear, prescriptive direction for future action and competitive advantage. Bliss requires Predictive Accuracy, with AI models and forecasts reliably anticipating market shifts, customer behavior, and operational needs, enabling proactive strategic maneuvering. Technical efficiency is achieved through Seamless Integration, meaning data flows effortlessly and instantaneously between all systems, eliminating manual transfer, latency, and unlocking cross-functional intelligence. A non-negotiable outcome is robust Data Security, ensuring a secure and compliant data environment that safeguards sensitive information and maintains public trust, mitigating reputation risk. The ultimate utility is AI-Powered Efficiency, where automation and machine learning enhance human work by handling repetitive tasks, driving overall productivity and augmenting human capabilities.

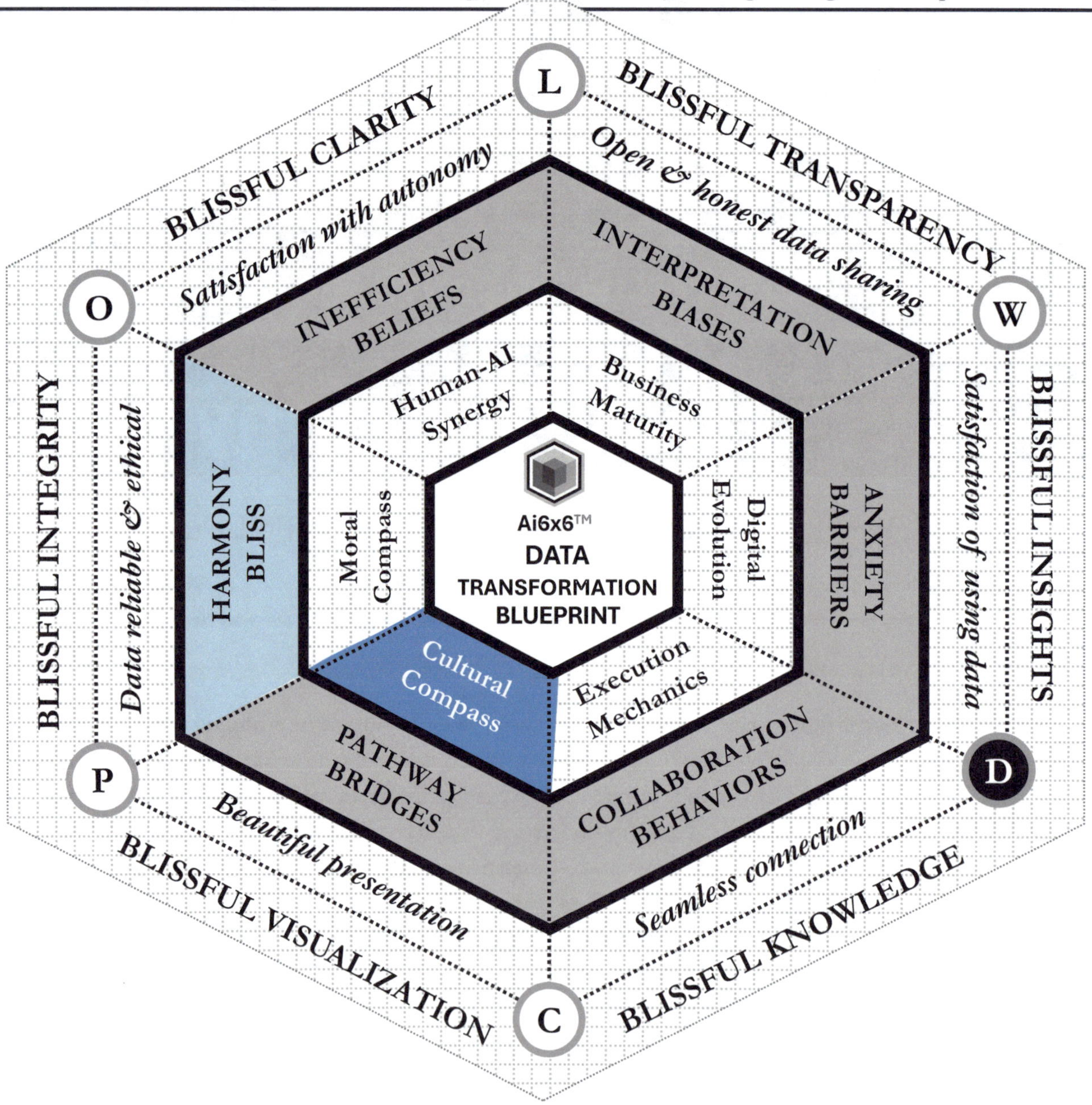

CULTURAL COMPASS FRAMEWORK FOR DATA LIFECYCLE

There is a profound sense of peace that comes from having transparent, accurate, and easy-to-understand data. This bliss is the "death of ambiguity." In your daily life, it means you never have to ask "is this number right?" again. You lead from a state of total awareness. This clarity reduces the mental load of management and allows you to focus 100% of your energy on creative strategy and high-level problem-solving. It's the feeling of "seeing through the noise" to the core truth of your business. This bliss is the reward for the hard work of building data integrity; it transforms the executive experience from a struggle for information into a mastery of insight and strategic foresight.

DO'S FOR THE LEADER	DON'TS FOR THE LEADER
Create a state of transparency.	Hide data from your teams.
Use accurate, easy-to-read data.	Use confusing or complex reports.
Connect data to clear business goals.	Let data have no purpose.
Ensure everyone understands the data.	Let people guess what the data means.
Provide clear, objective feedback.	Give vague and unhelpful feedback.
Link tasks to a larger purpose.	Make work seem meaningless.

SITUATION	SUGGESTED ACTION
Teams make decisions confidently because data is consistently clear, structured, and easy for everyone to interpret	Provide clarity standards, simplify visual structures, and reinforce shared understanding through transparent explanations and consistent communication
Employees trust insights quickly because information is presented without ambiguity, confusion, or unnecessary complexity	Use structured summaries, highlight essential context, and reinforce disciplined communication practices that support rapid comprehension
Managers resolve issues faster because transparent data reveals root causes without confusion or misinterpretation	Maintain clear lineage views, document assumptions thoroughly, and reinforce straightforward interpretation across all functions
Teams collaborate effectively because shared clarity eliminates misunderstandings and strengthens alignment across diverse roles	Use unified dashboards, align terminology, and reinforce consistent framing to support smooth collaboration
Employees adopt new tools easily because intuitive data design reduces cognitive load and confusion	Provide guided walkthroughs, simplify interfaces, and reinforce user-centered design principles across platforms
Leaders communicate direction confidently because clear data strengthens alignment, trust, and organizational coherence	Share concise insights, explain trends clearly, and reinforce transparency in strategic communication

The pleasure of a culture where data is shared openly and honestly is the bliss of "radical alignment." It's the feeling of an organization where there are "no secrets" and "no surprises." This bliss creates a high-trust environment where politics are replaced by performance. Daily, this means your meetings are focused on "how to improve" rather than "who to blame." It provides a deep sense of psychological safety and collective purpose. This bliss is the realization that when everyone sees the same reality, they can all move in the same direction with total confidence, turning the organization into a unified, high-velocity machine where the truth is everyone's most powerful ally in achieving market dominance and internal harmony.

DO'S FOR THE LEADER	DON'TS FOR THE LEADER
Share data openly and honestly.	Keep data a secret from employees.
Be honest about data limitations.	Overstate your data's accuracy.
Use clear, open data processes.	Use a black box data process.
Encourage open sharing of insights.	Let departments hoard their insights.
Communicate key decisions with data.	Keep data out of key decisions.
Be a leader in data ethics.	Be passive about data ethics.

SITUATION	SUGGESTED ACTION
Teams share information openly because transparency strengthens alignment and reduces unnecessary friction across functions	Establish open-access norms, clarify expectations, and reinforce collaborative interpretation practices
Employees communicate assumptions clearly because openness builds trust and strengthens shared understanding across groups	Use context templates, encourage clarification dialogue, and reinforce transparent communication
Managers publish metrics consistently because transparent reporting strengthens accountability and organizational credibility	Implement reporting rhythms, provide interpretation guidance, and reinforce constructive use of insights
Teams resolve conflicts quickly because transparent data eliminates speculation and hidden assumptions	Maintain shared dashboards, align definitions, and reinforce evidence-based dialogue across teams
Employees contribute insights confidently because transparency ensures fair recognition and shared ownership of ideas	Highlight contributions, encourage open discussion, and reinforce inclusive participation norms
Leaders model transparency because openness strengthens trust, alignment, and organizational progress	Share strategic context, explain rationale clearly, and reinforce open communication practices

3. BLISSFUL INSIGHTS | The Calculated Leap

The satisfaction of using clean, high-integrity data to generate a "game-changing" idea is the pinnacle of the AI era. This bliss is the "Eureka!" moment fueled by math. It's the joy of seeing a pattern that no human could have spotted on their own. Daily, this bliss makes work feel like an act of discovery. It validates your investment in AI and data transformation. This feeling of being "smarter than the market" because of your data capabilities provides a lasting sense of professional fulfillment and excitement. It's the bliss of knowing that your organization is "innovating on purpose" rather than by accident, guided by the precision of data-driven genius that few competitors can match or even comprehend.

DO'S FOR THE LEADER	DON'TS FOR THE LEADER
Use data to generate new ideas.	Only use data for reporting.
Encourage bold data innovation.	Punish people who try new things.
Provide resources for data experiments.	Keep all resources locked up.
Create a safe place for new ideas.	Create a culture of fear.
Reward data-driven innovation.	Reward conformity and old ideas.
Use data to test new theories.	Rely on intuition over data.

SITUATION	SUGGESTED ACTION
Teams generate creative solutions because insights reveal opportunities previously hidden or overlooked across datasets	Provide exploration tools, encourage hypothesis testing, and reinforce curiosity-driven analysis practices
Employees propose new ideas confidently because data supports experimentation and informed risk-taking	Highlight successful innovations, provide safe testing environments, and reinforce learning-focused culture
Managers identify emerging trends early because insights illuminate patterns shaping future organizational performance	Maintain trend dashboards, share scenario models, and reinforce proactive planning behaviors
Teams collaborate effectively because shared insights spark cross-functional creativity and alignment	Facilitate ideation sessions, integrate diverse perspectives, and reinforce open exploration practices
Employees refine concepts quickly because data provides clear feedback on potential value and feasibility	Use iterative testing, document learnings, and reinforce evidence-based refinement processes
Leaders champion insight-driven innovation because data strengthens strategic imagination and organizational agility	Highlight insight impact, align investments, and reinforce forward-looking thinking across teams

The benefit of having data seamlessly connected across the organization is the bliss of "total intelligence." It's the feeling of a company that "knows what it knows." This bliss eliminates the frustration of redundant work and "blind spot" errors. Daily, it means you can answer any complex question in seconds because the bridge between departments is permanent and perfect. It provides a sense of immense capability and scale. This bliss is the realization that your company is a "living brain," where every part is informed by the whole. It allows you to lead with a sense of effortless control, knowing that the entire weight of the organization's collective knowledge is backing every strategic move you make, with no information gaps.

DO'S FOR THE LEADER	DON'TS FOR THE LEADER
Connect data seamlessly across org.	Use data that is siloed and separate.
Create a single, complete view of data.	Have a fragmented view of the data.
Use a central data warehouse.	Let each team have its own data.
Promote a culture of data sharing.	Hoard data within departments.
Use tools to connect systems.	Let systems operate in isolation.
Create a single source of truth.	Let different systems have different data.

SITUATION	SUGGESTED ACTION
Teams work efficiently because connected data eliminates duplication and accelerates shared understanding across functions	Build unified platforms, align schemas, and reinforce cross-system consistency to support collaboration
Employees solve problems faster because integrated information reveals relationships across multiple organizational domains	Provide linked datasets, highlight dependencies, and reinforce holistic analysis practices
Managers coordinate effectively because shared knowledge strengthens alignment and reduces operational friction	Maintain integrated dashboards, clarify context, and reinforce collaborative planning behaviors
Teams innovate confidently because connected data provides a complete view of organizational dynamics	Use cross-source insights, encourage exploration, and reinforce comprehensive evaluation practices
Employees trust insights more because unified data reduces contradictions and confusion	Conduct reconciliation checks, document assumptions, and reinforce consistent interpretation across teams
Leaders make stronger decisions because connected knowledge strengthens foresight and strategic clarity	Share integrated summaries, explain implications clearly, and reinforce enterprise-wide visibility

The joy of seeing complex data presented in a beautiful, intuitive, and instantly understandable way is the bliss of "cognitive harmony." It's the feeling when a thousand rows of data become a single, elegant insight. This bliss makes communication effortless. Daily, it turns your "dashboard time" into a moment of inspiration rather than a chore. It ensures that your vision is shared perfectly by everyone who sees the data. This bliss is the aesthetic reward for valuing the representation of truth. It allows you to lead with "narrative power," using the beauty of data to inspire, persuade, and align your entire organization around the clear, bright future that the data reveals, making complexity feel simple and actionable.

DO'S FOR THE LEADER	DON'TS FOR THE LEADER
Present data in a beautiful way.	Create ugly or confusing charts.
Use visuals that are easy to read.	Use complex, messy data visuals.
Standardize your company's reports.	Let every team make their own charts.
Tell a clear story with your data.	Show data without context or narrative.
Get feedback on your visualizations.	Assume your charts are easy to read.
Link visuals to a business objective.	Show data just for the sake of it.

SITUATION	SUGGESTED ACTION
Teams understand insights quickly because visuals communicate complex information clearly and intuitively	Use clean design principles, standardize formats, and reinforce simplicity across all visual outputs
Employees engage deeply because compelling visuals make data exploration enjoyable and accessible	Provide visualization templates, encourage experimentation, and reinforce user-centered design practices
Managers communicate effectively because visuals highlight essential patterns without overwhelming detail	Use layered views, explain design choices, and reinforce clarity in communication
Teams collaborate smoothly because shared visuals create common understanding across diverse roles	Maintain consistent palettes, align scales, and reinforce interpretation norms across teams
Employees adopt dashboards readily because intuitive visuals reduce learning curves and confusion	Provide guided walkthroughs, simplify interactions, and reinforce usability principles
Leaders inspire confidence because strong visuals strengthen storytelling and strategic communication	Share visual summaries, highlight trends clearly, and reinforce narrative clarity across the organization

The deep satisfaction of knowing your organization's data is reliable, ethical, and "clean" is the bliss of "moral leadership." This is the peace of mind that comes from knowing your AI isn't hiding any biased "ghosts" or ethical "landmines." It's the feeling of pride in building a "clean-tech" company for the digital age. Daily, this bliss provides the ultimate professional security; you know your foundation is solid. This bliss is the realization that you haven't just built a successful company, but a responsible and trustworthy one. It's the feeling of "integrity in action," where your data reflects your highest values, ensuring that your organization's success is as sustainable as it is significant, building a legacy of excellence.

DO'S FOR THE LEADER	DON'TS FOR THE LEADER
Ensure data is reliable and ethical.	Use bad data to make decisions.
Implement a data quality framework.	Have no plan for data quality.
Correct bad data at the source.	Let data errors compound over time.
Use a single source of truth.	Allow multiple versions of the same data.
Be transparent about data quality.	Hide data problems from your staff.
Invest in tools to check for integrity.	Rely on manual data checks.

SITUATION	SUGGESTED ACTION
Teams trust insights fully because data is consistently accurate, validated, and ethically maintained	Implement quality controls, document lineage, and reinforce responsible stewardship across all functions
Employees act confidently because ethical data practices ensure fairness across decisions and outcomes	Provide governance guidelines, highlight ethical expectations, and reinforce accountability for responsible behavior
Managers support initiatives strongly because reliable data strengthens operational performance and organizational credibility	Monitor quality trends, allocate resources, and reinforce continuous improvement practices
Teams collaborate openly because integrity ensures shared confidence in every dataset used collectively	Maintain transparent standards, align definitions, and reinforce trust across teams
Employees escalate issues quickly because they know integrity concerns will be addressed responsibly	Provide reporting channels, communicate resolutions, and reinforce openness and accountability
Leaders champion integrity because ethical data strengthens long-term trust and organizational reputation	Establish oversight structures, monitor compliance, and reinforce sustained commitment to ethical practices

BONUS CHAPTER

Ai6x6™ Transformation Blueprint

Data Diversity

"Diversity creates change, which is the engine for innovation."

Chad E. Foster, Motivational Speaker and Author. This concisely links diversity as the essential spark that forces new ways of thinking, directly powering the innovation needed to drive successful digital change.

In **Data Transformation**, diversity acts as the essential ethical and technical safeguard against the pervasive risk of algorithmic bias, which can undermine customer trust and violate regulatory standards. As organizations build their new data architectures and deploy Artificial Intelligence (AI) and Machine Learning (ML), these systems are trained on historical data that often contains inherent human prejudices. Without an intentionally **diverse and multidisciplinary team** of data scientists, engineers, and ethicists, these biases will be amplified and coded directly into the new digital intelligence, leading to unfair or discriminatory outcomes—from faulty product recommendations to biased hiring decisions. A diverse team is vital for applying the necessary critical scrutiny to data provenance, challenging fairness metrics, and actively developing mitigation strategies for subtle biases. This broad perspective ensures a more **transparent**, accurate, and reliable data output, which is the foundational resource for all successful digital transformation efforts. By embedding diversity into the data lifecycle, the company secures its future by building trust, ensuring ethical compliance, and guaranteeing its new systems work equitably.

In today's interconnected world, diversity is a fundamental pillar of strength and innovation. Often simplified to visible differences, true diversity is a rich tapestry of attributes shaping how individuals perceive and contribute. Understanding this multifaceted nature is crucial for fostering environments where every voice thrives. This essay explores diversity through six critical dimensions: **demographic, cognitive, experiential, geographic, functional, and perspective diversity**. By examining these distinct facets, we can better appreciate how a holistic approach to diversity enriches teams and communities, driving robust solutions.

First, **Demographic Diversity** encompasses varied inherent and acquired characteristics. This includes age, bringing diverse generational insights, and gender, race, ethnicity, sexual orientation, ensuring a wide range of lived experiences. Socioeconomic status, religion, disability, and nationality further contribute to a richer collective understanding. Valuing demographic diversity ensures equitable representation and leverages unique insights from varied personal journeys.

Secondly, **Cognitive Diversity** highlights how people think, process information, and approach problems. This includes varied thinking styles like analytical vs. intuitive approaches, and different problem-solving or learning methods. When a group embraces cognitive diversity, it tackles problems from multiple angles, leading to more innovative solutions and a thorough understanding of complex issues, free from a single dominant thought process.

Next, **Experiential Diversity** focuses on the breadth of life and work experiences individuals bring. This covers varied work histories across industries and roles, providing a rich pool of professional knowledge. Diverse educational backgrounds, from different disciplines or institutions, also offer distinct frameworks. Beyond formal credentials, personal life experiences like travel and cultural immersion contribute significantly, equipping teams with a wide array of skills and adaptable approaches.

Geographic Diversity emphasizes the impact of place on an individual's perspective. This includes differences from regional or local backgrounds, shaping values and priorities. National backgrounds introduce distinct cultural norms, while urban versus rural upbringings offer varying insights into community dynamics and societal needs. These varied geographical lenses provide a broader understanding of global and local markets and behaviors.

Furthermore, **Functional Diversity** pertains to the variety of professional roles, specialized areas, and departmental expertise within a team. This dimension recognizes contributions from different functions like engineering, marketing, or finance, each bringing specific domain knowledge. Variations in hierarchical roles and industry experience also inform cross-pollination of best practices. Leveraging functional diversity ensures all necessary technical skills are present, fostering comprehensive problem-solving.

Finally, **Perspective Diversity** encompasses the fundamental differences in viewpoints and mental models individuals hold, often influenced by other diversity forms. This includes varied ideological stances, cultural interpretations, and distinct value systems. Understanding diverse stakeholder perspectives—from customers to employees—is crucial for holistic decision-making. Ultimately, perspective diversity allows for problems to be framed in multiple ways, challenging assumptions and encouraging empathetic, effective outcomes.

By intentionally cultivating and integrating these six dimensions—demographic, cognitive, experiential, geographic, functional, and perspective—organizations and societies can harness the profound power of true diversity. This comprehensive approach not only fosters equity but also enhances creativity, resilience, and innovation, demonstrating that a multitude of viewpoints is the clearest path to collective success.

Embracing demographic diversity within data teams and across data-related functions means intentionally cultivating a talent pool that brings varied life experiences to data collection, analysis, and interpretation. This goes beyond mere representation; it's about leveraging the inherent value that comes from data professionals with varied ages, genders, racial and ethnic backgrounds, physical abilities, socio-economic upbringings, and sexual orientations. These visible and non-visible differences are crucial for identifying and mitigating biases in data, ensuring ethical data practices, and developing data solutions that truly serve diverse user groups. A demographically diverse data team brings distinct insights into market needs, cultural sensitivities, and potential blind spots in algorithms or data models, which are crucial for effective data transformation. When data teams lack this breadth, they risk perpetuating biases, misinterpreting data from diverse populations, or developing data products that fail to resonate with broad customer bases. True demographic diversity fosters more accurate, ethical, and ultimately, more impactful data outcomes, better equipped to address complex data challenges and connect authentically with diverse data sources and users.

Topic	Tangible steps	Challenges you may face
AGE DIVERSITY IN DATA SCIENCE TEAMS		
Varied age groups bring different technical skills (legacy vs. modern) and perspectives on data usage. Enhances innovation and knowledge transfer in data projects.	Implement cross-generational data project teams. Facilitate mentorship for data seniors/juniors. Provide continuous learning on evolving data tools accessible to all ages.	Resistance to new data technologies from older talent. Overcoming biases favoring specific age groups in hiring data roles. Managing different work styles in data collaboration.
GENDER DIVERSITY IN DATA MODELING		
Balanced gender representation influences data model design, bias detection, and ethical AI development. Diverse gender perspectives lead to more robust and fair algorithms.	Set targets for gender balance in data modeling/AI teams. Implement bias detection tools in data pipelines. Conduct inclusive data literacy training.	Unconscious biases in data interpretation. Lack of female role models in data leadership. Resistance to scrutinizing existing biased data sets.
RACIAL & ETHNIC DIVERSITY IN DATA ETHICS		
Diverse racial & ethnic backgrounds foster critical insights into data fairness, privacy, and responsible AI. Improves ethical data practices and builds trust.	Develop cultural awareness training for data professionals. Establish ERGs focused on data ethics and inclusion. Prioritize diverse representation in data governance committees.	Systemic bias in historical data. Difficulty fostering psychological safety to discuss data biases. Insufficient understanding of cultural implications of data use.

2. COGNITIVE DIVERSITY

Cognitive diversity within data teams is paramount for navigating the inherent complexities of data transformation. It acknowledges that true innovation and robust problem-solving in data stem not just from varied experiences or demographics, but fundamentally from differences in how people think, process data, and approach analytical challenges. A data team composed solely of analytical thinkers might excel at optimizing existing data pipelines but could struggle with identifying novel data sources or designing truly disruptive algorithms. Conversely, a team too reliant on intuitive thinking might miss critical data quality issues or overlook practical constraints in data implementation. By fostering a mix of analytical, creative, systemic, and neurodiverse cognitive styles across all data roles, organizations are better equipped to challenge assumptions in data models, identify unseen patterns, predict unforeseen data risks, and design more comprehensive and resilient data strategies. It's about building a collective data intelligence where different cognitive strengths complement each other, leading to more thorough data evaluations, more innovative data solutions, and a greater capacity for adaptive learning in the face of evolving data landscapes. This richness in thought ensures that data-driven decisions are thoroughly vetted from multiple angles, preventing blind spots and fostering intellectual agility crucial for sustained data transformation success.

Topic	Tangible steps	Challenges you may face
ANALYTICAL VS. INTUITIVE THINKING IN DATA ANALYTICS		
Balancing data-driven & gut-feeling approaches in data analysis shapes insight generation & problem-solving. Leads to nuanced decision-making, combining factual rigor with insightful foresight.	Design data analysis workflows requiring diverse cognitive input. Encourage data scientists to articulate both data-backed and intuitive reasoning. Provide training on leveraging different thinking styles in data projects.	Friction between data rigor & rapid intuition. Difficulty quantifying intuitive data insights. Culture valuing one data analysis style over another hinders collaboration.
SYSTEMIC THINKERS IN DATA ARCHITECTURE		
Data professionals seeing interconnectedness of data systems & long-term implications drive comprehensive data architecture improvement. Ensures holistic, sustainable data infrastructure.	Provide systems thinking training for data architects & engineers. Assign systemic thinkers to cross-functional data integration initiatives. Use "data lineage mapping" to identify impacts.	Difficulty identifying/developing systemic thinking in data roles. Resistance to long-term data infrastructure views. Overwhelm from too much interconnected data system information.
CREATIVE PROBLEM-SOLVERS IN DATA VISUALIZATION		
Data professionals with imaginative thinking inspire novel data visualization & storytelling. Crucial for making complex data accessible & actionable.	Establish data visualization "hackathons" or ideation labs. Implement brainstorming sessions for unconventional data presentation ideas. Create dedicated time for "blue-sky" data exploration.	Risk aversion stifling new data visualization ideas. Difficulty translating creative data visuals into actionable insights. Perceived lack of "seriousness" for certain data roles.

3. EXPERIENTIAL DIVERSITY

Experiential diversity across data professionals is a powerful catalyst for successful data transformation, providing a rich tapestry of practical insights drawn from varied career paths, industries, educational backgrounds, and personal life stages. Data professionals who have navigated different organizational cultures, worked across diverse sectors (e.g., startup to corporate, public to private), or gained international exposure bring a broader repertoire of data solutions and a more nuanced understanding of complex data challenges. This breadth of experience prevents insular thinking in data strategy, encourages the adoption of "best practices" from seemingly unrelated data fields, and fosters resilience by drawing on lessons learned from past data successes and failures in varied contexts. For instance, a data professional with experience in financial fraud detection can bring invaluable composure and clear-headedness during periods of intense data security breaches. By intentionally cultivating a data workforce with diverse professional and personal journeys, an organization gains a formidable collective wisdom, enhancing its ability to foresee data opportunities, mitigate data risks, and execute data transformation initiatives with greater adaptability and effectiveness. It promotes a learning culture where varied practical data wisdom is valued and applied to new strategic data challenges.

Topic	Tangible steps	Challenges you may face
CROSS-INDUSTRY INSIGHTS IN DATA ANALYTICS		
Data professionals from different industries bring fresh perspectives & transferable solutions to data challenges. Broadens organizational data problem-solving.	Recruit data talent with diverse industry backgrounds. Organize peer-learning forums for cross-industry data insights. Encourage participation in external data science communities.	Difficulty translating data insights to new industry contexts. Resistance to "outsider" data approaches. Lack of understanding of new industry data nuances for integration.
STARTUP VS. CORPORATE DATA GOVERNANCE		
Impact of data professionals from agile startup vs. structured corporate environments on data governance & innovation. Blending these fosters agility & stability in data systems.	Integrate startup data talent into corporate data innovation projects. Create joint ventures pairing corporate/startup data teams. Foster internal "intrapreneurship" programs for data solutions.	Culture clash between agile & traditional data governance. Difficulty integrating fast data decisions into slower processes. High expectations for immediate data ROI.
GLOBAL VS. LOCAL DATA STRATEGY		
Data professionals with diverse international & local market exposure balance global data standardization & regional adaptation. Optimizes data operational efficiency & market relevance.	Implement global data talent rotation programs. Establish regional data knowledge hubs with local market data insights. Encourage dual reporting lines for international data projects.	Complex balance of global vs. local data needs. Resistance from local data teams to global data mandates. Misinterpretation of cultural nuances in data collection/use.

Geographic diversity within data professionals is indispensable for organizations operating in a globalized world, especially during periods of significant data transformation. Data professionals deeply rooted in varied regions, countries, or even urban versus rural settings bring critical local data market understanding, cultural nuances, and insights into unique regulatory environments and logistical challenges of data collection and use. Without this localized data perspective, global data strategies risk being homogenous, culturally insensitive, or simply impractical in execution. For example, a data professional experienced in emerging markets might bring crucial insights into scalable, low-cost data solutions, while someone from a highly regulated economy understands the complexities of data compliance (e.g., GDPR, CCPA). This diversity ensures that data transformation initiatives are not just conceptualized centrally but are designed with built-in adaptability for regional data variations. It empowers the organization to navigate geopolitical data risks, understand localized customer data behaviors, and foster stronger relationships with diverse regional data stakeholders. Data professionals with varied geographic exposure contribute to data strategies that are globally coherent yet locally relevant, maximizing impact and minimizing missteps in a fragmented data world.

Topic	Tangible steps	Challenges you may face
REGIONAL DATA MARKET UNDERSTANDING IN ANALYTICS		
Data professionals from different regions offer unique insights into local data market dynamics, consumer data behaviors, & competition. Vital for effective localized data analytics and strategy.	Recruit data talent with local market data knowledge. Mandate global data strategies be reviewed/adapted by regional data teams. Support employee immersion in diverse regions for data context.	Overlooking data nuances for standardization. Difficulty adapting to rapid changes in local data markets. Resistance from central data teams to local adaptations.
NAVIGATING GLOBAL DATA REGULATORY LANDSCAPES		
Data professionals familiar with diverse international legal/compliance frameworks are crucial for multi-national data operations & transformations. Ensures adherence & mitigates data risks.	Create global data compliance training programs for relevant data roles. Invest in continuous education on international data regulations (e.g., GDPR, CCPA). Consult local legal experts for every data project.	Keeping up with rapidly changing data privacy laws. Reconciling conflicting data requirements across jurisdictions. Risk of non-compliance due to inadequate local data understanding.
LOCAL INFRASTRUCTURE AND DATA RESOURCE KNOWLEDGE		
Data professionals understanding local infrastructure (digital, connectivity) & data resources impact operational data transformation. Optimizes local data implementation & data pipelines.	Empower local data teams for infrastructure investments. Develop centralized knowledge of regional data capabilities. Foster local data service provider partnerships.	Underestimating varied local data infrastructure impact. Inaccurate data resource assumptions. Difficulty scaling global data solutions to diverse local conditions.

5. FUNCTIONAL DIVERSITY

Functional diversity within data teams is critical for executing comprehensive and interconnected data transformations. It ensures that data strategic and operational decisions are informed by the deep expertise and unique perspectives of various professional disciplines, rather than being confined to a single departmental silo. A data team comprising experts from finance, technology, human resources, operations, marketing, and legal functions can collectively identify data opportunities, anticipate data challenges, and design data solutions that are holistic, practical, and sustainable. For instance, a finance analyst ensures fiscal viability of data investments, while a tech specialist champions digital data adoption, and an HR generalist focuses on people-centric change management around data. Without this blend, data transformation efforts might optimize one area at the expense of another, leading to fragmented implementation or unintended negative consequences in data usage. By fostering cross-functional collaboration and valuing diverse functional expertise, an organization can break down traditional barriers, enhance teamwork, and create truly integrated data strategies that leverage the strengths of every part of the business, ensuring that the data transformation is not just conceptually sound but also executable across all organizational facets.

Topic	Tangible steps	Challenges you may face
FINANCE-LED DATA ROI MEASUREMENT FOR TEAMS		
Data professionals with strong financial backgrounds prioritize cost optimization & ROI as primary data transformation drivers. Ensures fiscal responsibility in data projects.	Appoint finance professionals to data project teams to define ROI. Establish clear financial metrics & ROI targets for data initiatives. Implement rigorous cost-benefit analyses for all data investments.	Over-emphasis on short-term data cuts. Resistance if financial goals overshadow data innovation. Difficulty projecting ROI for intangible benefits in specific data functions.
TECHNOLOGY-DRIVEN DATA INNOVATION ACROSS WORKFLOWS		
Data professionals with deep tech expertise champion digital adoption & disruptive data innovation. Essential for driving tech-enabled data change throughout the organization.	Elevate technologists to strategic roles in data projects. Foster "data tech literacy" for all employees. Create cross-functional innovation teams combining data tech/business experts.	Resistance to new data technologies. Rapid evolution challenges data skills. Difficulty integrating new data tech with legacy systems and traditional data workflows.
HR'S ROLE IN DATA LITERACY & CULTURAL ADOPTION		
HR professionals focus on cultural shifts, talent management, & employee well-being during data transformation. Ensures human capital is central to data success.	Position HR teams as strategic partners in all data change initiatives. Develop robust change management/communication plans for data adoption. Implement well-being/reskilling programs for all data users.	Business units resist prioritizing people over data technology. Difficulty measuring HR ROI in data initiatives. Challenges addressing employee anxiety/resistance to new data processes.

Perspective diversity within data teams is arguably one of the most subtle yet profound drivers of successful data transformation. It goes beyond demographics, experience, or function, delving into the varied worldviews, core values, ethical frameworks, and risk tolerances that shape how data professionals interpret information and prioritize data objectives. A data team with diverse perspectives challenges inherent biases in data, avoids groupthink in analysis, and considers a wider spectrum of data possibilities and potential consequences. For example, balancing data professionals who prioritize short-term data gains (e.g., quick wins from dashboards) with those focused on long-term data sustainability (e.g., robust data governance), or those who are risk-seeking with data experimentation with those who are risk-averse with data privacy, leads to more robust and balanced data strategic decisions. This diversity ensures that data transformation is not just about efficiency or growth in data volume, but also about ethical responsibility, data security, employee well-being, and genuine stakeholder value from data. It encourages data professionals to approach data problems from fundamentally different angles, leading to more resilient data strategies, better data stakeholder engagement, and a more adaptive and morally grounded organization capable of navigating complex and ambiguous future data landscapes. It cultivates a culture of constructive challenge and holistic data-driven decision-making.

Topic	Tangible steps	Challenges you may face
CUSTOMER-CENTRIC VS. DATA-CENTRIC VIEWS		
Data professionals balance internal data needs with external customer data insights. Ensures data solutions solve customer problems & resonate with market.	Implement "Voice of Customer" programs for all data professionals. Conduct customer immersion days for data teams. Establish KPIs balancing data quality metrics with customer satisfaction.	Internal resistance to shifting from data-centric to customer-centric. Difficulty capturing diverse customer data needs. Data professionals struggle balancing data desires vs. feasibility.
SHORT-TERM VS. LONG-TERM DATA STRATEGY		
Data professionals' differing time horizons influence project pace, scope, & success of data initiatives. Balancing these ensures both immediate data gains & future data viability.	Develop multi-year data roadmaps with short-term data milestones. Assign data teams with complementary focuses. Use scenario planning to visualize impacts of data strategies.	Pressure for immediate data results. Data professionals may prioritize short-term data gains. Difficulty articulating long-term data value to all stakeholders.
RISK-SEEKING VS. RISK-AVERSE APPROACHES IN DATA INNOVATION		
Varied risk tolerance impacts pursuit of disruptive vs. incremental data improvements. A healthy balance drives calculated data growth and controlled data experimentation.	Foster culture where calculated data risks are encouraged. Establish "fail fast, learn faster" practices for data projects. Create data project portfolios balancing high-risk/reward with safer bets.	Decision paralysis when data risk tolerances conflict. Dominance of extreme data risk views. Difficulty setting clear, unified data risk parameters across teams.

SECTION G

BLUEPRINT SERIES

Moral Compass

Ai6x6™ Transformation Blueprint

Story 1 — The Nuremberg Trials and Ethical Accountability (1945–1946): After World War II, the world confronted atrocities committed under the guise of legal authority. The Nuremberg Trials established that individuals and institutions are accountable for actions that violate fundamental human rights. Legal frameworks were strengthened by moral reasoning, not replaced by it. Judges balanced evidence with ethical principles to define new global norms. The trials demonstrated that morality must guide power. They became a foundation for modern ethical governance.

Story 2 — The Tuskegee Syphilis Study Exposure (1932–1972): For decades, researchers withheld treatment from Black men to observe disease progression, violating consent and dignity. Public exposure triggered national outrage and forced reforms in medical ethics. Oversight boards, informed consent, and patient rights became mandatory. Machines and data had played roles in tracking outcomes, but human ethics had been absent. The scandal showed how systems fail when morality is excluded. It reshaped global standards for ethical research.

Story 3 — The Environmental Movement and Earth Day (1970): Industrial growth created pollution that threatened ecosystems and public health. Scientists provided data, but moral arguments mobilized action. Citizens demanded accountability, and governments introduced regulations to protect air, water, and wildlife. The movement reframed environmental protection as a shared ethical duty. It demonstrated how moral clarity can transform policy, industry, and public behavior. The cultural shift established sustainability as a long-term societal obligation.

Story 4 — The Anti-Apartheid Global Solidarity Movement (1950s–1994: Apartheid enforced racial segregation and oppression in South Africa. Activists, governments, and institutions worldwide united to challenge the regime through sanctions, protests, and moral pressure. Human stories amplified the injustice, while global networks coordinated action. The movement showed how collective moral conviction can influence political systems. It proved that ethical solidarity can reshape international norms. The end of apartheid became a landmark in moral mobilization.

Story 5 — The Rise of Corporate Social Responsibility (1990s–2000s): Globalization exposed labor abuses, environmental harm, and opaque supply chains. Consumers demanded transparency, and companies adopted codes of conduct, sustainability reporting, and ethical sourcing. Machines tracked compliance, but humans defined the standards. CSR shifted from public relations to strategic responsibility. Organizations learned that long-term value depends on ethical behavior. It marked a turning point where morality became integral to corporate identity.

Story 6 — The AI Ethics Movement (2015–2025): As AI systems influenced decisions in hiring, policing, healthcare, and finance, concerns grew about bias, transparency, and accountability. Researchers, policymakers, and civil society pushed for ethical frameworks to guide development. Machines provided capability, but humans defined boundaries. The movement emphasized fairness, explainability, and human oversight. It demonstrated that emerging technologies require moral governance as much as technical innovation. AI ethics became a global priority shaping the future of intelligent systems.

LESSONS FROM A DATA TRANSFORMATION PERSPECTIVE

Moral maturity in data practices emerges when organizations recognize that information carries ethical obligations. Over time, institutions learned that data can empower or harm depending on how it is collected, interpreted, and used. Mature cultures prioritize consent, transparency, and fairness, ensuring that data serves human dignity rather than undermining it. Machines process information at scale, but humans define boundaries that protect rights and prevent misuse. Maturity is achieved when data governance integrates ethical reasoning into every stage—collection, storage, analysis, and decision-making. Ethical data practices become essential to trust, accountability, and long-term legitimacy.

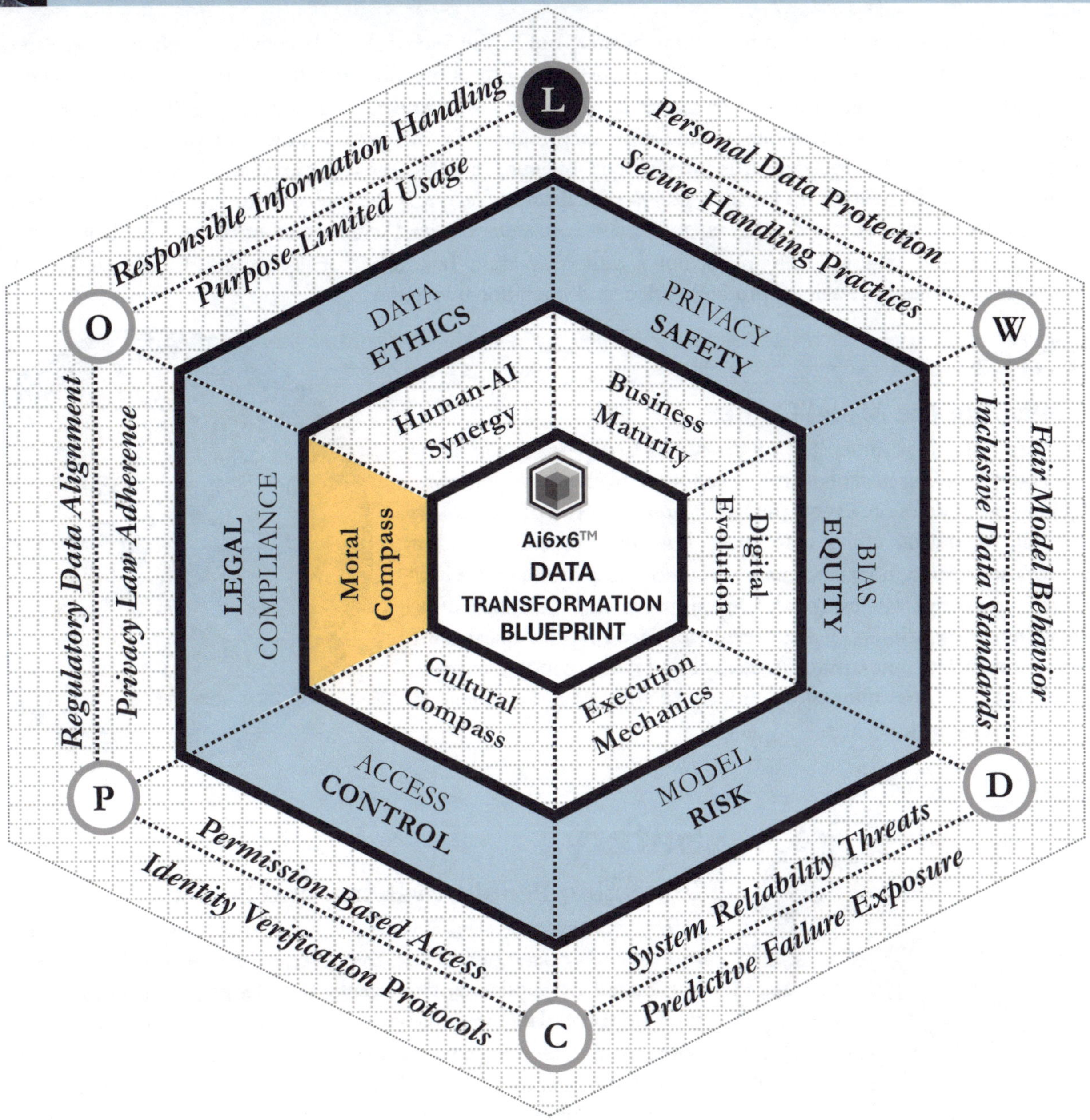

The **Ai6x6 Research Lab** identified the Moral Compass as the most vital lens because it serves as the final guardian of information honor and individual privacy. This is the ultimate filter for every data decision, ensuring the drive for intelligence never outpaces the bedrock of **Data Ethics** and **Minimization**. We chose this because an organization's legacy in the AI era is built on how it protected the sanctity of information, not just how much it collected. It is vital because it provides the moral authority required to manage the high responsibility of **Global Data Compliance** and **Bias Detection** in an automated world.

This lens is the heart's final check—a rigorous audit of **Model Risk**, **Drift**, and **Privilege Control**. It ensures the transformed enterprise is not only a titan of intelligence but a beacon of public trust and equity. The Lab believes that in the 2026 landscape, the greatest risk to a data strategy is not a system failure, but a moral breach. By focusing on privacy safety and impact assessments, we provide the tools to navigate the complex ethical social contract between the organization and the individuals it serves. This lens is about the courage to be honorable and transparent when the pressure for predictive power is highest. Ultimately, it ensures that data is treated with the respect, equity, and protection that its human origin demands.

Morality has always shaped how organizations behave, but in the age of AI, it becomes a structural requirement rather than an aspirational ideal. The **Moral Compass** lens forces organizations to confront a truth that is often ignored: intelligence—whether human or machine—does not create ethical behavior. It amplifies whatever moral foundation already exists. When an organization is grounded in integrity, clarity, and principled action, AI strengthens it. When the foundation is inconsistent, opaque, or misaligned, AI exposes those weaknesses with unforgiving speed. Morality is no longer a philosophical preference; it is the architecture that determines whether accelerated decisions, automated systems, and scaled consequences lead to progress or harm. This lens reframes morality as an enterprise-wide discipline, not a leadership trait. It applies equally to workforce practices, data governance, customer interactions, product design, and operational systems.

Ethics

Establishing the Moral Center

Ethics forms the first dimension of the Moral Compass. Ethics is not simply about avoiding misconduct; it is about establishing a moral center that guides every decision, every interaction, and every use of technology. Across all six pillars, ethics determines how data is collected, how employees are treated, how customers are served, how products are built, and how operations are executed. Organizations that prioritize ethics create cultures where integrity is expected, not encouraged. Ethical clarity becomes the anchor that ensures AI is deployed with intention, restraint, and respect. In this dimension, principled behavior becomes the foundation of moral authority.

Safety

Protecting People, Systems, and Society

Safety represents the second dimension. As AI accelerates decision-making and expands operational reach, the potential for harm increases. Safety is not limited to physical environments; it includes psychological safety for the workforce, data safety for customers, model safety for AI systems, and operational safety for mission-critical processes. Safety ensures that no individual, team, or automated system can cause harm through unchecked influence or unintended consequences. It is not about limiting innovation; it is about ensuring that innovation does not outpace responsibility. In this dimension, responsible protection becomes essential to moral operation.

Equity

Ensuring Fairness Across Stakeholders

Equity forms the third dimension. Influence in modern organizations is no longer tied solely to hierarchy; it flows through data pipelines, algorithmic recommendations, and automated workflows. Equity ensures that these systems do not privilege certain groups, perspectives, or outcomes at the expense of others. It applies to workforce advancement, customer treatment, product accessibility, and data representation. Equity is not about equalizing results; it is about ensuring that the system itself is not biased. In this dimension, fairness becomes a structural requirement rather than a social aspiration.

Risk

Navigating Uncertainty With Discipline

Risk represents the fourth dimension. AI-enabled decisions carry greater speed, greater reach, and greater uncertainty. Risk is not about avoiding uncertainty; it is about navigating it with discipline. Organizations must anticipate consequences, stress-test assumptions, and understand the risks embedded in both human and machine judgment. This applies to operational risks, product risks, customer risks, workforce risks, and data risks. Organizations that embrace this dimension act decisively without acting blindly. In this dimension, risk-aware decision-making becomes a moral responsibility.

Control

Defining Governance and Accountability

Control forms the fifth dimension. As organizations adopt AI, authority becomes more complex: who decides, who approves, who monitors, and who intervenes when systems drift? Control ensures that governance structures are clear, decision rights are defined, and accountability is transparent. It applies to data governance, product governance, operational governance, and workforce governance. Control is not about bureaucracy; it is about ensuring that power—human or machine—is exercised with clarity and accountability. In this dimension, structured oversight becomes the backbone of moral operation.

Legal

Upholding Lawful and Responsible Conduct

Legal completes the Moral Compass. In a world where data, automation, and intelligence shape every action, legal obligations become inseparable from organizational obligations. Compliance is not a constraint; it is a safeguard. Legal integrity ensures that decisions align with regulatory expectations, societal norms, and ethical standards across all six pillars. It protects the organization from missteps and reinforces public trust. In this dimension, lawful stewardship becomes the final test of moral credibility.

The Moral Compass lens reminds organizations that moral authority is not a byproduct of good intentions; it is the result of deliberate, disciplined choices. Morality must be engineered into systems, not assumed in people. When organizations prioritize ethics, enforce safety, ensure equity, manage risk, establish control, and uphold legal accountability, they create environments capable of navigating complexity with confidence and integrity. They build systems where people feel protected, customers feel respected, and society feels considered. They create the moral conditions for transformation to take root and endure. In this maturity, morality becomes an enterprise capability—one that strengthens trust, accelerates adoption, and stabilizes transformation. It becomes the differentiator that determines whether AI amplifies value or amplifies harm. Organizations that master the Moral Compass do not merely deploy intelligent systems; they deploy them responsibly. They create futures where intelligence and integrity advance together, ensuring that progress is not only possible but principled.

Family Values — ETHICS

The foundational moral code—religious, cultural, and social—that dictates the ground rules for "Right vs. Wrong." Establishing the core integrity of the home so that all who belong understand the non-negotiable standards of conduct.

Home Security — SAFETY

Integrated protection—cameras, sensors, and smoke detectors—that provide total awareness of the premises. Deploying proactive warning systems and delegation guardrails to protect the inhabitants from "hidden fires."

Balanced Contribution — EQUITY

Ensuring everyone helps with the cooking and upkeep, with expenses shared equitably by each earning member. Maintaining participation and representation so that no member feels excluded from the collective success.

The Insurance Shield — RISK

Comprehensive fire, flood, and auto coverage to provide long-term peace of mind for the entire household. Using stress testing and "what-if" scenarios to protect the family assets from unforeseen market storms.

Access & Keys — CONTROL

Ensuring all adults have keys and codes, with clear emergency rules and a defined line of command for disasters. Establishing rigid RACI controls to ensure that a human always has the final word on the most critical locks.

The Estate Will — LEGAL

Recording Wills, Trusts, and Medical Mandates with a legal custodian to safeguard the valuables for the future. Securing the organization's legacy through regulatory accountability, ensuring the estate is legally unassailable for generations.

CHAPTER

G1

BLUEPRINT SERIES

Ai6x6™ Transformation Blueprint

Data Ethics

"A sustainable future requires intelligent machines and wiser humans."
Elon Musk, entrepreneur known for Tesla, SpaceX, and AI commentary.

Data Ethics requires principled information practices, and Responsible Information Handling defines how data must be collected, stored, and used with integrity. It establishes the expectations that prevent misuse, manipulation, or careless exposure. This matters because data decisions shape trust across customers, employees, and partners. Responsible Information Handling prevents ethical violations by ensuring data is treated as a protected asset. It also reinforces fairness by ensuring all data is managed under the same ethical rules. When ignored, data becomes vulnerable to exploitation or inappropriate use. Teams experience uncertainty about what is permissible. Over time, Data Ethics deteriorates as boundaries weaken and misuse becomes normalized. Purpose Limited Usage ensures data is used only for legitimate, clearly defined objectives.

Data is the "land" of the AI era, and its value is entirely dependent on its integrity. Leaders use Purpose Limitation Statements and Consent Capture Records to focus on the intent behind collection, ensuring the "Data Land" isn't poisoned by unconsented noise. By sitting with "Data Minimization Reviews," the leader ensures the organization only holds what it truly needs. This topic treats data as a sacred trust rather than a raw commodity. Ethical data use is the clinical protocol that ensures your intelligence remains a valid asset, preventing the "rot" of misinformation from corrupting your decision logic. Engaging with these logs in print allows you to focus on the long-term ethical implications of your data strategy. You can trace the path of consent across the framework, ensuring that your organization's "Data Ethics" are as robust as its business logic. This protects you from the "Kill Zone" of public mistrust and regulatory scrutiny.

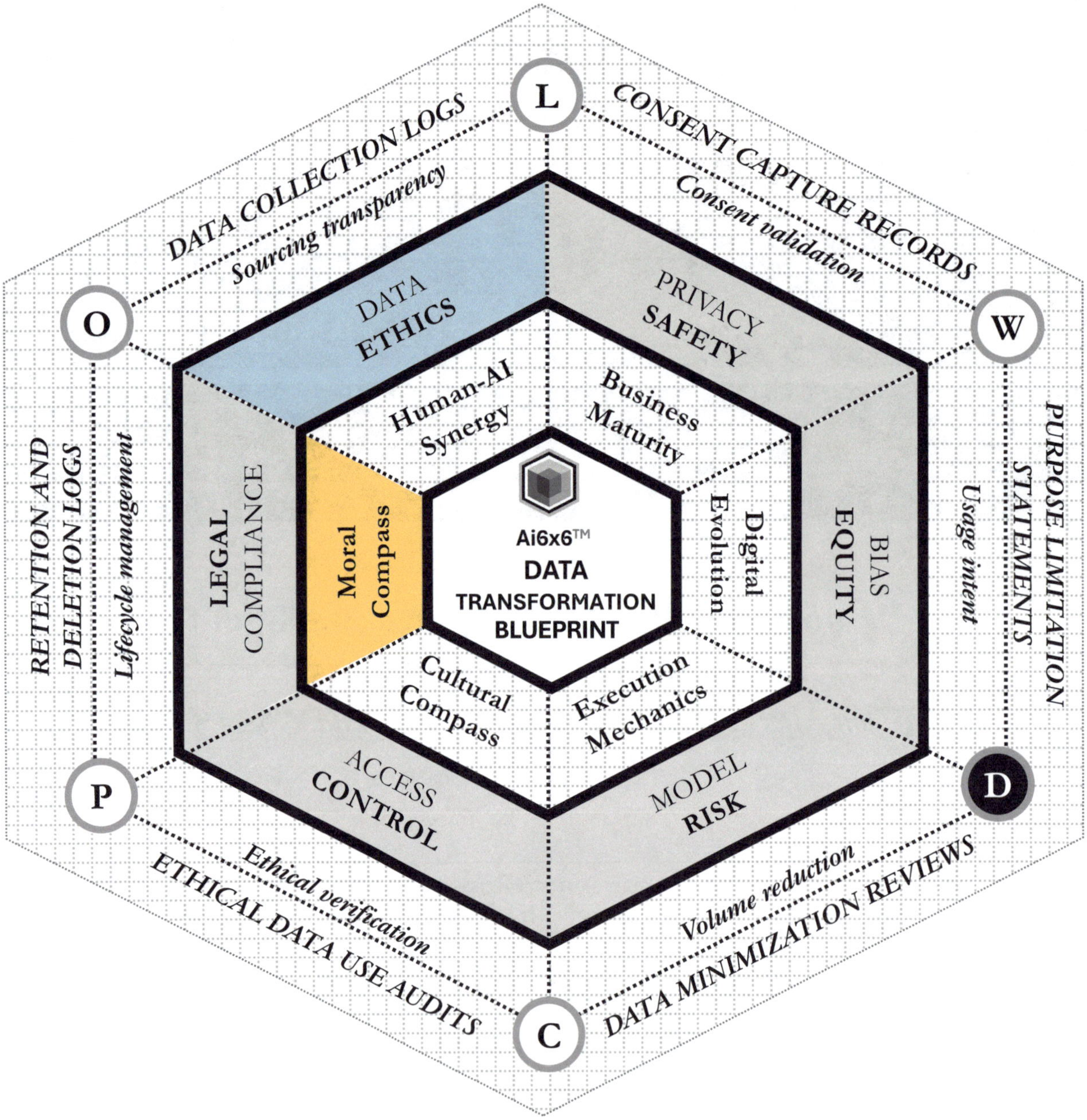

Ai6x6 DATA ETHICS FOR MORAL COMPASS | DATA TRANSFORMATION

1. DATA COLLECTION LOGS

Data Ethics requires transparency in how information is gathered, and Data Collection Logs document the sources, methods, and purposes behind every data intake. They create a traceable record that ensures data is collected lawfully, ethically, and with clear intent. This structure ensures that Data Ethics is grounded in visibility rather than assumption. It matters because undisclosed or poorly governed collection practices expose organizations to privacy violations and ethical breaches. Logs prevent ambiguity by clarifying what data was collected, why it was collected, and who authorized it. They also reinforce fairness by ensuring individuals are not unknowingly subjected to inappropriate data capture. When ignored, data collection becomes opaque and vulnerable to misuse. Stakeholders lose trust in the organization's integrity. Over time, Data Ethics deteriorates into risk, inconsistency, and reputational harm.

When done well, Data Collection Logs strengthen Data Ethics by creating transparency and disciplined governance. People trust that data is gathered responsibly and with respect for individual rights. This improves compliance, reduces risk, and enhances ethical credibility. Logs reinforce the principle that ethical data practices require documentation, not intuition. They remind leaders that responsible collection is the foundation of trustworthy analytics. Ultimately, they build a data environment where integrity begins at the point of capture.

MORAL COMPASS SCENARIO	LEADERSHIP ACTION
Maribel realizes her team is collecting far more customer data than they actually use. She worries the excess creates unnecessary exposure.	The leader instructs Maribel to update the collection log and remove non-essential fields. They also require quarterly reviews to ensure data collection stays aligned with business need.
Darius notices that several teams are collecting the same data independently. He sees duplication and inconsistency forming.	The leader consolidates all collection logs into a unified registry. They also assign data stewards to maintain accuracy across teams.
Leena discovers that field agents are capturing handwritten notes that never make it into official logs. She fears important data is being lost.	The leader standardizes digital capture methods and integrates them into the log. They also train field agents on compliant documentation.
Omar finds that timestamps in the collection log are missing or inaccurate. He knows this undermines auditability.	The leader enforces mandatory timestamping and updates the logging system to automate it. They also run a full audit to correct historical gaps.
Serena sees that third-party vendors are submitting incomplete collection logs. She worries compliance gaps are widening.	The leader requires vendors to follow the same logging standards as internal teams. They also add compliance clauses to vendor contracts.
Hugo notices that sensitive data is being logged without classification labels. He fears mishandling.	The leader updates the log to include mandatory sensitivity tags. They also train teams on proper classification practices.

2. CONSENT CAPTURE RECORDS

Data Ethics requires respect for individual autonomy, and Consent Capture Records document when, how, and under what terms individuals agreed to data use. They verify that permissions were granted knowingly and voluntarily. This structure ensures that Data Ethics is upheld through explicit, traceable consent. It matters because consent is the cornerstone of ethical data handling. Records prevent unauthorized use by proving that individuals understood and accepted the terms. They also reinforce fairness by ensuring consent is not assumed or implied. When ignored, organizations risk violating privacy rights and eroding trust. Individuals feel exploited when their data is used without permission. Over time, Data Ethics collapses as consent becomes inconsistent or unverifiable.

When done well, Consent Capture Records strengthen Data Ethics by ensuring data use aligns with individual expectations. People trust that their choices are honored and respected. This improves transparency, compliance, and ethical alignment. Consent documentation reinforces the principle that autonomy must be protected. It reminds leaders that ethical data practices require explicit permission. Ultimately, it builds a data ecosystem where trust is earned through respect for individual agency.

MORAL COMPASS SCENARIO	LEADERSHIP ACTION
Aanya discovers that a marketing team launched a campaign using customer emails collected years ago without renewed consent. She recognizes the compliance risk.	The leader halts the campaign and verifies consent records before resuming. They also implement automated consent-expiration alerts.
Rowan notices that consent checkboxes on a form are pre-selected. He knows this invalidates consent.	The leader updates the form to require explicit opt-in and logs the correction. They also audit all active forms for similar issues.
Serena sees that verbal consents from support calls are not being recorded anywhere. She fears they won't hold up in an audit.	The leader implements call-recording consent capture and links it to customer profiles. They also train agents on proper disclosure scripts.
Omar learns that a product team is using data from minors without parental consent. He recognizes the legal exposure.	The leader deletes the improperly collected data and implements age-verification steps. They also require parental-consent logging for all youth interactions.
Lila notices that consent withdrawals are not being reflected in downstream systems. She sees a dangerous disconnect.	The leader synchronizes consent records across all systems and validates the fix. They also implement real-time consent propagation.
Andre discovers that consent records are stored in spreadsheets with no access controls. He fears unauthorized edits.	The leader migrates all records to a secure system with audit trails. They also restrict editing rights to authorized personnel.

Data Ethics requires clarity of intent, and Purpose Limitation Statements define exactly how collected data will be used. They articulate boundaries that prevent repurposing or expanding data use without justification. This structure ensures that Data Ethics is upheld through disciplined purpose control. It matters because data misuse often begins with subtle expansions beyond the original intent. Statements prevent scope creep by anchoring data use to declared objectives. They also reinforce fairness by ensuring individuals are not subjected to unexpected or intrusive applications. When ignored, data becomes vulnerable to exploitation and mission drift. Stakeholders question whether the organization respects their privacy. Over time, Data Ethics erodes as purpose boundaries blur.

When done well, Purpose Limitation Statements strengthen Data Ethics by creating transparency and disciplined use. People trust that their data will not be repurposed without consent. This improves ethical credibility and reduces regulatory risk. Purpose clarity reinforces the principle that data must serve defined, legitimate goals. It reminds leaders that ethical use requires honoring original intent. Ultimately, it builds a data environment where boundaries are respected, and trust is protected.

MORAL COMPASS SCENARIO	LEADERSHIP ACTION
Keisha realizes her team is using customer data collected for support purposes to inform marketing decisions. She knows this violates purpose limitations.	The leader stops the secondary use immediately and documents the breach. They also require a purpose-review checkpoint before any new data use.
Damon sees that a product team wants to repurpose behavioral data for employee monitoring. He senses ethical overreach.	The leader denies the request and reinforces the original purpose boundaries. They also update the purpose statement to prevent misinterpretation.
Liora discovers that a partner company is using shared data for analytics not covered in the agreement. She recognizes the contractual risk.	The leader halts data sharing until the partner complies with the stated purpose. They also revise the agreement to tighten usage restrictions.
Mateo notices that purpose statements are vague and open to interpretation. He fears teams will stretch boundaries.	The leader rewrites the statements with precise, narrow definitions. They also require legal review for all future purpose declarations.
Serena sees that employees don't understand why purpose limitation matters. She senses a cultural gap.	The leader runs a training session explaining purpose-based governance. They also add purpose prompts to all data-access workflows.
Omar finds that purpose statements haven't been updated after a major product pivot. He worries they no longer reflect reality.	The leader refreshes all statements and aligns them with the new operating model. They also schedule annual reviews to maintain accuracy.

4. DATA MINIMIZATION REVIEWS

Data Ethics requires restraint, and Data Minimization Reviews evaluate whether the organization collects only what is necessary. They assess relevance, necessity, and proportionality across data categories. This structure ensures that Data Ethics is upheld through disciplined limitation. It matters because excessive data collection increases risk, cost, and ethical exposure. Reviews prevent overreach by challenging assumptions about what data is truly required. They also reinforce fairness by reducing unnecessary intrusion into personal information. When ignored, organizations accumulate data they cannot justify or protect. Individuals feel surveilled rather than served. Over time, Data Ethics deteriorates as data volume grows without purpose.

When done well, Data Minimization Reviews strengthen Data Ethics by promoting efficiency, privacy, and responsible stewardship. People trust that the organization collects only what is essential. This improves compliance, reduces storage risk, and enhances ethical alignment. Minimization reinforces the principle that less data often means better ethics. It reminds leaders that restraint is a form of protection. Ultimately, it builds a data ecosystem where necessity guides every decision.

MORAL COMPASS SCENARIO	LEADERSHIP ACTION
Nadia notices her team is collecting full birthdates when only age ranges are needed. She sees unnecessary risk.	The leader removes the birthdate field and updates the minimization review. They also require justification for every data element collected.
Trevor discovers that a legacy system stores years of unused customer data. He fears regulatory exposure.	The leader initiates a minimization review and deletes non-essential historical data. They also implement retention limits to prevent future buildup.
Aisha sees that engineers are logging full user IDs in debug logs. She recognizes this as excessive.	The leader replaces full IDs with hashed or truncated identifiers. They also update logging standards to enforce minimization.
Marco notices that surveys ask for demographic details irrelevant to the analysis. He suspects the fields were added without purpose.	The leader removes the unnecessary fields and updates the survey template. They also require review of all data-collection instruments.
Evelyn finds that a vendor is requesting more data than needed for their service. She worries they're over-collecting.	The leader renegotiates the data-sharing agreement to limit fields. They also require vendors to justify every requested attribute.
Jonah discovers that internal dashboards display sensitive data that isn't needed for decision-making. He sees a visibility risk.	The leader masks or removes sensitive fields and updates dashboard permissions. They also implement a minimization review for all analytics tools.

5. ETHICAL DATA USE AUDITS

Data Ethics requires accountability, and Ethical Data Use Audits evaluate whether data is being applied in ways that align with stated values, policies, and legal obligations. They review decisions, processes, and outcomes to ensure ethical consistency. This structure ensures that Data Ethics is verified through objective oversight. It matters because misuse often occurs subtly, through shortcuts or misaligned incentives. Audits prevent drift by identifying where data use deviates from ethical expectations. They also reinforce fairness by ensuring all teams follow the same standards. When ignored, unethical practices persist unnoticed and become normalized. Stakeholders lose confidence in the organization's integrity. Over time, Data Ethics collapses under unchecked misuse.

When done well, Ethical Data Use Audits strengthen Data Ethics by creating transparency and corrective action. People trust that data is used responsibly and with oversight. This improves credibility, compliance, and cultural alignment. Audits reinforce the principle that ethics must be verified, not assumed. They remind leaders that responsible data use requires continuous scrutiny. Ultimately, they build a data environment where integrity is monitored and maintained.

MORAL COMPASS SCENARIO	LEADERSHIP ACTION
Rhea learns that a team is using customer data to predict personal traits not disclosed in the privacy policy. She recognizes the ethical breach.	The leader halts the analysis and logs the violation. They also require an ethics review before any predictive modeling.
Omar discovers that employee data is being used to rank "team loyalty." He sees clear ethical overreach.	The leader stops the project and deletes the derived data. They also retrain the team on ethical boundaries.
Lila notices that a machine-learning model is using sensitive attributes indirectly through proxy variables. She fears hidden bias.	The leader orders a full audit of model inputs and removes proxy features. They also implement fairness checks for all models.
Andre finds that a partner company is using shared data to build unrelated commercial products. He recognizes the misuse.	The leader suspends data sharing and initiates a contractual review. They also implement stricter monitoring of partner data use.
Serena sees that analysts are accessing raw customer data when aggregated data would suffice. She worries about unnecessary exposure.	The leader restricts access to aggregated datasets and updates access policies. They also require justification for raw-data access.
Hugo discovers that a team is using scraped public data without verifying ethical sourcing. He senses reputational risk.	The leader pauses the project and conducts a sourcing audit. They also implement ethical-sourcing guidelines for all external data.

6. RETENTION AND DELETION LOGS

Data Ethics requires lifecycle discipline, and Retention and Deletion Logs document how long data is kept and when it is removed. They ensure that information is not stored indefinitely or without justification. This structure ensures that Data Ethics is upheld through responsible data lifecycle management. It matters because excessive retention increases risk, cost, and privacy exposure. Logs prevent neglect by enforcing timely deletion and compliance with retention policies. They also reinforce fairness by ensuring individuals' data is not held longer than necessary. When ignored, data accumulates without purpose and becomes a liability. Stakeholders question whether the organization respects privacy boundaries. Over time, Data Ethics deteriorates as retention practices become inconsistent.

When done well, Retention and Deletion Logs strengthen Data Ethics by creating discipline and reducing unnecessary exposure. People trust that their data is handled responsibly throughout its lifecycle. This improves compliance, reduces risk, and enhances ethical credibility. Lifecycle governance reinforces the principle that data stewardship includes knowing when to let go. It reminds leaders that ethical responsibility extends beyond collection and use. Ultimately, it builds a data environment where retention is intentional, and deletion is accountable.

MORAL COMPASS SCENARIO	LEADERSHIP ACTION
Keisha notices that expired customer data is still stored in multiple systems. She fears regulatory penalties.	The leader initiates a full deletion sweep and updates retention logs. They also automate deletion triggers to prevent recurrence.
Damon discovers that deletion requests from customers are not being propagated to backup systems. He sees a compliance gap.	The leader updates the deletion workflow to include backups and verifies the fix. They also document the new process in the retention log.
Liora finds that teams are extending retention periods without approval. She recognizes the governance breach.	The leader resets retention periods to policy standards and logs the violations. They also require approval for any future extensions.
Mateo notices that deletion logs are incomplete and missing timestamps. He worries audits will fail.	The leader enforces mandatory timestamping and updates the logging system. They also run a historical audit to fill gaps.
Serena sees that sensitive data is being retained "just in case." She knows this violates minimization principles.	The leader deletes the unnecessary data and reinforces retention rules. They also require justification for any exceptions.
Omar discovers that a vendor is retaining data beyond the contractual period. He recognizes the legal risk.	The leader demands immediate deletion and updates the vendor's compliance status. They also add retention-verification clauses to future contracts.

Ai6x6™ Transformation Blueprint

Privacy Safety

"Morality must scale with technology, or technology will scale without morality."
Yuval Noah Harari, historian and author known for insights on humanity's future.

Privacy Safety requires disciplined protection, and Personal Data Protection defines the safeguards that keep sensitive information secure from unauthorized access or exposure. It establishes the standards that govern how personal data is collected, stored, and processed. This matters because privacy breaches create immediate harm for individuals and significant risk for organizations. Personal Data Protection prevents violations by ensuring data is shielded from unnecessary visibility. It also reinforces fairness by ensuring all individuals receive equal privacy safeguards. When ignored, privacy risks escalate rapidly. Teams experience confusion about data boundaries. Over time, Privacy Safety deteriorates as protective practices weaken. Secure Handling Practices ensure data is managed with rigor throughout its lifecycle.

This is the quest for decision integrity through protection. Mapping Breach Response Logs and Encryption Compliance Checks on paper allows a leader to reclaim the "Armor" of their data from the threat of unauthorized friction. It involves using "Privacy Impact Assessments" to identify vulnerabilities before they are exploited. For the leader, privacy isn't just a legal requirement—it's a competitive moat. By protecting the identity of the data, you protect the sovereignty of the organization. Privacy safety ensures that the data driving your Ai6x6™ framework remains clean, secure, and entirely under your strategic control. The print series helps you visualize the "Identity Protection Reports" in context with your broader infrastructure. This clinical focus on "Privacy Safety" allows you to build a "Safe Zone" of data integrity, ensuring that as you scale your intelligence, you are not simultaneously scaling your risk of a catastrophic, brand-destroying breach or legal liability.

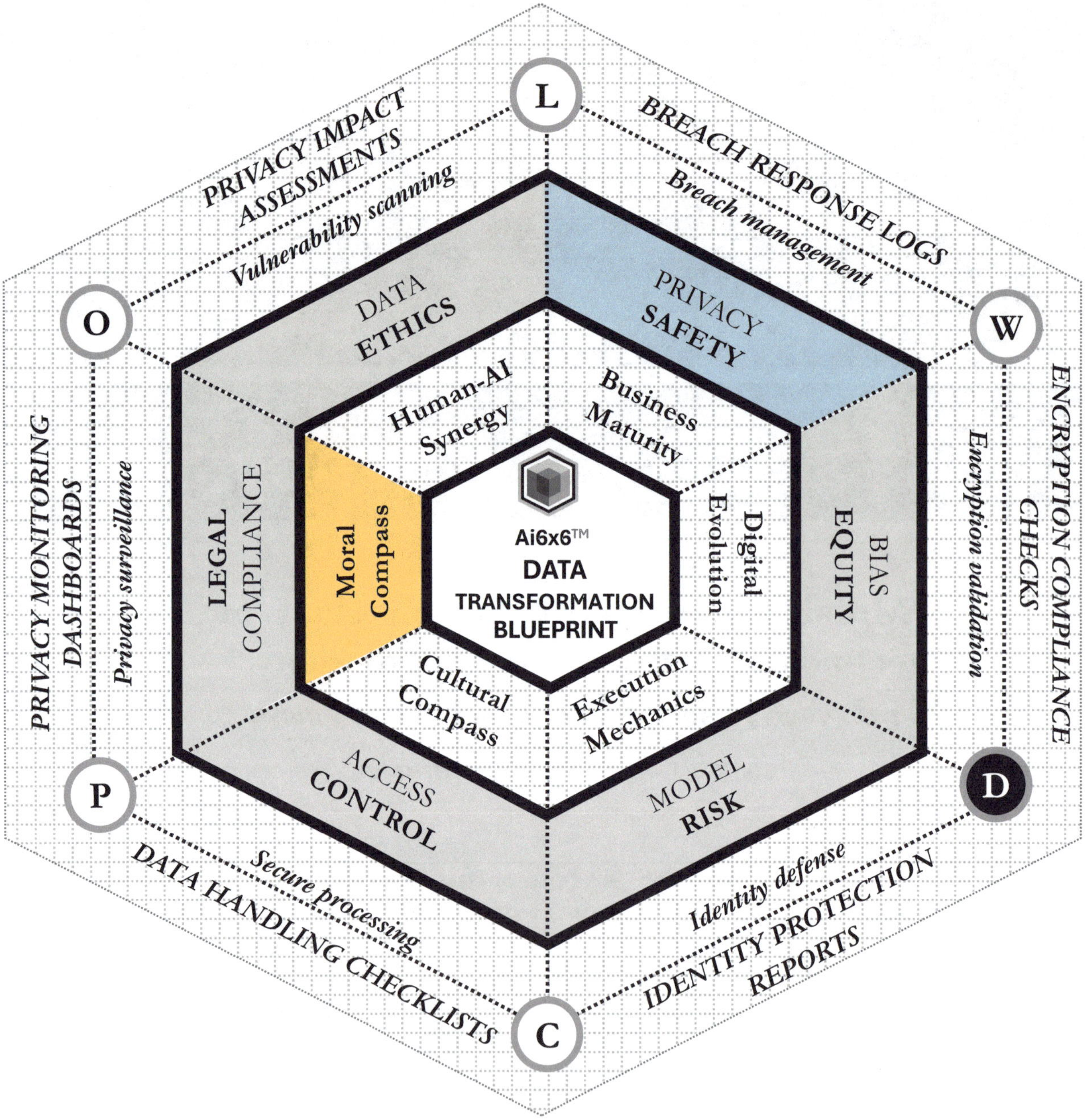

Ai6x6 PRIVACY SAFETY FOR MORAL COMPASS | DATA TRANSFORMATION

1. PRIVACY IMPACT ASSESSMENTS

Privacy Safety requires proactive evaluation, and Privacy Impact Assessments analyze how data practices affect individual privacy rights. They examine collection methods, processing flows, and potential exposure points. This structure ensures that Privacy Safety is grounded in foresight rather than reaction. It matters because privacy risks often emerge from design decisions made without considering downstream consequences. Assessments prevent harm by identifying vulnerabilities before systems are deployed. They also reinforce fairness by ensuring individuals are not subjected to hidden or excessive data use. When ignored, privacy risks escalate unnoticed and expose the organization to legal, ethical, and reputational damage. Stakeholders lose trust in the organization's ability to protect their information. Over time, Privacy Safety deteriorates into systemic vulnerability.

When done well, Privacy Impact Assessments strengthen Privacy Safety by embedding privacy considerations into every stage of design and operation. People trust that their data is handled responsibly and with respect. This improves compliance, reduces risk, and enhances ethical credibility. Assessments reinforce the principle that privacy must be engineered, not assumed. They remind leaders that protection begins with anticipation. Ultimately, they build a data environment where privacy is safeguarded through disciplined evaluation.

MORAL COMPASS SCENARIO	LEADERSHIP ACTION
Talia realizes her team is launching a new feature that collects location data without assessing privacy risks. She worries the rollout could violate regional laws.	The leader pauses the launch and requires a full privacy impact assessment. They also add a mandatory PIA checkpoint to all future product workflows.
Damon sees that a vendor integration will expose customer identifiers to a third party. He senses the team hasn't evaluated the privacy implications.	The leader orders an immediate PIA and blocks data sharing until risks are mitigated. They also require legal review for all third-party integrations.
Liora discovers that a new analytics tool processes sensitive data in the cloud. She fears the team hasn't assessed cross-border transfer risks.	The leader initiates a privacy impact assessment focused on data residency. They also update vendor contracts to include regional compliance clauses.
Mateo notices that a machine-learning model uses inferred attributes. He realizes the privacy impact hasn't been evaluated.	The leader mandates a PIA specifically for inferred data. They also require model teams to document all derived attributes.
Serena sees that a customer-support workflow exposes unnecessary personal data to agents. She worries about over-access.	The leader conducts a PIA and redesigns the workflow to minimize exposure. They also implement role-based access controls.
Omar finds that a new internal dashboard displays sensitive employee data. He knows the privacy impact was never assessed.	The leader requires a PIA and masks sensitive fields. They also enforce privacy-by-design standards for all dashboards.

2. BREACH RESPONSE LOGS

Privacy Safety requires accountability during incidents, and Breach Response Logs document every action taken when data exposure occurs. They capture timelines, decisions, communications, and remediation steps. This structure ensures that Privacy Safety is upheld through transparent and disciplined incident management. It matters because breaches create immediate harm and long term trust erosion. Logs prevent confusion by providing a clear record of how the organization responded. They also reinforce fairness by ensuring affected individuals receive timely and accurate information. When ignored, breach responses become chaotic, inconsistent, or concealed. Stakeholders feel misled or abandoned. Over time, Privacy Safety collapses as incident handling becomes unreliable.

When done well, Breach Response Logs strengthen Privacy Safety by creating visibility, accountability, and structured learning. People trust that incidents are managed with rigor and honesty. This improves recovery, reduces harm, and reinforces ethical responsibility. Logs reinforce the principle that privacy protection includes transparent response. They remind leaders that accountability must be documented, not implied. Ultimately, they build a data environment where breaches are handled with integrity and discipline.

MORAL COMPASS SCENARIO	LEADERSHIP ACTION
Keisha discovers that a minor data incident wasn't logged because the team didn't think it was "serious enough." She knows this undermines breach readiness.	The leader logs the incident retroactively and reinforces that all breaches—large or small—must be recorded. They also retrain the team on breach-classification rules.
Damon notices that timestamps in the breach log are inconsistent. He fears the organization won't be able to reconstruct events accurately.	The leader standardizes timestamping and updates the logging system. They also audit past entries to correct gaps.
Liora sees that breach logs don't include containment actions. She worries critical steps are being lost.	The leader updates the log template to require containment details. They also assign a breach coordinator to verify completeness.
Mateo finds that a breach involving a vendor wasn't logged internally. He recognizes the governance failure.	The leader logs the breach and notifies legal and security teams. They also update vendor contracts to require immediate breach reporting.
Serena discovers that breach logs are stored in unsecured folders. She sees a dangerous irony.	The leader moves all logs to a secure, access-controlled system. They also restrict permissions to essential personnel.
Omar notices that breach logs are reviewed only annually. He fears slow detection of patterns.	The leader implements monthly breach-log reviews. They also assign owners to address recurring issues.

3. ENCRYPTION COMPLIANCE CHECKS

Privacy Safety depends on strong technical safeguards, and Encryption Compliance Checks verify that sensitive data is protected at rest and in transit. They ensure encryption standards are applied consistently across systems, devices, and workflows. This structure ensures that Privacy Safety is upheld through verifiable security controls. It matters because unencrypted data is vulnerable to interception, theft, and unauthorized access. Checks prevent oversight by confirming that encryption is implemented correctly and continuously. They also reinforce fairness by ensuring all individuals' data receives equal protection. When ignored, sensitive information becomes exposed to preventable risks. Stakeholders lose confidence in the organization's security posture. Over time, Privacy Safety deteriorates as vulnerabilities accumulate.

When done well, Encryption Compliance Checks strengthen Privacy Safety by ensuring data is shielded from unauthorized access. People trust that their information is protected by robust safeguards. This improves security, compliance, and ethical alignment. Encryption oversight reinforces the principle that privacy requires technical rigor. It reminds leaders that protection must be verified, not assumed. Ultimately, it builds a data environment where confidentiality is consistently maintained.

MORAL COMPASS SCENARIO	LEADERSHIP ACTION
Nadia realizes that several internal tools transmit data without encryption. She worries sensitive information is exposed.	The leader mandates immediate encryption updates and documents the gaps. They also add automated encryption checks to deployment pipelines.
Trevor discovers that encryption keys are shared informally between engineers. He recognizes the security risk.	The leader rotates all keys and implements a secure key-management system. They also retrain engineers on proper handling.
Aisha sees that mobile devices used by field staff lack encryption. She fears data loss if devices are stolen.	The leader enforces device-level encryption and updates compliance checks. They also require remote-wipe capabilities.
Marco notices that backups are stored unencrypted in a legacy system. He knows this violates policy.	The leader encrypts all backups and migrates them to a secure environment. They also add backup-encryption checks to the compliance schedule.
Evelyn finds that a vendor claims to encrypt data but won't provide evidence. She senses a compliance gap.	The leader demands encryption attestation and pauses data sharing until verified. They also update vendor-risk scoring.
Jonah discovers that expired certificates are still in use. He fears encrypted channels are no longer secure.	The leader replaces all expired certificates and automates renewal reminders. They also audit certificate usage quarterly.

Privacy Safety requires safeguarding individuals from misuse of their personal information, and Identity Protection Reports monitor risks related to impersonation, credential theft, and unauthorized access. They analyze patterns that indicate identity compromise. This structure ensures that Privacy Safety is upheld through continuous vigilance. It matters because identity misuse can cause severe personal and organizational harm. Reports prevent escalation by detecting anomalies early. They also reinforce fairness by ensuring all individuals receive equal protection from identity based threats. When ignored, identity risks go undetected and lead to breaches or fraud. Stakeholders feel unprotected and vulnerable. Over time, Privacy Safety erodes as identity threats proliferate.

When done well, Identity Protection Reports strengthen Privacy Safety by enabling rapid detection and response. People trust that their identities are monitored and safeguarded. This improves confidence, reduces harm, and enhances organizational integrity. Reports reinforce the principle that privacy includes protecting identity, not just data. They remind leaders that vigilance is essential to security. Ultimately, they build a data environment where identity protection is continuous and reliable.

MORAL COMPASS SCENARIO	LEADERSHIP ACTION
Rhea notices a spike in failed login attempts for a group of employees. She worries their accounts are being targeted.	The leader triggers an identity-protection review and forces password resets. They also enable multi-factor authentication for the affected group.
Omar sees that identity-protection reports show unusual access patterns from foreign IPs. He suspects credential compromise.	The leader blocks suspicious IPs and investigates the access logs. They also require geo-fencing for sensitive systems.
Lila discovers that privileged accounts are not included in identity-protection monitoring. She recognizes the oversight.	The leader adds all privileged accounts to the monitoring scope. They also implement real-time alerts for high-risk activity.
Andre notices that identity-protection alerts are being ignored because of alert fatigue. He fears real threats are slipping through.	The leader tunes alert thresholds and consolidates duplicate alerts. They also assign a dedicated analyst to triage identity risks.
Serena finds that identity-protection reports are not shared with team leads. She sees a communication gap.	The leader distributes reports to relevant managers and provides interpretation guidance. They also schedule monthly identity-risk reviews.
Hugo discovers that employees reuse passwords across systems. He knows this increases breach risk.	The leader enforces password-uniqueness rules and deploys a password-manager tool. They also run a training session on credential hygiene.

5. DATA HANDLING CHECKLISTS

Privacy Safety requires disciplined execution, and Data Handling Checklists define the required steps for securely managing personal information. They outline procedures for access, storage, transfer, and disposal. This structure ensures that Privacy Safety is upheld through consistent, repeatable practices. It matters because privacy failures often stem from human error or inconsistent handling. Checklists prevent oversight by standardizing how data is managed across teams. They also reinforce fairness by ensuring all individuals' data is treated with equal care. When ignored, handling practices become inconsistent and expose data to unnecessary risk. Employees become uncertain about proper procedures. Over time, Privacy Safety deteriorates into operational inconsistency.

When done well, Data Handling Checklists strengthen Privacy Safety by embedding protection into daily workflows. People trust that data is managed with discipline and care. This improves compliance, reduces risk, and enhances ethical reliability. Checklists reinforce the principle that privacy requires operational rigor. They remind leaders that protection is built through consistent action. Ultimately, they build a data environment where secure handling is the norm, not the exception.

MORAL COMPASS SCENARIO	LEADERSHIP ACTION
Keisha notices employees are emailing spreadsheets with sensitive data. She realizes they're bypassing secure channels.	The leader updates the handling checklist to prohibit email transfers. They also train employees on secure file-sharing tools.
Damon sees that teams skip verification steps when transferring data between systems. He fears data corruption.	The leader enforces mandatory verification and updates the checklist with validation steps. They also assign owners for each transfer.
Liora discovers that printed documents with sensitive data are left on desks. She recognizes the physical-security risk.	The leader adds secure-disposal steps to the checklist and installs locked bins. They also conduct a clean-desk audit.
Mateo notices that employees are not labeling sensitive data properly. He worries mishandling will occur.	The leader updates the checklist with classification requirements. They also train teams on labeling standards.
Serena finds that contractors are not following internal data-handling procedures. She sees inconsistent practices.	The leader requires contractors to follow the same checklist and signs them into compliance agreements. They also assign a contractor-compliance monitor.
Omar sees that data-handling checklists are outdated and missing cloud-specific steps. He knows this creates gaps.	The leader updates the checklist to include cloud-security requirements. They also schedule semi-annual reviews to keep it current.

6. PRIVACY MONITORING DASHBOARDS

Privacy Safety requires real time visibility, and Privacy Monitoring Dashboards track compliance, access patterns, and emerging risks across systems. They consolidate key indicators that reveal whether privacy controls are functioning as intended. This structure ensures that Privacy Safety is upheld through continuous oversight. It matters because privacy risks evolve quickly and require timely detection. Dashboards prevent blind spots by highlighting anomalies and deviations. They also reinforce fairness by ensuring all data subjects receive consistent protection. When ignored, privacy issues remain hidden until they escalate into breaches. Stakeholders lose trust in the organization's ability to safeguard information. Over time, Privacy Safety collapses under unmanaged risk.

When done well, Privacy Monitoring Dashboards strengthen Privacy Safety by enabling rapid insight and intervention. People trust that privacy controls are actively monitored. This improves responsiveness, compliance, and ethical alignment. Dashboards reinforce the principle that privacy protection requires constant awareness. They remind leaders that oversight must be continuous, not periodic. Ultimately, they build a data environment where privacy is safeguarded through real time intelligence.

MORAL COMPASS SCENARIO	LEADERSHIP ACTION
Nadia notices a spike in access to sensitive customer records. She worries unauthorized browsing is occurring.	The leader investigates the spike and restricts access where needed. They also add anomaly-detection alerts to the dashboard.
Trevor sees that the dashboard shows repeated access attempts from deprecated accounts. He fears dormant accounts are still active.	The leader disables all deprecated accounts and updates the access-review process. They also add automated alerts for dormant-account activity.
Aisha discovers that privacy alerts are going unreviewed for weeks. She senses a monitoring breakdown.	The leader assigns a dedicated reviewer and sets daily alert-review expectations. They also track alert-resolution times.
Marco notices that the dashboard lacks visibility into third-party data use. He worries blind spots exist.	The leader integrates vendor-usage data into the dashboard. They also require vendors to provide usage logs regularly.
Evelyn sees that the dashboard doesn't track data-retention violations. She recognizes a compliance gap.	The leader adds retention-compliance indicators and links them to deletion logs. They also schedule monthly retention reviews.
Jonah finds that dashboard metrics are inconsistent across regions. He fears data-quality issues.	The leader standardizes data-collection methods and recalibrates the dashboard. They also assign regional data stewards.

Ai6x6™ Transformation Blueprint

Bias Equity

"AI ethics is not a debate for tomorrow; it is a responsibility for today."
Sundar Pichai, CEO of Alphabet/Google, known for global AI leadership.

Bias Equity requires fairness in data driven systems, and Fair Model Behavior ensures algorithms operate without favoring or disadvantaging specific groups. It establishes the expectations that prevent discriminatory outcomes. This matters because biased models can amplify inequity at scale. Fair Model Behavior prevents harm by ensuring decisions are grounded in balanced, representative data. It also reinforces trust by demonstrating that automated systems treat all individuals fairly. When ignored, bias becomes embedded and difficult to detect. Teams experience uncertainty about the reliability of model outputs. Over time, Bias Equity deteriorates as disparities widen. Inclusive Data Standards ensure datasets reflect diverse populations and reduce structural bias.

AI is a force multiplier for cultural and operational rot if not checked. Model Fairness Dashboards and Bias Detection Reports are the diagnostics required to ensure the automated future doesn't inherit the prejudices of the past. This topic involves using "Representative Sampling Logs" to verify the integrity of the data sets fueling your algorithms. For a leader, "Bias Equity" is about ensuring "Decision Integrity." By clinically auditing models for fairness, you prevent the organization from automating discrimination, which would otherwise lead to massive legal liability and the erosion of customer trust. Engaging with "Bias Mitigation Plans" in print allows for the engrossed reflection needed to catch subtle logical flaws. This ensures that your "Data Transformation" is not just technically fast, but morally sound, preventing your AI from creating "Kill Zones" in your own market. Bias equity is the tool that ensures your algorithms are as fair as your leadership intent.

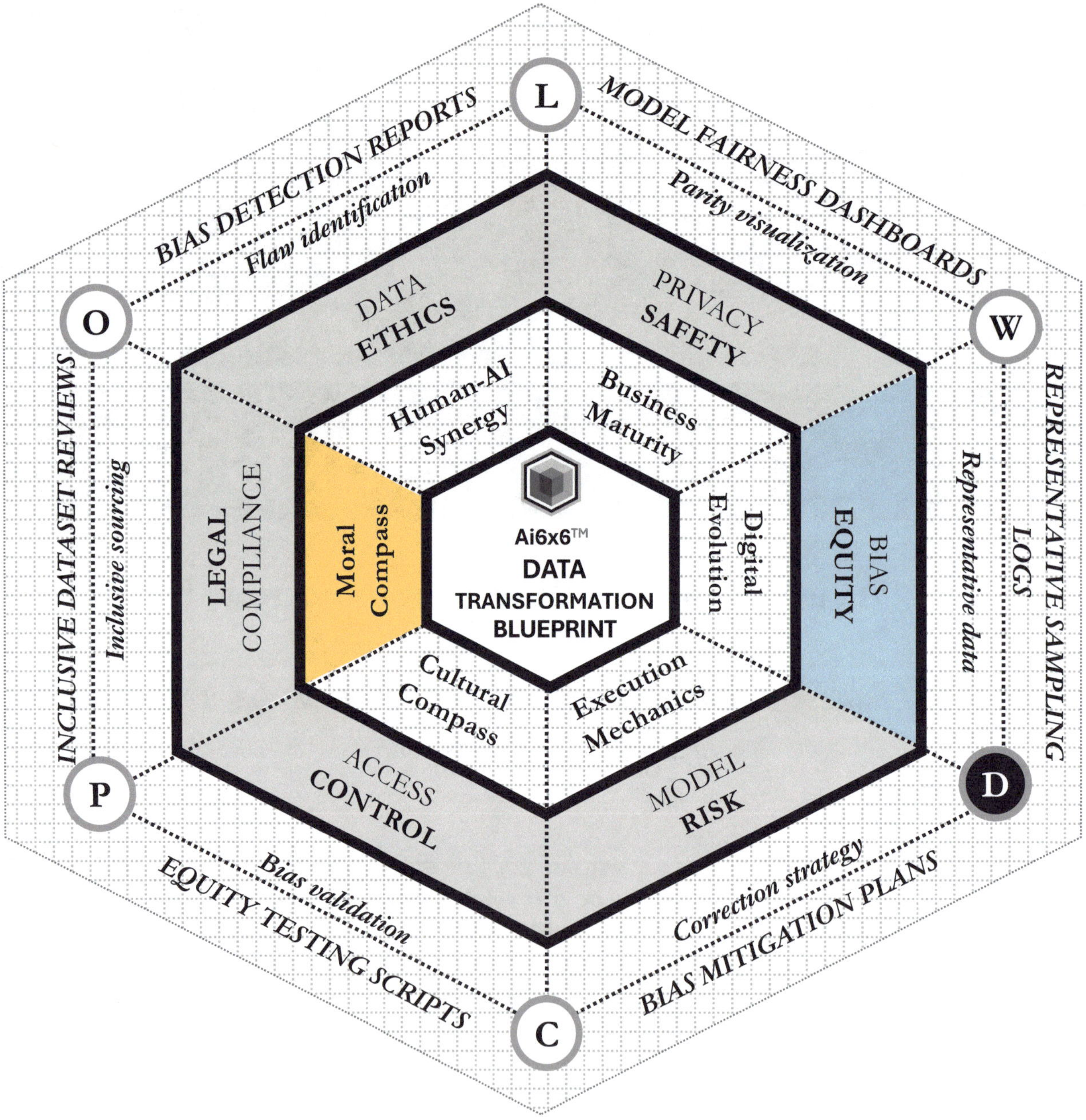

Ai6x6 BIAS EQUITY FOR MORAL COMPASS | DATA TRANSFORMATION

1. BIAS DETECTION REPORTS

Bias Equity requires systematic visibility into unfair patterns, and Bias Detection Reports identify where models, datasets, or decisions produce unequal outcomes. They analyze demographic performance differences, error rates, and distributional impacts. This structure ensures that Bias Equity is grounded in measurable evidence rather than intuition. It matters because bias often emerges subtly and becomes embedded in automated systems. Reports prevent blind spots by revealing where inequities are forming. They also reinforce fairness by ensuring all groups receive equal consideration in evaluation. When ignored, biased outcomes persist unchallenged and harm affected individuals. Stakeholders lose trust in the organization's data practices. Over time, Bias Equity deteriorates into systemic discrimination.

When done well, Bias Detection Reports strengthen Bias Equity by enabling targeted correction and continuous improvement. People trust that inequities are identified and addressed with rigor. This improves model reliability, ethical alignment, and organizational credibility. Reports reinforce the principle that fairness must be measured, not assumed. They remind leaders that equity requires ongoing vigilance. Ultimately, they build a data environment where bias is surfaced early and resolved responsibly.

MORAL COMPASS SCENARIO	LEADERSHIP ACTION
Rina sees the latest bias report showing the model under-predicts creditworthiness for first-generation applicants. She realizes the team has normalized the discrepancy as "expected variance."	The leader halts model use for that segment and assigns a cross-functional team to investigate root causes. They also require monthly bias-report reviews with documented remediation steps.
Damon notices the bias report flags a sudden spike in false positives for Spanish-language users. He suspects a recent preprocessing change introduced linguistic bias.	The leader orders an immediate rollback and a targeted linguistic-bias audit. They also add language-specific fairness metrics to future reports.
Liora discovers that the bias report shows consistent disparities across age groups, but no one has taken ownership. She senses a governance vacuum.	The leader assigns a bias-owner role and creates a remediation timeline. They also integrate bias-report accountability into quarterly performance reviews.
Mateo sees that the report highlights proxy variables correlating with protected attributes. He realizes the team overlooked indirect discrimination.	The leader removes the proxy features and retrains the model. They also require a proxy-variable scan for all future feature sets.
Serena notices the report shows fairness improvements in one segment but regressions in another. She worries the team is optimizing in silos.	The leader mandates multi-segment fairness evaluation before any model approval. They also create a cross-segment fairness dashboard.
Omar finds that the bias report is technically accurate but unreadable to non-experts. He fears leaders will miss critical risks.	The leader requires a plain-language summary for every report. They also train analysts on executive-ready bias communication.

2. MODEL FAIRNESS DASHBOARDS

Bias Equity requires real time oversight, and Model Fairness Dashboards visualize how algorithms perform across demographic groups. They track disparities, drift, and fairness metrics in operational systems. This structure ensures that Bias Equity is monitored continuously rather than periodically. It matters because fairness can degrade over time as data shifts or conditions change. Dashboards prevent unnoticed inequities by highlighting deviations immediately. They also reinforce fairness by ensuring all groups receive equal protection from algorithmic harm. When ignored, models evolve in ways that disadvantage certain populations. Stakeholders question the legitimacy of automated decisions. Over time, Bias Equity collapses as fairness becomes inconsistent.

When done well, Model Fairness Dashboards strengthen Bias Equity by providing transparency and rapid insight. People trust that fairness is actively monitored and maintained. This improves accountability, compliance, and ethical reliability. Dashboards reinforce the principle that equity requires constant visibility. They remind leaders that fairness is dynamic and must be managed continuously. Ultimately, they build a data environment where equitable performance is sustained through real time intelligence.

MORAL COMPASS SCENARIO	LEADERSHIP ACTION
Aavya realizes her team cannot explain how a recommendation model makes decisions. She knows this violates transparency expectations.	The leader requires a full transparency sheet documenting inputs, logic, and limitations. They also prohibit deployment until the sheet is complete.
Rowan sees that a model's risk scores are being used in operations without any explanation of confidence levels. He senses misuse.	The leader updates the transparency sheet to include confidence ranges and usage guidelines. They also train teams on interpreting model outputs.
Serena discovers that a vendor-supplied model has no documentation at all. She worries the company is relying on a black box.	The leader demands a transparency sheet from the vendor and pauses use until provided. They also update procurement rules to require transparency documentation.
Omar notices that transparency sheets haven't been updated after major model changes. He recognizes the compliance gap.	The leader enforces version-controlled updates for every model change. They also assign a transparency owner for each model.
Lila finds that transparency sheets are written in highly technical language. She fears non-technical teams can't understand them.	The leader rewrites the sheets in plain language and adds visual explanations. They also require readability checks before approval.
Andre sees that transparency sheets don't include known limitations. He worries teams will over-trust the model.	The leader mandates a "limitations and risks" section for every sheet. They also require teams to review limitations before deployment.

3. REPRESENTATIVE SAMPLING LOGS

Bias Equity requires balanced data foundations, and Representative Sampling Logs document whether datasets reflect the populations they aim to serve. They track demographic composition, sampling methods, and inclusion criteria. This structure ensures that Bias Equity is upheld through intentional dataset design. It matters because unrepresentative data produces biased models and unfair outcomes. Logs prevent skew by revealing gaps in population coverage. They also reinforce fairness by ensuring all groups are adequately represented in training data. When ignored, datasets amplify historical inequities and distort predictions. Stakeholders lose confidence in the validity of model outputs. Over time, Bias Equity deteriorates as biased data shapes biased systems.

When done well, Representative Sampling Logs strengthen Bias Equity by ensuring datasets reflect real world diversity. People trust that models are built on inclusive foundations. This improves accuracy, fairness, and ethical alignment. Sampling oversight reinforces the principle that equity begins with data quality. It reminds leaders that representation is a prerequisite for fairness. Ultimately, it builds a data environment where inclusion is engineered into the dataset itself.

MORAL COMPASS SCENARIO	LEADERSHIP ACTION
Rhea reviews the sampling log and sees that older adults are severely underrepresented in the dataset. She realizes the model will fail for a key demographic.	The leader mandates targeted sampling to fill the gap and documents the correction. They also require demographic-coverage thresholds for all datasets.
Omar notices that the sampling log shows heavy reliance on urban data sources. He worries rural behaviors are missing entirely.	The leader expands sampling to rural regions and updates the log with new sources. They also assign a sampling steward to monitor coverage.
Lila discovers that the log shows minimal representation from non-native English speakers. She fears linguistic bias will emerge downstream.	The leader adds multilingual sampling requirements and validates coverage. They also integrate language-diversity checks into the log.
Andre sees that sampling was done during a single season. He realizes seasonal bias will distort predictions.	The leader expands sampling across multiple seasons and updates the log. They also require temporal-diversity checks for future datasets.
Serena notices that the sampling log includes outdated sources from a deprecated product. She worries the data no longer reflects current behavior.	The leader removes obsolete sources and refreshes the sampling plan. They also require annual sampling-source reviews.
Hugo finds that sampling logs are incomplete for third-party datasets. He senses a transparency gap.	The leader requires vendors to provide full sampling documentation. They also add vendor-sampling audits to procurement.

4. BIAS MITIGATION PLANS

Bias Equity requires corrective action, and Bias Mitigation Plans outline the strategies used to reduce unfair outcomes in data systems. They define interventions such as reweighting, resampling, constraint application, or model redesign. This structure ensures that Bias Equity is operationalized through deliberate remediation. It matters because identifying bias is insufficient without action. Plans prevent stagnation by specifying how inequities will be addressed. They also reinforce fairness by ensuring mitigation is systematic rather than ad hoc. When ignored, bias persists even after detection. Stakeholders question whether the organization is committed to equitable outcomes. Over time, Bias Equity collapses as issues remain unresolved.

When done well, Bias Mitigation Plans strengthen Bias Equity by creating structured pathways to improvement. People trust that fairness concerns lead to concrete action. This improves model performance, ethical credibility, and stakeholder confidence. Mitigation planning reinforces the principle that equity requires intervention, not observation. It reminds leaders that fairness must be engineered into system behavior. Ultimately, it builds a data environment where bias reduction is continuous and accountable.

MORAL COMPASS SCENARIO	LEADERSHIP ACTION
Keisha sees that the mitigation plan relies solely on post-processing adjustments. She realizes the team is avoiding deeper structural fixes.	The leader mandates upstream data and feature corrections before post-processing. They also require multi-layer mitigation strategies for all high-risk models.
Damon notices the plan proposes reducing model complexity to "smooth out" bias. He fears this will degrade performance without solving root causes.	The leader rejects the shortcut and requires targeted bias-source analysis. They also assign a fairness engineer to redesign the mitigation approach.
Liora discovers that the mitigation plan ignores intersectional groups. She recognizes the risk of hidden disparities.	The leader expands the plan to include intersectional fairness testing. They also require intersectional reporting for all mitigation outcomes.
Mateo sees that the plan lacks timelines and owners. He senses it will never be executed.	The leader assigns named owners and sets milestone deadlines. They also track progress in the governance dashboard.
Serena notices the plan focuses on one protected group while worsening outcomes for another. She worries about fairness trade-offs.	The leader requires multi-group optimization and trade-off documentation. They also run a fairness-impact simulation before approval.
Omar finds that the plan relies on synthetic data without validation. He fears it may introduce new bias.	The leader mandates validation of synthetic samples and restricts their use. They also require provenance documentation for all synthetic data.

5. EQUITY TESTING SCRIPTS

Bias Equity requires repeatable evaluation, and Equity Testing Scripts automate fairness checks across models and datasets. They run standardized tests that measure disparities, error patterns, and demographic impacts. This structure ensures that Bias Equity is embedded into the development lifecycle. It matters because manual fairness testing is inconsistent and prone to oversight. Scripts prevent gaps by ensuring every model undergoes the same rigorous evaluation. They also reinforce fairness by making equity testing a non negotiable step. When ignored, fairness becomes optional and inconsistently applied. Models reach production with untested biases. Over time, Bias Equity deteriorates as inequities propagate across systems.

When done well, Equity Testing Scripts strengthen Bias Equity by creating consistency and automation. People trust that fairness checks are systematic and reliable. This improves quality, compliance, and ethical alignment. Scripts reinforce the principle that equity must be built into the pipeline. They remind leaders that fairness requires disciplined, repeatable testing. Ultimately, they build a data environment where equitable performance is validated continuously.

MORAL COMPASS SCENARIO	LEADERSHIP ACTION
Nadia discovers that the equity testing script only evaluates accuracy, not false-positive or false-negative rates. She knows this hides critical disparities.	The leader expands the script to include error-rate parity checks. They also require subgroup-specific reporting for every test run.
Trevor sees that the script uses outdated demographic categories. He realizes it no longer reflects the current population.	The leader updates the categories and aligns them with regulatory standards. They also schedule annual script reviews.
Aisha notices that the script doesn't test for drift-induced inequity. She fears fairness will degrade over time.	The leader adds drift-fairness tests and integrates them into CI/CD pipelines. They also require fairness gating before deployment.
Marco discovers that the script fails silently when data is missing. He recognizes the risk of false "all clear" results.	The leader adds error-handling and mandatory completeness checks. They also require test-run logs for auditability.
Evelyn sees that the script only tests one model version. She worries teams are skipping fairness checks for updates.	The leader enforces version-wide testing and ties script execution to release workflows. They also track compliance in the governance dashboard.
Jonah finds that the script doesn't test for intersectional fairness. He senses blind spots.	The leader adds intersectional test cases and documents the results. They also require intersectional coverage for all future scripts.

6. INCLUSIVE DATASET REVIEWS

Bias Equity requires intentional inclusion, and Inclusive Dataset Reviews evaluate whether datasets reflect diverse perspectives, contexts, and populations. They examine sources, coverage, and potential exclusions. This structure ensures that Bias Equity is upheld through comprehensive dataset evaluation. It matters because datasets often reflect historical biases or narrow viewpoints. Reviews prevent exclusion by identifying missing groups or underrepresented contexts. They also reinforce fairness by ensuring data reflects the full spectrum of real world diversity. When ignored, models inherit and amplify systemic inequities. Stakeholders question the legitimacy of outcomes derived from incomplete data. Over time, Bias Equity collapses as exclusion becomes embedded.

When done well, Inclusive Dataset Reviews strengthen Bias Equity by ensuring datasets are broad, representative, and ethically constructed. People trust that data reflects the populations it affects. This improves accuracy, fairness, and cultural relevance. Reviews reinforce the principle that equity requires intentional inclusion. They remind leaders that fairness begins with what data is chosen. Ultimately, they build a data environment where diversity is foundational, not incidental.

MORAL COMPASS SCENARIO	LEADERSHIP ACTION
Rhea discovers that the dataset excludes people with disabilities because the collection process wasn't accessible. She realizes the exclusion is structural.	The leader redesigns the data-collection workflow to ensure accessibility. They also add disability-representation metrics to dataset reviews.
Omar notices that cultural groups are unevenly represented due to biased sampling locations. He fears cultural misclassification.	The leader expands sampling to culturally diverse regions and updates the dataset review. They also require cultural-competency checks for annotators.
Lila sees that the dataset overrepresents heavy users and underrepresents casual users. She worries the model will skew toward power-user behavior.	The leader rebalances the dataset and documents the correction. They also add usage-pattern diversity checks.
Andre discovers that annotators lacked training on sensitive categories. He recognizes annotation bias.	The leader retrains annotators and adds calibration sessions. They also require annotation-quality audits.
Serena notices that the dataset excludes older devices and browsers. She fears the model will fail for users with older tech.	The leader expands data collection to include legacy environments. They also add device-diversity metrics to the review.
Hugo finds that dataset reviews are happening only at the end of projects. He senses issues are being caught too late.	The leader mandates dataset reviews at every project milestone. They also integrate review checkpoints into the development lifecycle.

Ai6x6™ Transformation Blueprint

Model Risk

"If we automate decisions, we must automate fairness too."

Michelle Obama, former First Lady known for advocacy and social leadership.

Model Risk requires awareness of system vulnerabilities, and System Reliability Threats identify where models may behave unpredictably under stress or changing conditions. They establish visibility into weaknesses that compromise accuracy or stability. This matters because unreliable models create operational, financial, and ethical risk. System Reliability Threats prevent failure by revealing where recalibration or redesign is needed. They also reinforce fairness by ensuring all model users receive dependable outputs. When ignored, errors multiply and become harder to correct. Teams experience confusion as predictions lose consistency. Over time, Model Risk increases as reliability gaps widen. Predictive Failure Exposure highlights where models are most likely to produce incorrect or harmful results.

Managing the "Black Box" requires constant clinical oversight. Leaders must be the primary auditors of Model Drift, Hallucination Error Logs, and Adversarial Exposure Tests. This topic focuses on "Robustness Assessments" to ensure that the AI engine remains reliable under stress. By engaging with a "Model Risk Register" in a print format, the leader can spot the patterns of failure that digital dashboards often obscure. Model risk management is about maintaining sovereignty over algorithmic logic, ensuring that the machine's output always aligns with the leader's intent and the organization's broader transformation goals. In the print blueprint, you can compare "Model Drift Reports" with "Reliability Stress Tests" side-by-side. This clinical focus prevents you from being "owned" by your algorithms, ensuring that the intelligence you build remains an evidence-based certainty rather than a million-dollar guess. Model risk is the diagnostic that protects your decision integrity from the unpredictable nature of black-box machine logic.

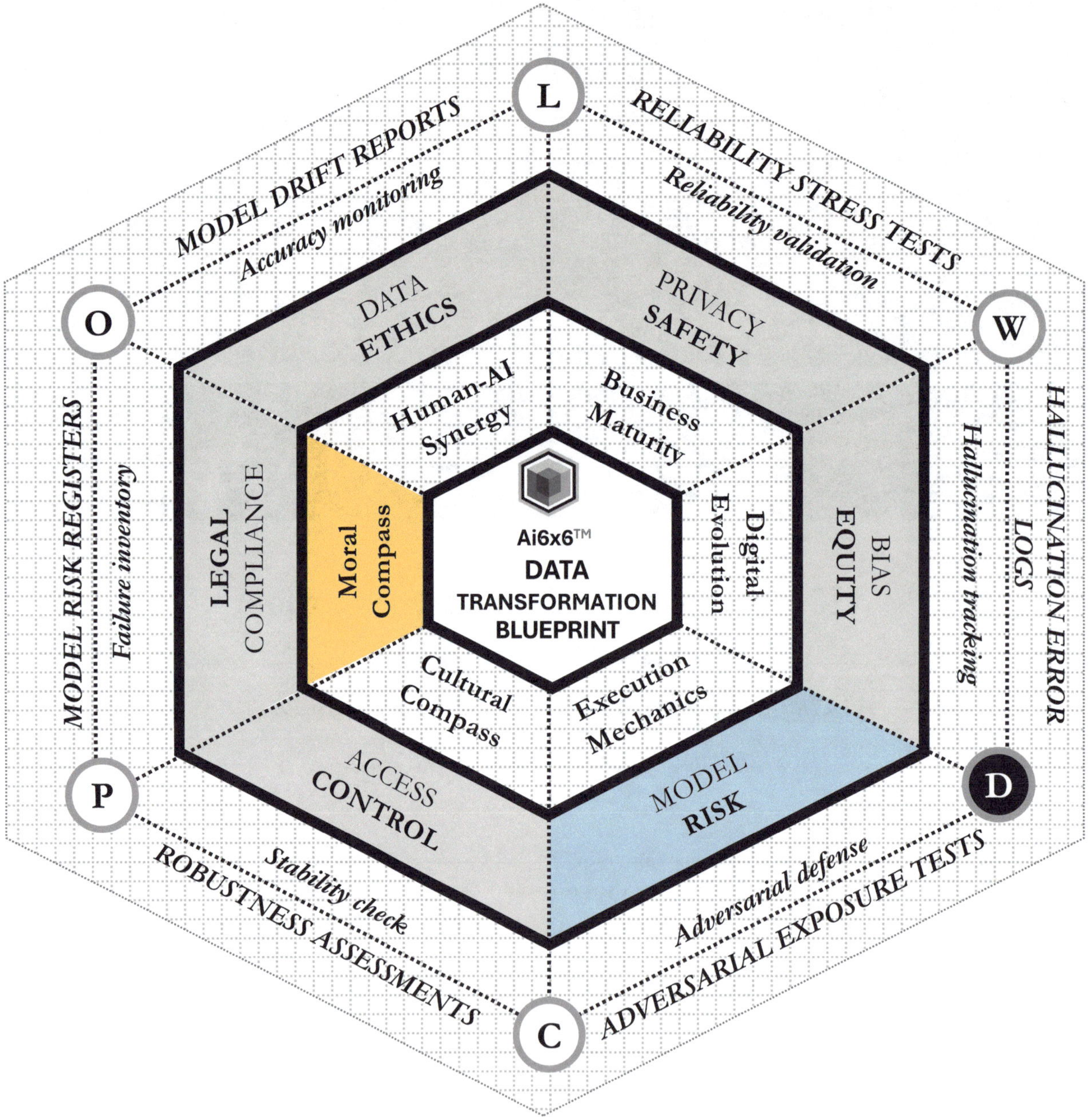

Ai6x6 MODEL RISK FOR MORAL COMPASS | DATA TRANSFORMATION

1. MODEL DRIFT REPORTS

Model Risk requires continuous vigilance, and Model Drift Reports track how model performance changes over time as data patterns evolve. They analyze shifts in accuracy, distribution, and behavior that signal degradation. This structure ensures that Model Risk is monitored through measurable indicators rather than intuition. It matters because even high performing models deteriorate when real world conditions shift. Reports prevent silent failure by revealing when outputs no longer reflect current realities. They also reinforce fairness by identifying when drift disproportionately affects specific groups. When ignored, models produce unreliable or biased results that undermine trust. Stakeholders question the validity of automated decisions. Over time, Model Risk escalates into operational and ethical instability.

When done well, Model Drift Reports strengthen Model Risk management by enabling timely recalibration and retraining. People trust that models remain aligned with real world conditions. This improves accuracy, fairness, and long term reliability. Drift monitoring reinforces the principle that models require maintenance, not blind confidence. It reminds leaders that performance must be validated continuously. Ultimately, it builds a data environment where model behavior remains stable, predictable, and trustworthy.

MORAL COMPASS SCENARIO	LEADERSHIP ACTION
Nira sees the drift report showing a steady decline in model accuracy after a major market shift. She realizes the model is now misaligned with real-world behavior.	The leader triggers an immediate retraining cycle and updates drift thresholds. They also assign a drift-response owner to monitor recovery.
Colton notices the report highlights a sudden spike in input-distribution drift for one customer segment. He suspects a data-pipeline issue.	The leader investigates the segment-specific pipeline and corrects the source. They also add segment-level drift alerts to prevent recurrence.
Eleni discovers that drift reports are being generated but never reviewed. She senses a governance breakdown.	The leader mandates weekly drift-review meetings and assigns remediation owners. They also track closure rates in the risk dashboard.
Mateo sees that drift is occurring only in multilingual inputs. He realizes the model was never validated for language diversity.	The leader expands training data to include multilingual samples. They also add language-specific drift metrics.
Soren notices that drift reports show stable accuracy but rising fairness disparities. He recognizes fairness drift.	The leader adds fairness-drift indicators and retrains the model with balanced data. They also require fairness checks in every drift review.
Hana finds that drift reports rely on outdated baselines. She worries the team is comparing against irrelevant benchmarks.	The leader recalibrates baselines and documents the update. They also schedule quarterly baseline reviews.

Model Risk requires understanding how systems behave under pressure, and Reliability Stress Tests evaluate model performance in extreme or adverse conditions. They simulate edge cases, resource constraints, and unexpected inputs. This structure ensures that Model Risk is assessed through rigorous scenario testing. It matters because models often fail when confronted with conditions outside their training distribution. Stress tests prevent overconfidence by revealing vulnerabilities that normal evaluation misses. They also reinforce fairness by identifying where stress disproportionately affects certain groups. When ignored, models collapse under real world volatility. Stakeholders lose confidence in the system's resilience. Over time, Model Risk becomes a structural weakness that threatens operational stability.

When done well, Reliability Stress Tests strengthen Model Risk management by exposing weaknesses before they cause harm. People trust that models can withstand unpredictable conditions. This improves resilience, safety, and ethical reliability. Stress testing reinforces the principle that robustness must be engineered. It reminds leaders that reliability is proven through pressure, not assumption. Ultimately, it builds a data environment where models remain dependable even under strain.

MORAL COMPASS SCENARIO	LEADERSHIP ACTION
Suruchi sees the stress test showing the model collapses under high-volume inputs. She realizes the system can't handle peak demand.	The leader mandates performance optimization and reruns the stress test. They also add load-testing gates to deployment workflows.
Rowan notices the model fails when inputs contain missing fields. He suspects brittle preprocessing.	The leader updates preprocessing to handle incomplete data gracefully. They also require robustness testing for all edge cases.
Serena discovers that stress tests only evaluate accuracy, not latency. She fears slow responses will degrade user experience.	The leader adds latency metrics to the stress test suite. They also require latency thresholds for production approval.
Omar sees that the model fails when exposed to noisy inputs. He recognizes a vulnerability to real-world variability.	The leader introduces noise-augmented training and reruns the stress test. They also add noise-robustness KPIs.
Lila notices that stress tests are run only once per quarter. She senses risks accumulating between cycles.	The leader increases test frequency and automates execution. They also assign a reliability steward to monitor results.
Andre finds that stress tests don't simulate multi-model interactions. He worries cascading failures are being missed.	The leader adds system-level stress scenarios. They also require cross-model dependency mapping.

3. HALLUCINATION ERROR LOGS

Model Risk requires visibility into incorrect or fabricated outputs, and Hallucination Error Logs document instances where models generate false information. They capture patterns, triggers, and contexts that lead to hallucinations. This structure ensures that Model Risk is addressed through systematic error tracking. It matters because hallucinations undermine credibility, distort decisions, and create ethical exposure. Logs prevent normalization of incorrect outputs by making failures explicit. They also reinforce fairness by identifying whether hallucinations disproportionately affect certain users or topics. When ignored, fabricated outputs spread unchecked and damage trust. Stakeholders question the reliability of the entire system. Over time, Model Risk escalates into widespread misinformation.

When done well, Hallucination Error Logs strengthen Model Risk management by enabling targeted mitigation and model refinement. People trust that errors are monitored and corrected. This improves accuracy, transparency, and ethical alignment. Error logging reinforces the principle that reliability requires confronting failure. It reminds leaders that hallucinations must be treated as critical defects. Ultimately, it builds a data environment where truthfulness is continuously protected.

MORAL COMPASS SCENARIO	LEADERSHIP ACTION
Rhea sees the hallucination log showing the model fabricates citations under pressure. She realizes users may trust false information.	The leader adds citation-validation rules and retrains the model on verified sources. They also require hallucination-rate monitoring.
Omar notices the model invents nonexistent product features when asked ambiguous questions. He senses a prompt-handling flaw.	The leader updates prompt-routing logic and adds ambiguity-detection safeguards. They also retrain the model on refusal patterns.
Lila discovers hallucinations spike during long conversational sessions. She suspects context-window degradation.	The leader adjusts context-management logic and tests long-session stability. They also add session-length hallucination metrics.
Andre finds that hallucination logs are stored without severity labels. He worries critical issues are buried.	The leader adds severity scoring and triage workflows. They also require weekly hallucination-review meetings.
Serena notices hallucinations occur mostly in specialized domains. She realizes the model lacks domain-specific grounding.	The leader integrates domain-verified datasets and adds expert-review checkpoints. They also require domain-specific hallucination tests.
Hugo sees that hallucination logs are not linked to user-impact assessments. He senses blind spots.	The leader connects hallucination logs to impact scoring. They also prioritize fixes based on user-risk levels.

4. ADVERSARIAL EXPOSURE TESTS

Model Risk requires defense against manipulation, and Adversarial Exposure Tests evaluate how models respond to intentionally deceptive or hostile inputs. They probe vulnerabilities that attackers could exploit. This structure ensures that Model Risk is assessed through proactive security testing. It matters because adversarial attacks can distort outputs, bypass safeguards, or compromise systems. Tests prevent exploitation by revealing weaknesses before they are targeted. They also reinforce fairness by ensuring all users are protected from manipulated outcomes. When ignored, models become susceptible to subtle or overt attacks. Stakeholders lose confidence in the system's integrity. Over time, Model Risk becomes a security liability.

When done well, Adversarial Exposure Tests strengthen Model Risk management by building resilience against manipulation. People trust that models can withstand malicious attempts to distort behavior. This improves safety, reliability, and ethical robustness. Adversarial testing reinforces the principle that security must be proactive. It reminds leaders that protection requires anticipating threats. Ultimately, it builds a data environment where models remain trustworthy even under attack.

MORAL COMPASS SCENARIO	LEADERSHIP ACTION
Keisha sees the model collapses when exposed to simple adversarial prompts. She realizes the system is easily manipulated.	The leader mandates adversarial-training augmentation. They also add adversarial-resilience KPIs.
Damon notices that adversarial tests reveal vulnerabilities in input-sanitization logic. He fears injection attacks.	The leader strengthens sanitization rules and reruns exposure tests. They also require security review for all input pathways.
Liora discovers that the model reveals internal reasoning when probed adversarially. She recognizes a confidentiality risk.	The leader patches the leakage vector and adds reasoning-protection layers. They also require red-team testing before deployment.
Mateo sees that adversarial tests are only run on text inputs. He suspects multimodal vulnerabilities are untested.	The leader expands tests to all input types. They also assign a multimodal-risk owner.
Serena notices that adversarial failures are not documented with reproducible steps. She worries issues will be hard to fix.	The leader standardizes adversarial-test documentation. They also require reproducibility checks.
Omar finds that adversarial exposure tests are run manually. He senses delays in detection.	The leader automates adversarial testing in CI/CD. They also add alerting for high-severity failures.

Model Risk requires evaluating stability across diverse conditions, and Robustness Assessments measure how consistently models perform when inputs vary. They test sensitivity to noise, perturbations, and distribution shifts. This structure ensures that Model Risk is understood through comprehensive performance evaluation. It matters because fragile models fail when confronted with even minor deviations. Assessments prevent overreliance on brittle systems by revealing where stability is lacking. They also reinforce fairness by identifying whether instability disproportionately affects certain groups. When ignored, models behave unpredictably and undermine decision quality. Stakeholders question the reliability of automated outputs. Over time, Model Risk becomes embedded in core operations.

When done well, Robustness Assessments strengthen Model Risk management by ensuring models remain stable across conditions. People trust that outputs are consistent and dependable. This improves accuracy, fairness, and operational confidence. Robustness evaluation reinforces the principle that reliability requires resilience. It reminds leaders that stability must be tested, not assumed. Ultimately, it builds a data environment where models perform reliably in real world variability.

MORAL COMPASS SCENARIO	LEADERSHIP ACTION
Nadia sees the robustness assessment showing the model fails when inputs contain slang. She realizes linguistic diversity wasn't considered.	The leader expands training data to include informal language. They also add slang-robustness tests.
Trevor notices the model performs poorly on edge-case numerical ranges. He suspects insufficient numeric stress testing.	The leader adds extreme-value tests and retrains the model. They also require numeric-robustness validation.
Aisha discovers that robustness assessments ignore rare but high-impact scenarios. She fears catastrophic failures.	The leader adds rare-event simulations. They also require scenario-based robustness scoring.
Marco sees that robustness tests don't include cross-platform variations. He worries the model behaves differently across environments.	The leader expands assessments to all deployment platforms. They also add environment-specific robustness metrics.
Evelyn notices robustness declines sharply when the model is throttled. She suspects resource-sensitivity issues.	The leader tests under constrained resources and optimizes model efficiency. They also add resource-robustness KPIs.
Jonah finds that robustness assessments are not tied to risk registers. He senses fragmented governance.	The leader links robustness findings to model-risk entries. They also require risk-aligned remediation plans.

Model Risk requires structured oversight, and Model Risk Registers document all known risks, vulnerabilities, and mitigation plans associated with each model. They consolidate information across lifecycle stages, from design to deployment. This structure ensures that Model Risk is managed through centralized visibility. It matters because fragmented risk knowledge leads to oversight gaps and unmanaged exposure. Registers prevent blind spots by cataloging risks in a single authoritative source. They also reinforce fairness by ensuring all models receive consistent scrutiny. When ignored, risks accumulate silently and become difficult to track or mitigate. Stakeholders lose confidence in governance discipline. Over time, Model Risk becomes unmanageable and systemic.

When done well, Model Risk Registers strengthen Model Risk management by creating transparency and accountability. People trust that risks are documented, monitored, and addressed. This improves governance, compliance, and ethical reliability. Registers reinforce the principle that risk must be cataloged to be controlled. They remind leaders that oversight requires structure. Ultimately, they build a data environment where model risks are visible, managed, and continuously reduced.

MORAL COMPASS SCENARIO	LEADERSHIP ACTION
Rhea discovers the risk register hasn't been updated since the last major model overhaul. She realizes risks are outdated.	The leader refreshes all entries and aligns them with the new architecture. They also schedule quarterly register reviews.
Omar notices that high-severity risks lack mitigation owners. He senses accountability gaps.	The leader assigns named owners and sets remediation deadlines. They also track progress in the governance dashboard.
Lila finds that the register lists risks but not user-impact scores. She worries prioritization is arbitrary.	The leader adds impact scoring and reorders priorities. They also require user-impact justification for each entry.
Andre sees that the register doesn't include third-party model risks. He recognizes a blind spot.	The leader expands the register to include vendor models. They also require vendor-risk attestations.
Serena discovers that risk entries are written in technical jargon. She fears executives can't interpret them.	The leader rewrites entries in executive-ready language. They also add a plain-language summary section.
Hugo notices that resolved risks remain in the register without closure notes. He sees clutter and confusion.	The leader adds closure documentation and archives resolved risks. They also enforce lifecycle management for all entries.

Ai6x6™ Transformation Blueprint

Access Control

"Sustainability is intelligence applied to survival."

David Attenborough, natural historian and environmental storyteller.

Access Control requires strict governance, and Permission Based Access defines who can view, modify, or use specific data assets. It establishes boundaries that prevent unauthorized or excessive access. This matters because uncontrolled access creates immediate security and privacy risk. Permission Based Access prevents misuse by ensuring data visibility aligns with role and responsibility. It also reinforces fairness by ensuring access is granted consistently rather than informally. When ignored, data becomes vulnerable to leakage or manipulation. Teams experience uncertainty about access rights. Over time, Access Control deteriorates as permissions drift and oversight weakens. Identity Verification Protocols ensure only authenticated individuals can access protected systems.

The boundary of sovereignty is defined by who can touch the data. Engaging with Identity and Credential Logs and Data Segmentation Maps ensures that decision integrity is never compromised by unauthorized exposure. This topic involves setting "Privilege Boundary Rules" that limit data access to the right people at the right time. By reviewing "Access Audit Trails" on a printed blueprint, the leader can physically trace the flow of information. Access control is the clinical procedure for eliminating internal "noise" and protecting the "Intellectual Property" that forms the backbone of the company's AI strategy. In the print series, you can map "Permission Change Records" against your "RACI Charts," seeing the whole map of data ownership. This prevents the "Kill Zone" of internal data leaks or systemic corruption, ensuring that your organization's data moves only with the intent and approval of the leadership, protecting the "Armor" of your enterprise valuation.

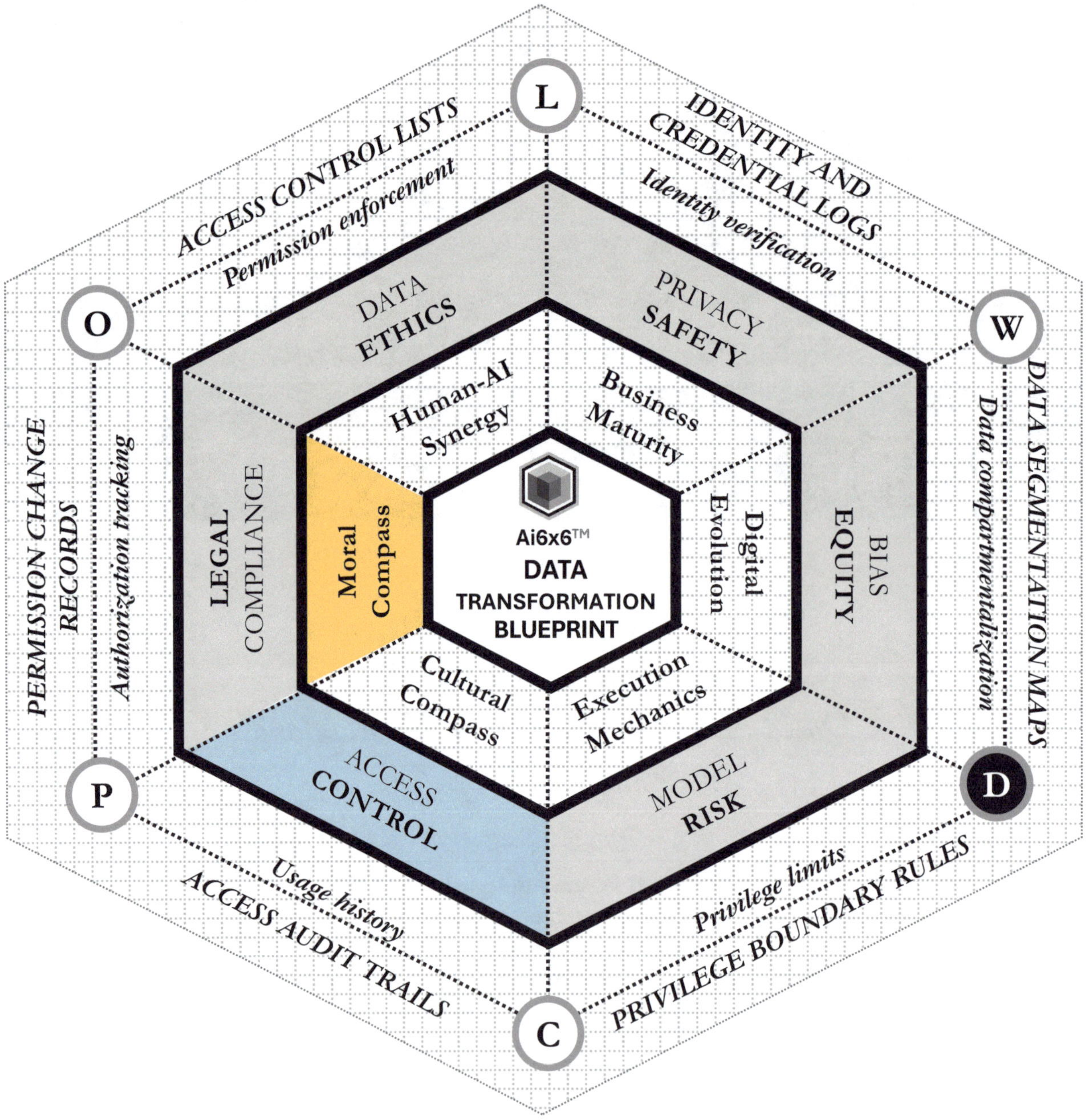

Ai6x6 ACCESS CONTROL FOR MORAL COMPASS | DATA TRANSFORMATION

1. ACCESS CONTROL LISTS

Access Control requires precise governance, and Access Control Lists define which individuals can view, modify, or interact with specific data assets. They map permissions to roles, responsibilities, and operational needs. This structure ensures that Access Control is enforced through explicit, rule based authorization. It matters because excessive or misaligned access creates privacy, security, and ethical vulnerabilities. Lists prevent unauthorized actions by restricting access to those with legitimate need. They also reinforce fairness by ensuring privileges are assigned based on role, not influence or hierarchy. When ignored, sensitive data becomes exposed to misuse or accidental disclosure. Stakeholders lose trust in the organization's ability to safeguard information. Over time, Access Control deteriorates into uncontrolled access and systemic risk.

When done well, Access Control Lists strengthen Access Control by creating disciplined, transparent permission structures. People trust that data is protected and access is granted responsibly. This improves compliance, security, and operational reliability. Lists reinforce the principle that authority must be limited to necessity. They remind leaders that protection requires intentional restriction. Ultimately, they build a data environment where access is controlled, accountable, and aligned with responsibility.

MORAL COMPASS SCENARIO	LEADERSHIP ACTION
Nira discovers that several interns have access to confidential customer files. She realizes permissions were copied from a senior analyst by mistake.	The leader revokes all inappropriate access immediately and rebuilds the ACL from role-based templates. They also implement a mandatory access review for all onboarding cases.
Colton notices that a deprecated team still appears on multiple ACLs. He fears dormant accounts could be exploited.	The leader removes all inactive groups and aligns ACLs with the current org structure. They also schedule quarterly ACL hygiene checks.
Eleni sees that a vendor was granted broad access "temporarily" months ago. She recognizes the risk of privilege creep.	The leader restricts the vendor to least-privilege access and documents the correction. They also add expiration dates to all vendor ACL entries.
Mateo finds that ACLs differ across identical systems. He suspects inconsistent governance.	The leader standardizes ACL templates and enforces centralized approval. They also assign system owners to maintain alignment.
Soren notices that ACL changes are being requested informally through chat messages. He sees a compliance gap.	The leader mandates formal ticket-based ACL changes and disables informal channels. They also train teams on proper request procedures.
Hana discovers that ACLs don't distinguish between read and write access. She worries about accidental data modification.	The leader updates ACLs with granular permissions and audits all write access. They also require justification for elevated privileges.

2. IDENTITY AND CREDENTIAL LOGS

Access Control requires traceability, and Identity and Credential Logs document who accessed systems, when they accessed them, and under what credentials. They create a verifiable record of authentication events and identity linked actions. This structure ensures that Access Control is upheld through continuous monitoring. It matters because unauthorized access often occurs through compromised or misused credentials. Logs prevent invisibility by making every identity interaction observable. They also reinforce fairness by ensuring all users are held to the same authentication standards. When ignored, suspicious activity goes undetected and breaches escalate. Stakeholders question whether the organization can protect sensitive information. Over time, Access Control collapses as identity misuse becomes unmanageable.

When done well, Identity and Credential Logs strengthen Access Control by enabling rapid detection of anomalies and misuse. People trust that identity based risks are monitored with rigor. This improves security, accountability, and ethical reliability. Logs reinforce the principle that access must be both granted and traceable. They remind leaders that identity governance is foundational to protection. Ultimately, they build a data environment where every access event is visible, auditable, and accountable.

MORAL COMPASS SCENARIO	LEADERSHIP ACTION
Priya sees multiple failed login attempts from a single employee account overnight. She suspects credential compromise.	The leader forces a password reset and investigates access logs. They also enable MFA enforcement for all high-risk accounts.
Rowan notices that credential logs show shared passwords across a team. He recognizes a major security violation.	The leader resets all shared credentials and retrains the team on identity hygiene. They also implement unique-credential enforcement.
Serena discovers that privileged accounts have no rotation history. She fears long-term exposure.	The leader rotates all privileged credentials and adds automated rotation schedules. They also restrict privileged access to essential personnel.
Omar finds that identity logs are missing IP metadata. He realizes investigations will be incomplete.	The leader updates logging configurations to capture full metadata. They also run a validation audit to ensure completeness.
Lila sees that offboarded employees still appear in identity logs. She senses a broken offboarding workflow.	The leader removes all inactive identities and integrates HR offboarding with identity systems. They also add automated deactivation triggers.
Andre notices that credential logs are stored in plaintext. He recognizes a severe security flaw.	The leader encrypts all logs and restricts access to security teams. They also implement secure-logging standards.

3. DATA SEGMENTATION MAPS

Access Control requires structured separation, and Data Segmentation Maps define how information is divided into zones based on sensitivity, purpose, and risk. They visualize which datasets belong in which protection tiers. This structure ensures that Access Control is upheld through intentional data partitioning. It matters because mixing sensitive and non sensitive data increases exposure and complicates governance. Maps prevent over access by ensuring users only reach the segments relevant to their role. They also reinforce fairness by applying consistent segmentation rules across teams. When ignored, data becomes intermixed and difficult to protect. Stakeholders lose confidence in the organization's ability to manage information responsibly. Over time, Access Control deteriorates into chaotic data sprawl.

When done well, Data Segmentation Maps strengthen Access Control by creating clarity and disciplined separation. People trust that sensitive information is isolated and protected. This improves compliance, reduces risk, and enhances operational efficiency. Segmentation reinforces the principle that protection requires boundaries. It reminds leaders that data must be organized before it can be secured. Ultimately, it builds a data environment where access is structured, intentional, and risk aligned.

MORAL COMPASS SCENARIO	LEADERSHIP ACTION
Rhea discovers that sensitive financial data is stored in the same segment as general analytics data. She worries about accidental exposure.	The leader separates sensitive datasets into isolated segments and updates the segmentation map. They also enforce strict access boundaries.
Omar sees that segmentation maps haven't been updated since a major data-warehouse migration. He suspects misaligned boundaries.	The leader rebuilds the segmentation map and validates it with engineering. They also require updates after every structural change.
Lila notices that customer data from different regions is mixed in one segment. She fears regulatory violations.	The leader re-segments data by region and documents compliance requirements. They also add region-specific access controls.
Andre finds that segmentation maps don't reflect third-party data flows. He recognizes a blind spot.	The leader integrates vendor data paths into the map. They also require vendors to provide segmentation documentation.
Serena sees that segmentation boundaries are too broad, allowing teams to access more data than needed. She senses privilege bloat.	The leader refines segmentation granularity and enforces least-privilege access. They also review segment boundaries quarterly.
Hugo discovers that segmentation maps are stored in outdated diagrams. He worries teams are using stale information.	The leader migrates maps to a dynamic, version-controlled system. They also assign ownership for continuous updates.

4. PRIVILEGE BOUNDARY RULES

Access Control requires limits on authority, and Privilege Boundary Rules define the maximum actions users can take within their assigned roles. They establish thresholds that prevent escalation beyond intended permissions. This structure ensures that Access Control is upheld through strict privilege containment. It matters because privilege creep—where users accumulate unnecessary permissions—creates major security and ethical risks. Rules prevent overreach by enforcing the principle of least privilege. They also reinforce fairness by ensuring no individual gains disproportionate control. When ignored, users inadvertently or intentionally exceed their authority. Stakeholders lose trust in the integrity of system governance. Over time, Access Control collapses as privilege boundaries blur.

When done well, Privilege Boundary Rules strengthen Access Control by creating disciplined, enforceable limits. People trust that authority is distributed responsibly and monitored. This improves security, compliance, and operational stability. Boundary rules reinforce the principle that power must be constrained. They remind leaders that access must be earned, justified, and controlled. Ultimately, they build a data environment where privilege is tightly governed and risk is minimized.

MORAL COMPASS SCENARIO	LEADERSHIP ACTION
Keisha notices engineers are granting themselves temporary admin rights during debugging. She recognizes a privilege-boundary breach.	The leader enforces strict admin-request workflows and revokes unauthorized privileges. They also implement just-in-time access controls.
Damon sees that privilege boundaries are defined inconsistently across teams. He fears uneven enforcement.	The leader standardizes privilege tiers and publishes clear boundary rules. They also train managers on enforcement expectations.
Liora discovers that a junior analyst has access to production systems. She realizes privilege boundaries were never applied.	The leader removes production access and updates boundary rules. They also require role-based privilege reviews.
Mateo finds that privilege boundaries don't account for cross-team collaboration. He sees teams bypassing rules to get work done.	The leader defines cross-team privilege pathways and updates boundary documentation. They also add approval gates for shared access.
Serena notices that privilege boundaries are not enforced during emergencies. She fears ad-hoc access will become normalized.	The leader creates emergency-access protocols with automatic expiration. They also audit all emergency access after incidents.
Omar discovers that privilege boundaries don't include data-export restrictions. He recognizes a major leak risk.	The leader adds export-control rules and updates privilege tiers. They also monitor export logs for violations.

5. ACCESS AUDIT TRAILS

Access Control requires accountability, and Access Audit Trails record every access event, modification, and data interaction across systems. They provide a chronological record that reveals who did what and when. This structure ensures that Access Control is upheld through verifiable oversight. It matters because unauthorized or inappropriate access often goes unnoticed without detailed logs. Audit trails prevent invisibility by making every action traceable. They also reinforce fairness by ensuring all users are subject to the same level of scrutiny. When ignored, misconduct becomes difficult to detect or investigate. Stakeholders lose confidence in the organization's governance. Over time, Access Control deteriorates into opacity and unmanaged risk.

When done well, Access Audit Trails strengthen Access Control by enabling investigation, accountability, and continuous improvement. People trust that access is monitored and misuse will be detected. This improves security, compliance, and ethical reliability. Audit trails reinforce the principle that transparency is essential to protection. They remind leaders that oversight must be documented, not assumed. Ultimately, they build a data environment where every access event is visible and accountable.

MORAL COMPASS SCENARIO	LEADERSHIP ACTION
Nadia sees that audit trails show unexplained access to sensitive HR files. She suspects unauthorized browsing.	The leader investigates the access pattern and restricts permissions. They also add anomaly-detection alerts to the audit system.
Trevor notices that audit trails are missing entries for certain systems. He realizes logging was never enabled.	The leader enables full logging and backfills where possible. They also require logging verification during system onboarding.
Aisha discovers that audit trails are stored without integrity checks. She fears tampering.	The leader adds cryptographic integrity controls and restricts write access. They also audit historical logs for anomalies.
Marco finds that audit trails are reviewed only during incidents. He senses risks accumulating unnoticed.	The leader mandates weekly audit-trail reviews. They also assign owners for recurring issues.
Evelyn sees that audit trails don't include user-agent data. She worries investigations will lack context.	The leader updates logging configurations to capture device and browser metadata. They also validate the new fields across systems.
Jonah notices that audit trails are siloed across departments. He recognizes fragmented visibility.	The leader centralizes audit logs into a unified platform. They also enforce cross-department access governance.

6. PERMISSION CHANGE RECORDS

Access Control requires disciplined evolution, and Permission Change Records document every modification to user access rights. They track who changed permissions, why they changed them, and when the changes occurred. This structure ensures that Access Control is upheld through transparent governance of privilege adjustments. It matters because unauthorized or poorly justified permission changes create major vulnerabilities. Records prevent privilege drift by ensuring all modifications are reviewed and auditable. They also reinforce fairness by ensuring changes follow consistent criteria. When ignored, permissions expand unpredictably and expose systems to misuse. Stakeholders lose trust in the integrity of access governance. Over time, Access Control collapses as privilege changes become opaque.

When done well, Permission Change Records strengthen Access Control by creating visibility and disciplined oversight. People trust that access rights evolve responsibly and transparently. This improves security, compliance, and operational integrity. Change tracking reinforces the principle that privilege must be controlled throughout its lifecycle. It reminds leaders that access governance requires continuous documentation. Ultimately, it builds a data environment where permissions are managed with precision and accountability.

MORAL COMPASS SCENARIO	LEADERSHIP ACTION
Rhea discovers that permission changes are being made without documentation. She realizes no one can trace who approved what.	The leader enforces mandatory change-record logging and blocks undocumented changes. They also add approval signatures to every record.
Omar sees that permission changes spike during late-night hours. He suspects unauthorized activity.	The leader investigates the changes and restricts after-hours access. They also add time-based alerts for unusual modifications.
Lila notices that permission changes are not tied to ticket numbers. She fears changes are being made outside formal workflows.	The leader integrates change records with the ticketing system. They also require ticket references for every modification.
Andre finds that permission changes are not reviewed after implementation. He senses hidden errors.	The leader adds post-change validation steps. They also assign reviewers to confirm accuracy.
Serena discovers that permission changes persist long after temporary needs expire. She recognizes privilege creep.	The leader adds expiration dates to all temporary permissions. They also run monthly cleanup sweeps.
Hugo sees that permission change records are stored in spreadsheets. He worries about tampering.	The leader migrates records to a secure, immutable system. They also restrict editing rights to authorized administrators.

Ai6x6™ Transformation Blueprint

Legal Compliance

"The measure of AI is not its power, but its humanity."
Dalai Lama, global spiritual leader known for compassion and ethics.

Legal Compliance requires disciplined alignment with data regulations, and Regulatory Data Alignment defines the legal standards organizations must follow when collecting and processing information. It establishes the minimum requirements for lawful data practices. This matters because noncompliance exposes organizations to penalties, litigation, and reputational damage. Regulatory Data Alignment prevents violations by ensuring data operations follow applicable laws. It also reinforces integrity by demonstrating that compliance is a core obligation. When ignored, legal exposure increases rapidly. Teams experience uncertainty as decisions drift outside regulatory boundaries. Over time, Legal Compliance deteriorates as adherence becomes inconsistent. Privacy Law Adherence ensures data practices remain aligned with evolving legal requirements.

The strategic moat of data is built on a foundation of law. By mastering GDPR, CCPA, and HIPAA in a unified, engrossed view, the leader ensures the organization's "Intelligence" remains a compliant asset. This topic treats "Legal Compliance" as a tool for "Opportunity Capture" rather than a hurdle. It involves using "Audit Trails" and "Permission Records" to prove the integrity of the data ecosystem. In the AI era, being legally unassailable in your data practices is the only way to avoid the catastrophic fines and reputational damage that follow "Data Blindness" or ethical lapses. The print series allows you to sit with these complex regulations, seeing how they anchor your "Moral Compass." This clinical mastery ensures that your "Data Transformation" is legally sound and morally unshakeable, providing the "Self-Healing" insurance policy that protects your operations from being dismantled by the regulatory sledgehammers of the modern digital landscape.

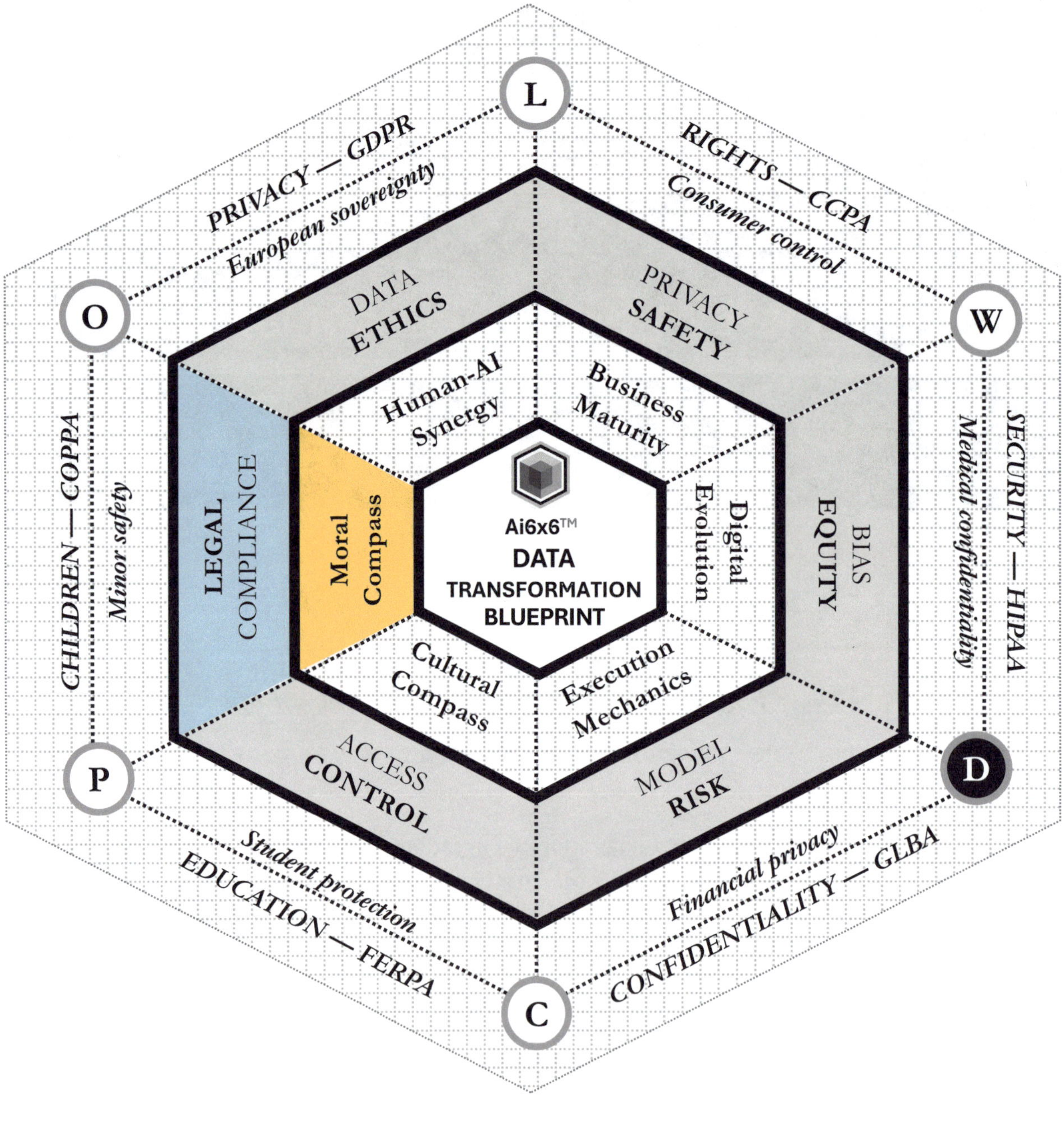

Ai6x6 LEGAL COMPLIANCE FOR MORAL COMPASS| DATA TRANSFORMATION

1. PRIVACY — GENERAL DATA PROTECTION REGULATION (GDPR)

Legal Compliance requires strict adherence to global privacy standards, and the General Data Protection Regulation establishes the most comprehensive framework for protecting personal data. It defines lawful bases for processing, mandates transparency, and enforces individual rights such as access, correction, and erasure. This structure ensures that Legal Compliance is grounded in accountability and explicit consent. It matters because GDPR violations expose organizations to severe penalties and reputational harm. Compliance prevents misuse by requiring organizations to justify every data action. It also reinforces fairness by ensuring individuals maintain control over their information. When ignored, data practices become opaque, intrusive, and legally vulnerable. Stakeholders lose trust in the organization's ethical posture. Over time, Legal Compliance deteriorates into systemic privacy risk.

When done well, GDPR alignment strengthens Legal Compliance by embedding privacy into every operational layer. People trust that their rights are respected and protected. This improves transparency, reduces risk, and enhances global credibility. GDPR compliance reinforces the principle that privacy is a fundamental right. It reminds leaders that lawful processing requires discipline, documentation, and respect for individual autonomy. Ultimately, it builds a data environment where privacy is treated as a core organizational obligation.

MORAL COMPASS SCENARIO	LEADERSHIP ACTION
Nira discovers that a product team is processing EU customer data without a lawful basis. She realizes the team assumed "legitimate interest" without documentation.	The leader halts all processing and requires a lawful-basis assessment. They also mandate GDPR compliance checks before any new data workflow.
Colton notices that data-subject access requests are taking over 60 days to fulfill. He fears the company is violating GDPR timelines.	The leader restructures the DSAR workflow and assigns a dedicated response team. They also implement SLA tracking for all future requests.
Eleni finds that EU data is being transferred to a non-compliant region. She recognizes a cross-border transfer violation.	The leader suspends transfers and implements SCCs and regional controls. They also require legal review for all international data flows.
Mateo sees that consent banners are vague and bundled. He knows this invalidates consent.	The leader redesigns consent flows with granular options. They also require explicit opt-in for all sensitive processing.
Soren notices that retention periods for EU data are undefined. He senses a compliance gap.	The leader establishes GDPR-aligned retention schedules. They also enforce automated deletion triggers.
Hana discovers that privacy notices are outdated and incomplete. She worries customers are being misled.	The leader updates all notices and adds transparency requirements to product workflows. They also schedule annual notice reviews.

2. RIGHTS — CALIFORNIA CONSUMER PRIVACY ACT (CCPA)

Legal Compliance requires honoring individual rights, and the California Consumer Privacy Act grants consumers control over how their personal data is collected, used, and sold. It establishes rights to access, delete, and opt out of data sharing. This structure ensures that Legal Compliance is grounded in transparency and consumer empowerment. It matters because modern data ecosystems often obscure how information flows. CCPA compliance prevents exploitation by requiring organizations to disclose practices clearly. It also reinforces fairness by giving individuals meaningful control over their digital footprint. When ignored, consumers lose visibility into how their data is monetized or distributed. Trust erodes as privacy expectations are violated. Over time, Legal Compliance collapses as regulatory exposure increases.

When done well, CCPA alignment strengthens Legal Compliance by ensuring consumer rights are respected and operationalized. People trust that their choices influence how their data is handled. This improves ethical credibility, reduces risk, and enhances customer relationships. CCPA compliance reinforces the principle that individuals deserve agency over their information. It reminds leaders that transparency is both a legal requirement and a competitive advantage. Ultimately, it builds a data environment where consumer rights are protected with rigor.

MORAL COMPASS SCENARIO	LEADERSHIP ACTION
Priya sees that opt-out requests for data sale are not being propagated to downstream systems. She realizes customers' rights are being ignored.	The leader synchronizes opt-out flags across all systems and validates propagation. They also require automated enforcement for future requests.
Rowan notices that the "Do Not Sell My Info" link is buried deep in the footer. He fears the company is non-compliant.	The leader moves the link to a prominent location and updates UI guidelines. They also audit all customer-facing pages for compliance.
Serena discovers that CCPA deletion requests are being partially fulfilled. She recognizes a systemic gap.	The leader enforces full-stack deletion and updates the deletion workflow. They also assign owners for each system involved.
Omar finds that customer identity verification steps are inconsistent. He worries about fraudulent requests.	The leader standardizes verification procedures and documents them. They also train support teams on secure identity checks.
Lila sees that CCPA disclosures are outdated and missing categories. She senses transparency risk.	The leader updates disclosures and aligns them with current data practices. They also require quarterly disclosure reviews.
Andre notices that employees don't understand what counts as "data sale." He fears accidental violations.	The leader runs a training session on CCPA definitions and examples. They also add a pre-launch compliance checklist.

3. SECURITY — HEALTH INSURANCE PORTABILITY AND ACCOUNTABILITY ACT (HIPAA)

Legal Compliance requires safeguarding sensitive information, and the Health Insurance Portability and Accountability Act establishes strict standards for protecting health data. It defines security controls, privacy rules, and breach notification requirements. This structure ensures that Legal Compliance is upheld through disciplined protection of medical information. It matters because health data is deeply personal and highly sensitive. HIPAA compliance prevents unauthorized access, disclosure, or misuse. It also reinforces fairness by ensuring all patients receive equal protection regardless of provider or system. When ignored, health information becomes vulnerable to exposure and exploitation. Patients lose trust in healthcare institutions. Over time, Legal Compliance deteriorates into ethical and legal crisis.

When done well, HIPAA alignment strengthens Legal Compliance by ensuring health data is handled with exceptional care. People trust that their medical information remains confidential and secure. This improves patient confidence, regulatory alignment, and ethical integrity. HIPAA compliance reinforces the principle that sensitive data requires heightened protection. It reminds leaders that privacy in healthcare is both a legal mandate and a moral duty. Ultimately, it builds a data environment where health information is safeguarded with uncompromising rigor.

MORAL COMPASS SCENARIO	LEADERSHIP ACTION
Rhea discovers that PHI is being stored in an unencrypted database. She recognizes a major HIPAA violation.	The leader encrypts all PHI immediately and documents the remediation. They also enforce encryption checks for all health-related systems.
Omar notices that access logs for PHI are incomplete. He fears unauthorized access is going undetected.	The leader enables full audit logging and assigns a compliance monitor. They also run a retrospective access review.
Lila finds that PHI is being emailed between departments. She knows this violates secure-transmission rules.	The leader bans email transmission of PHI and deploys a secure messaging tool. They also retrain all staff on HIPAA communication rules.
Andre sees that a vendor handling PHI lacks a Business Associate Agreement. He recognizes contractual non-compliance.	The leader halts data sharing and executes a BAA. They also update procurement rules to require BAAs before onboarding.
Serena discovers that PHI is visible on shared screens during meetings. She senses a privacy breach.	The leader enforces screen-privacy protocols and updates meeting guidelines. They also train teams on PHI handling etiquette.
Hugo notices that backup files containing PHI are stored without access controls. He fears exposure.	The leader secures all backups and restricts access. They also add PHI-specific backup policies.

Legal Compliance requires protecting financial information, and the Gramm–Leach–Bliley Act mandates safeguards for consumer financial data. It establishes requirements for privacy notices, data security, and information sharing limitations. This structure ensures that Legal Compliance is grounded in confidentiality and responsible stewardship. It matters because financial data misuse can cause severe personal and economic harm. GLBA compliance prevents unauthorized access and ensures institutions disclose how information is shared. It also reinforces fairness by giving consumers transparency into financial data practices. When ignored, financial information becomes vulnerable to fraud, identity theft, and unethical use. Stakeholders lose trust in financial institutions. Over time, Legal Compliance collapses as confidentiality breaches escalate.

When done well, GLBA alignment strengthens Legal Compliance by embedding security and transparency into financial operations. People trust that their financial information is protected and handled responsibly. This improves credibility, reduces risk, and enhances regulatory alignment. GLBA compliance reinforces the principle that confidentiality is essential to financial integrity. It reminds leaders that consumer trust depends on responsible data stewardship. Ultimately, it builds a data environment where financial information is safeguarded with discipline and clarity.

MORAL COMPASS SCENARIO	LEADERSHIP ACTION
Keisha sees that customer financial data is being shared with a partner without proper safeguards. She recognizes a GLBA violation.	The leader halts data sharing and implements safeguard requirements. They also update vendor contracts with confidentiality clauses.
Damon notices that financial-data access is not role-restricted. He fears unnecessary exposure.	The leader enforces least-privilege access and updates access controls. They also audit all financial-data permissions.
Liora discovers that privacy notices required under GLBA are outdated. She senses regulatory risk.	The leader updates notices and aligns them with current practices. They also schedule annual GLBA notice reviews.
Mateo finds that encryption is inconsistent across financial systems. He recognizes a confidentiality gap.	The leader standardizes encryption requirements and validates compliance. They also add encryption checks to deployment workflows.
Serena sees that incident-response plans don't include GLBA-specific requirements. She worries about incomplete reporting.	The leader updates the plan with GLBA reporting steps. They also train incident-response teams on financial-data protocols.
Omar notices that customer-service agents can see full account numbers. He fears unnecessary exposure.	The leader masks sensitive fields and updates UI rules. They also restrict full-number access to authorized roles.

5. EDUCATION — FAMILY EDUCATIONAL RIGHTS AND PRIVACY ACT (FERPA)

Legal Compliance requires protecting student information, and the Family Educational Rights and Privacy Act governs access to educational records. It grants students and parents rights to review, correct, and control disclosure of academic data. This structure ensures that Legal Compliance is upheld through transparency and respect for educational privacy. It matters because student records contain sensitive academic, behavioral, and personal information. FERPA compliance prevents unauthorized disclosure and ensures institutions honor student rights. It also reinforces fairness by ensuring all students receive equal protection. When ignored, educational data becomes vulnerable to misuse or inappropriate sharing. Students and families lose trust in academic institutions. Over time, Legal Compliance deteriorates into privacy violations and reputational harm.

When done well, FERPA alignment strengthens Legal Compliance by ensuring educational data is handled with respect and precision. People trust that their academic information is protected and accessible only to authorized parties. This improves institutional credibility and ethical alignment. FERPA compliance reinforces the principle that privacy is essential to educational integrity. It reminds leaders that student rights must be protected consistently. Ultimately, it builds a data environment where educational records are managed responsibly and transparently.

MORAL COMPASS SCENARIO	LEADERSHIP ACTION
Nadia discovers that student records are accessible to non-authorized staff. She recognizes a FERPA breach.	The leader restricts access and updates role-based permissions. They also train staff on FERPA access rules.
Trevor notices that student data is being used for analytics without consent. He fears non-compliance.	The leader halts the analysis and verifies consent requirements. They also implement a FERPA-aligned data-use review.
Aisha finds that parents' requests for record access are delayed. She senses a rights-fulfillment gap.	The leader streamlines the request workflow and assigns a response coordinator. They also track turnaround times.
Marco sees that student identifiers are included in public dashboards. He recognizes a privacy risk.	The leader anonymizes all dashboards and updates publishing rules. They also require privacy checks before release.
Evelyn discovers that student records are stored in unsecured folders. She worries about unauthorized access.	The leader secures all storage locations and adds access logs. They also enforce encryption for sensitive files.
Jonah notices that staff are unaware of what counts as "directory information." He fears accidental disclosure.	The leader clarifies directory-information rules and updates training materials. They also require annual FERPA certification.

6. CHILDREN — CHILDREN'S ONLINE PRIVACY PROTECTION ACT (COPPA)

Legal Compliance requires heightened protection for minors, and the Children's Online Privacy Protection Act regulates how organizations collect and use data from children under 13. It mandates parental consent, limits data collection, and enforces strict transparency requirements. This structure ensures that Legal Compliance is grounded in safeguarding vulnerable populations. It matters because children cannot fully understand or consent to data practices. COPPA compliance prevents exploitation by restricting how organizations engage with young users. It also reinforces fairness by ensuring minors receive stronger protections than adults. When ignored, children's data becomes exposed to misuse, profiling, or unauthorized sharing. Parents lose trust in digital platforms. Over time, Legal Compliance collapses as violations trigger legal and ethical consequences.

When done well, COPPA alignment strengthens Legal Compliance by ensuring children's data is handled with exceptional care. Families trust that digital environments respect and protect young users. This improves platform credibility, reduces risk, and enhances ethical responsibility. COPPA compliance reinforces the principle that vulnerable groups require elevated safeguards. It reminds leaders that children's privacy is a legal and moral priority. Ultimately, it builds a data environment where minors are protected through rigorous, transparent controls.

MORAL COMPASS SCENARIO	LEADERSHIP ACTION
Rhea discovers that a product collects behavioral data from children under 13 without parental consent. She recognizes a COPPA violation.	The leader deletes the data and implements verified parental-consent flows. They also add age-screening to onboarding.
Omar notices that ads are being personalized for child users. He fears unlawful profiling.	The leader disables personalization for minors and updates ad-serving rules. They also audit all child-directed experiences.
Lila finds that the privacy policy for a child-focused feature is written in complex language. She realizes it's not child-readable.	The leader rewrites the policy in plain, age-appropriate language. They also require readability checks for all child-facing disclosures.
Andre sees that third-party SDKs collect data from child users. He recognizes a compliance risk.	The leader removes non-compliant SDKs and verifies vendor COPPA adherence. They also add SDK-screening to development workflows.
Serena discovers that parental-consent records are stored inconsistently. She fears gaps in verification.	The leader centralizes consent records and enforces standardized logging. They also add periodic verification checks.
Hugo notices that child accounts are not segmented from adult accounts. He worries about accidental data mixing.	The leader creates a dedicated child-data segment and updates access rules. They also enforce strict separation in all pipelines.

BONUS CHAPTER

Ai6x6™ Transformation Blueprint

The Intelligence Paradox

"The digital world is not a cloud; it is a physical resource that requires a physical landscape." -- **Gerry McGovern**, author of *Worldwide Waste* and a leading voice on digital sustainability, identifies the core of the **Intelligence Paradox**. His work exposes the massive physical energy and e-waste costs of our data-driven lives, urging leaders to move away from "digital obesity" and toward a lean, "Green Data" strategy that treats every byte as a carbon-heavy resource.

Traditional IT strategies have operated on the "more is better" principle, hoarding vast amounts of data in hopes of future insights. However, this has created a hidden environmental crisis: a massive digital carbon footprint fueled by "Dark Data" and energy-hungry AI models. Traditional thinking fails to see that our most powerful tool for saving the planet is also its fastest-growing energy consumer. To gain traction, leaders must embrace the **Intelligence Paradox**, moving from data hoarding to data minimalism. This multi-dimensional approach requires balancing "compute power" with "climate cost," prioritizing "Green Coding" over easy offsets, and securing transparent data flows without compromising safety. True transformation occurs when the digital engine is as lean as the world it seeks to protect.

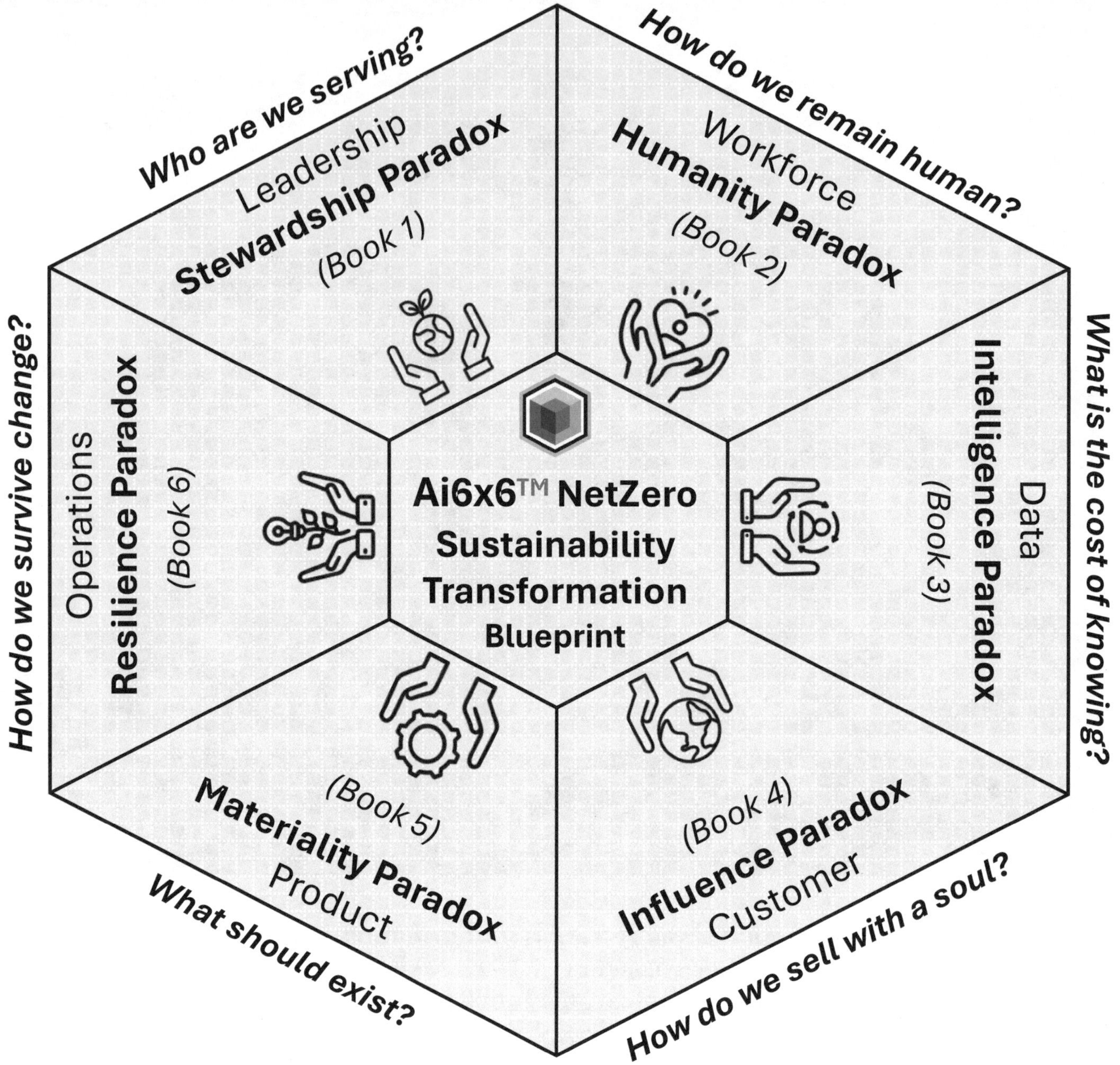

The **Ai6x6 NetZero Sustainability Transformation Blueprint** redefines the path to Net Zero by replacing the traditional "check-the-box" compliance mindset with a rigorous framework of **paradoxical leadership**. In a world accelerated by AI, sustainability is rarely a choice between a clear right and wrong; instead, it is a constant navigation of thirty-six competing tensions across six organizational lenses. By framing the journey through these paradoxes—such as the conflict between **Digital Intelligence and Energy Consumption** or **Global Targets and Local Infrastructure**—the framework prepares leaders for the messy, non-linear realities of systemic change.

Rather than viewing these tensions as obstacles to be avoided, the **Ai6x6 methodology** utilizes them as engines for innovation. It guides executives to move past the paralysis of "either/or" thinking, teaching them to embrace a "both/and" strategy that maintains fiduciary health while securing long-term planetary survival. This approach ensures that a company's moral compass remains calibrated even as technology and climate volatility shift the ground beneath them. Ultimately, the series provides a diagnostic roadmap that turns systemic friction into strategic advantage, empowering leaders to achieve radical Net Zero goals without sacrificing the operational resilience or human dignity required to thrive in the post-AI era.

THE INTELLIGENCE PARADOX is the hidden friction of the digital age: the more we use data to solve the climate crisis, the more energy we consume to power the solution. We are in a race to build "smarter" systems, yet the physical infrastructure of AI—the servers, cooling, and chips—is placing a massive new burden on the planet's resources. This creates a circular struggle where digital growth risks canceling out environmental gains. To transform, we must move beyond simply "collecting more" and learn to build a digital engine that is as lean and responsible as the world it seeks to save.

1. THE COMPUTE-CLIMATE PARADOX (Smart AI vs. High Energy)

We are relying on massive AI models to predict weather patterns and optimize energy grids, yet training these very models creates a staggering carbon footprint. The paradox is that our most powerful tool for saving the planet is also one of its fastest-growing energy consumers. Leaders must navigate the tension between the "intelligence gain" of a large model and the "carbon cost" of running it. The goal is to move toward "Green AI," where the environmental benefit of the algorithm's insight clearly outweighs the electricity and water used to keep the processors running and cool.

2. THE DARK DATA PARADOX (Saving Everything vs. Digital Waste)

In the AI era, the instinct is to save every byte of data to feed future machine learning models. However, this creates a mountain of "Dark Data"—vast amounts of digital waste that sits unused in data centers, consuming energy 24/7. The paradox lies in the fear of deleting a future insight versus the reality of current environmental damage. Data transformation requires the discipline to distinguish between valuable knowledge and high-carbon digital clutter. It is a shift from "data hoarding" to "data minimalism," ensuring that our digital storage doesn't become the landfill of the future.

3. THE TRANSPARENCY-SECURITY PARADOX (Being Open vs. Staying Safe)

To prove sustainability progress, companies must share granular, real-time data across the entire supply chain. Yet, the paradox is that the more "open" and transparent this data becomes for the sake of the planet, the more "vulnerable" the company becomes to cyberattacks and industrial espionage. Sharing high-resolution energy or logistics data can reveal corporate secrets or weak points. Navigating this requires a "Zero-Trust" approach to green data, where information is shared to build trust with stakeholders without compromising the fundamental security and safety of the organization's digital borders.

4. THE HARDWARE-REFRESH PARADOX (New Gadgets vs. Electronic Waste)

There is a constant push to upgrade to the latest, most energy-efficient chips and servers to reduce operational carbon. However, the paradox is that the manufacturing of this new hardware creates a massive spike in e-waste and consumes rare minerals. We often "save" energy at the plug while "spending" it at the factory and the landfill. A true moral compass in data transformation requires looking at the full lifecycle of hardware, choosing to extend the life of existing machines through better software rather than constantly chasing the marginal gains of the newest silicon.

5. THE PREDICTIVE-OVERLOAD PARADOX (Checking Everything vs. Too Much Power)

Using billions of sensors to monitor every pipe and wire can prevent resource leaks and save tons of carbon. But a paradox emerges when the energy needed to transmit, store, and process that "mountain of data" exceeds the energy saved by fixing the leak. We risk building a surveillance system that is more resource-heavy than the problem it is trying to solve. Transformation means finding the "sweet spot" of monitoring—collecting only the data that is truly actionable and avoiding the trap of high-energy, low-value digital "noise."

6. THE DATA ETHICS PARADOX (Fixing the Code vs. Buying Credits)

When a company's digital footprint grows too large, it is often easier to buy carbon offsets than to re-engineer complex legacy code. This paradox represents the easy way out versus the ethical way forward. Offsets might balance the books on paper, but they don't solve the underlying problem of inefficient, "heavy" software. The moral choice in data transformation is to invest in "Green Coding"—writing, efficient algorithms that require less compute power to execute. It is about fixing the engine from the inside rather than just paying for the smoke.

1. THE COMPUTE-CLIMATE PARADOX (Smart AI vs. High Energy)

- **Cloud Computing:** A major provider moves all non-urgent LLM training to "Cold Regions" like Iceland, utilizing 100% local geothermal energy and free-air cooling.
- **Social Media:** A platform implements "Energy-Aware Streaming," automatically throttling auto-play video resolution during peak hours when the local energy grid is most carbon-intensive.
- **Bio-Tech:** A research firm adopts "DNA Data Storage" for archival genomic sequences, eliminating the perpetual electricity and cooling costs of traditional silicon-based servers.
- **Search Engines:** A global tech giant builds on-site "Micro-grids" and massive battery arrays to ensure its AI search queries are powered exclusively by surplus solar and wind.
- **Crypto:** A blockchain network completes a shift from "Proof of Work" to "Proof of Stake," reducing its global energy consumption by 99.9% while maintaining transaction security.
- **Semiconductors:** A chipmaker prioritizes the development of "Neuromorphic Chips" that mimic the human brain's efficiency, using 1,000x less power for AI inference than current GPUs.

2. THE DARK DATA PARADOX (Saving Everything vs. Digital Waste)

- **Telecommunications:** A global telco implements "Data Sunsetting" protocols that automatically delete 40% of unused network logs after six months to save massive storage energy.
- **Healthcare:** A hospital system moves historical patient scans to "Cold Storage" that remains unpowered until a specific request is made, reducing the carbon footprint of its digital archives.
- **Marketing:** A global retailer purges all inactive email subscribers every 90 days, reducing the server-load and "ghost carbon" associated with sending ignored digital communications.
- **Security:** A smart-city project implements "Selective Retention," where AI deletes 99% of empty street-view footage daily, keeping only data that contains actionable safety events.
- **Manufacturing:** An industrial plant deletes high-frequency sensor noise every 24 hours, keeping only the aggregated "Trend Reports" required for long-term predictive maintenance.
- **Government:** A city agency deduplicates 15 years of overlapping municipal records, shrinking its physical data center footprint and reducing active server maintenance costs.

3. THE TRANSPARENCY-SECURITY PARADOX (Being Open vs. Staying Safe)

- **Logistics:** A freight company uses "Zero-Knowledge Proofs" to verify its Scope 3 emissions to auditors without revealing sensitive, proprietary shipping routes or client volumes.
- **Energy Grid:** Regional utility competitors share real-time grid-load data on a private blockchain to balance renewable surges without exposing trade secrets.
- **Financial Services:** A bank uses "Federated Learning" to improve its green-loan fraud detection models, training on local data so that sensitive customer information never leaves the device.
- **Public Health:** A national health service shares anonymized "Environmental Health" trends with researchers while using "Differential Privacy" to ensure individual patient identities are mathematically unrecoverable.
- **Oil & Gas:** A major firm discloses real-time methane leak data via satellite to a public dashboard, using the transparency to validate its proactive repair efforts while masking exact site layouts.
- **Aviation:** Airlines share "Black Box" fuel efficiency data through a neutral third-party trust, allowing the industry to optimize flight paths without compromising individual corporate safety protocols.

4. THE HARDWARE-REFRESH PARADOX (New Gadgets vs. Electronic Waste)

- **Higher Education:** A university implements "Thin Client" virtualization, allowing 10,000 students to run high-performance software on older laptops, extending device life by four years.
- **Financial Services:** A global bank partners with a "Circular IT" vendor to refurbish and redeploy 50,000 employee monitors, choosing high-quality reuse over purchasing new energy-efficient models.
- **IT Services:** A firm shifts to "Modular Servers" where only the CPU and RAM are upgraded, while the heavy steel chassis and power supplies are kept for a 12-year lifespan.
- **Public Sector:** A city government mandates that all municipal hardware must have a "Reparability Score" of 8/10, ensuring that devices are fixed locally rather than scrapped.
- **Consumer Tech:** A hardware manufacturer launches a "Second-Life" program, turning trade-in smartphones into low-power IoT sensors for industrial environmental monitoring.
- **Defense:** A contractor uses software optimization and "Retro-fitting" to extend the life of field-service tablets, avoiding the embodied carbon of manufacturing new ruggedized gear.

5. THE PREDICTIVE-OVERLOAD PARADOX (Checking Everything vs. Too Much Power)

- **Smart Cities:** A municipality moves to "Edge Processing" for its traffic lights, analyzing data locally at the pole to reduce the energy cost of transmitting billions of bytes to the cloud.
- **Agriculture:** A large-scale farm replaces 24/7 moisture sensors with "Triggered Sampling," where sensors only activate when local weather stations predict a high-evaporation event.
- **Utilities:** A water company uses "Acoustic Sampling" for leak detection, processing only the sound frequency of a burst pipe locally to minimize data transmission energy.
- **Logistics:** A warehouse uses "Passive RFID" tags that only transmit data when scanned, avoiding the need for millions of battery-powered sensors that require constant charging.
- **Retail:** A fashion brand uses "Predictive Stocking" based on existing trends to eliminate the need for real-time tracking of every individual garment in the supply chain.
- **Oil & Gas:** A pipeline operator uses "Fibre-Optic Sensing" that uses light to monitor 100 miles of pipe simultaneously, replacing thousands of individual, high-energy electronic sensors.

6. THE DATA ETHICS PARADOX (Fixing the Code vs. Buying Credits)

- **FinTech:** A payment processor invests in "Green Refactoring," rewriting its entire transaction engine in more efficient code to reduce server-side CPU cycles by 40% per swipe.
- **Software-as-a-Service:** A CRM provider offers a "Low-Carbon Mode" for its dashboard, allowing users to disable energy-intensive animations and high-resolution images.
- **E-Commerce:** A retailer optimizes its website's "DOM" and CSS to reduce the data weight of its homepage, saving energy for every one of its 100 million annual visitors.
- **Web Hosting:** A provider offers "Carbon-Priority Routing," automatically hosting client websites in regions where the current renewable energy mix is highest.
- **AI Research:** A lab prioritizes "Model Distillation," creating smaller, leaner AI models that achieve 95% of the accuracy of giant models at 10% of the energy cost.
- **Architecture:** A firm creates a "Green Coding Standard" for its 3D-modeling software, mandating that all custom plugins must be audited for their energy consumption.

WE NEED YOU!

WELCOMING OUR NEXT SET OF THOUGHT LEADERS

The remarkable success of the Ai6x6 Organization Transformation Series is built upon the **hard-won wisdom** of executive leaders who have successfully navigated the AI transformation within their industries, turning theory into a deeply practical, actionable guide. As we expand the series, we are extending a highly selective invitation to a new cohort of pioneers—individuals whose professional journey is at the **leading edge of AI strategy and execution**. We are looking for executives who possess a proven track record of rewriting business models, mastering unforeseen ethical and competitive challenges, and offering a **clear blueprint** for the next 3-5 years of the AI-Era.

Contributing to the Ai6x6 Series is a unique opportunity to **cement your professional legacy**, joining an exclusive, curated network of global authorities whose insights are distilled into executive-level knowledge cards, directly shaping the governance and innovation decisions of the world's most influential companies. Our selection process is based strictly on **pure expertise and demonstrated leadership** in the AI-Era, reinforcing our commitment to editorial integrity and authentic thought leadership—there is no monetary exchange involved. Select contributors will also receive an invitation to blog on our expert network microsite, further amplifying their platform. To confidentially inquire about contributing your expertise, please contact us via our dedicated portal at **www.Ai6x6.com/contact** or by scanning the adjacent QR code.

Contact

SUBSCRIBE TO OUR NEWSLETTERS: To stay ahead of the curve and receive continuous, executive-level intelligence on the rapid evolution of the AI-Era, we invite you to register for our **Ai6x6 Weekly Newsletter**. Each edition delivers curated insights directly from our thought leaders and highlights critical shifts in AI strategy, governance, and provides relevant excepts from our website and books. Keep your finger on the pulse of the future and ensure your company remains competitive by signing up on the same page

PLEASE REVIEW US ON AMAZON!

Data Blueprint

We know your time is valuable, but if the Ai6x6 Organization Transformation Series delivered the real-world, actionable strategies you needed, please consider leaving a brief review on Amazon. If we missed the mark, we would also like to know, and that is the only we we can improve. Your honest feedback is crucial; it helps fellow executives discover this unique framework and benefit from the insights of our thought leaders. Thank you for supporting our mission!

THE Ai6x6™ PRACTITIONER PROGRAM

Your Authority. Our Infrastructure. A Shared Legacy.

The **Ai6x6™ Practitioner Program** is designed for the high-level strategist—with proven industry leadership experience of over thirty years—who seeks a modern, standardized engine to deploy their operational wisdom. We provide the institutional infrastructure of a global consulting firm, delivered as a white-label solution that honors your professional identity and independence.

The "Un-Franchise" Model

From $2000 per month membership

We have re-engineered the consulting partnership for the modern era. Think of this as a **franchise without the friction or the high fees to get started**:

- **No Long-Term Commitments:** Maintain your agility with a model built for the independent architect.
- **No Loyalty Fees or Royalties:** You keep 100% of your engagement fees. We don't tax your success.
- **A Partnership of Mutual Trust:** Our relationship is powered by a transparent monthly membership, ensuring our incentives are perfectly aligned with your practice's growth.

The Foundations of Practice

All Practitioners gain immediate access to the **Ai6x6™ "Firm-in-a-Box"**:

- **White-Label Ecosystem:** Seamlessly integrate your own brand into our proprietary presentations, client pitches, and diagnostic audit templates.
- **The Clinical Engine:** Augment your strategic intuition with fact-based, quantitative scoring systems that replace subjective brainstorming with measurable transformation.

The White Glove Partnership

For those who wish to move from independent advisor to institutional authority, our **White Glove** tier offers a deep-tissue integration into the Ai6x6™ ecosystem:

- **Monthly 1-on-1 Strategic Check-ins:** Direct access to the **Ai6x6™ leadership team** to stress-test your pipeline, refine high-stakes offerings, or audit active client delivery.
- **Institutional Visibility:** Gain a global platform as a **Co-Author or Featured Expert** in upcoming volumes. We showcase your success stories on **Ai6x6.com**, cementing your legacy as a master of the 6 Lens Sequence.
- **The Bulk Purchase Program:** Exclusive access to deep discounts on the **Ai6x6™ Book Series**, enabling you to provide every client stakeholder with the "source code" for orchestration.

APPLY TO OUR PRACTITIONER PROGRAM AT:

Ai6x6.com/practitioner

Consulting Practitioner

THE Ai6x6™ STRATEGIC ROADMAP SERVICE

Your 3-Year Vision, Stress-Tested for the AI Era

The **Ai6x6™** book series has given you the "source code"—the clinical lenses, the vulnerability audits, and the diagnostic frameworks. But strategic orchestration at the highest level requires a cold, objective eye. If you are ready to move from self-diagnosis to radical transformation, we invite you to engage the **Ai6x6™ Roadmap Service.**

The 21-Day Transformation Engine

We don't believe in "billable hours" or six-month discovery phases. We believe in high-fidelity data and rapid execution. Our domain experts take your raw diagnostic output and synthesize it into a personalized, holistic blueprint for the next three years of your organization's life.

How It Works

1. **Submit Your Data:** Upload your completed Discovery Questionnaires, Vulnerability Audits, and Radar Charts directly to our secure portal or via email or via courier.
2. **Contextualize:** Provide your current 3-year roadmap and any business documentation that can provide more context on your line of business, your mission and success KPIs.
3. **The Clinical Review:** Our experts analyze your organizational DNA against the **6 Lens Sequence**, identifying hidden frictions and untapped leverage points. Assign a single point within your organization for asking questions.
4. **The Delivery:** Within **2 to 3 weeks**, you receive a personalized **Holistic Transformation Roadmap**, including our proprietary **Self-Disruption Strategy**.

Why Self-Disruption?

The greatest threat to your organization isn't your competitor; it is your own legacy success. Our roadmap doesn't just show you how to grow; it shows you how to cannibalize your outdated models before the market does it for you.

The Commitment

- **Speed:** 14–21-day turnaround.
- **Security:** Full NDA-backed engagement.
- **Precision:** Fact-based, quantitative analysis. No "fluff".

Packages from $3000

SECURE YOUR DIAGNOSTIC REVIEW:

Ai6x6.com/roadmap

Consulting Roadmap

THE Ai6x6™ TRANSFORMATION RETREAT

One Playbook. Your Leadership. Total Orchestration.

The **Ai6x6™ Business Transformation Series** is designed to eliminate the need for traditional, slow-moving consulting teams. Using our "Clinical Blueprints," your organization can self-organize, self-diagnose, and self-disrupt. Everything you need to facilitate a high-stakes, 2.5-day executive offsite is contained within the books.

The Self-Guided Mastery (100% Internal & Free)

You have the "Source Code." Your internal Offsite Coordinator or Chief of Staff can lead the entire 2.5-day agenda end-to-end using the **Consultant Playbook (Part 2)** and our **Supplemental Workbooks**.

- **No External Fees:** Run the entire diagnostic without "billable hours."
- **Full Control:** Use the 6-book series to "mix and match" exercises that fit your specific departmental frictions.
- **Tactile Execution:** Move from whiteboards to prioritized MoSCoW cards and RACI matrices using the physical templates provided in the series. Our Advisory section has all the inspiration your team needs.

Packages from $4999

The Orchestrated Upgrade (White-Glove Support)

For leadership teams seeking institutional weight and an objective "Cold Eye" review without the overhead of travel, we offer the **Orchestrated Retreat** package. This provides three critical layers of external authority:

- **Phase 1: The 30-Minute Architect's Briefing (Pre-Retreat):** A high-velocity strategic sprint with your internal Facilitator. we help you select the exact modules from across the 6-book series to target your unique friction points and set the tactical "Rules of Engagement."
- **Phase 2: The 30-Minute Executive Keynote (Live Remote):** A virtual address delivered by the **Ai6x6™ Leadership Team** to kick off Day 1. This keynote provides the external authority required to break through internal political silos and align the room's energy.
- **Phase 3: The 14-Day Clinical Roadmap Audit (Post-Retreat) :** Submit your digital outputs—prioritized lists, RACI matrices, and visualized Roadmaps. Our team performs a rigorous review and delivers a **Comprehensive Strategic Report** within **two weeks**, stress-testing your plan for diagnostic integrity and self-disruption readiness.

UPGRADE TO AN ORCHESTRATED RETREAT:

Ai6x6.com/retreat

Consulting Retreat

THE Ai6x6™ C-SUITE GHOST PAIRING

Intelligence Off the Record and Under the Radar

The **Ai6x6™** series gave you the diagnostic frameworks. The Roadmap Service gave you the blueprint. But the reality of the C-Suite is a high-friction environment where internal politics often suppress the "Clinical Truth." If you are ready to operate with a dedicated, silent partner who provides the radical candor your inner circle cannot, we invite you to engage the **Ai6x6™ Strategic Ghost**.

The Shadow Cabinet

We do not believe in public accolades or "boardroom visibility." We believe in being the invisible intellectual infrastructure for the modern leader. This is a high-fidelity, secure partnership designed to be the objective "second brain" that your legacy organization is politically incapable of providing.

The Protocol: Subscribed Strategy

The Ghost Partnership is a precision-engineered monthly engagement. We don't appear on your org chart, your cap table, or your public calendar.

- **The Weekly Calibration:** 60 minutes of raw, unfiltered strategic synthesis—a "Cold Eye" review of your current friction points and tactical moves.
- **The Monthly Artifact:** One bespoke, high-stakes strategic asset—a Red Team analysis of your board deck, a Vulnerability Audit of a new initiative, or a competitor "Disruption Map."
- **The Direct Feed:** 24/7 access via a secure, encrypted WhatsApp channel for real-time decision support and "sanity checks."
- **The Annual Pivot:** A comprehensive year-end roadmap review to ensure your personal trajectory remains aligned with the 6 Lens Sequence.

The Shield: Total Secrecy

The Ghost operates under a dual-layer legal framework. A rigorous NDA and a specialized **Secrecy Agreement** ensure that our involvement remains entirely invisible. In a world of corporate noise, we are your only pure signal.

The Commitment

- **Access:** 4 hours of dedicated sync-up time per month.
- **Output:** 1 high-fidelity strategic artifact per month.
- **Security:** Full encryption and total anonymity.
- **Focus:** Your personal asymmetric advantage.

Subscriptions from $2500 per month

SECURE YOUR SHADOW PARTNER :

Ai6x6.com/ghost

Consulting Ghost

"Peace is its own reward."

Mahatma Gandhi

Ai6x6 Research Lab
Publishing Team

NEW YORK OFFICE
(Editorial Hub)

Editorial Lead: Marisa Thompson
Research Lead: Talia Rosen
Line Editor: Jordan Alvarez

LONDON OFFICE
(Design Hub)

Design Lead: Keiko Tanaka
Layout Specialist: Oliver Whitmore
Illustrator: Aisha Rahman
Design Team from Sumu Arts

KOLKATA OFFICE
(Support Hub)

Project Manager: Sohini Chatterjee
Production : Arindam Ghosh
Quality Lead: Priya Deshmukh
Fact Checker: Ravi Patel
Research Assistant: Ananya Sengupta
Research Assistant: Rohan Mukherjee
Project Assistant: Madhumita Roy
Production Assistant: Sagnik Dutta
Quality Assistant: Ishita Banerjee
Series Lead (Blueprint): Amit Verma
Series Lead (Playbook): Neha Reddy
Series Lead (Diagnostic): Karan Mehta

The End